MW00570742

APPLIED INORGANIC CHEMISTRY

The University of Calgary Press gratefully acknowledges the financial assistance of the Department of Chemistry of The University of Calgary and the Alberta Foundation for the Literary Arts for the publication of this book.

APPLIED INORGANIC CHEMISTRY

T. W. Swaddle
Professor of Chemistry
University of Calgary

University of Calgary Press

©1990 Thomas Wilson Swaddle. All rights reserved

ISBN 0-919813-58-5

The University of Calgary Press
2500 University Drive N.W.
Calgary, Alberta, Canada T2N 1N4

Canadian Cataloguing in Publication Data

Swaddle, Thomas W. (Thomas Wilson), 1937–
    Applied inorganic chemistry.

    Includes bibliographical references.
    ISBN 0-919813-58-5

    1. Chemistry, Inorganic.   I. Title
    QD151.2.S92 1990   546   C89-091628-2-

Printed and bound in Canada
Corrected printing, 1992

To May Swaddle

Cover photograph: View northwest from Heart Mountain of the Lafarge Canada Inc. plant at Exshaw, Alberta, where the Bow River flows (left to right, in the picture) through the easternmost ranges of the Canadian Rockies. The plant produces 1.1 million tonnes of finished Portland cement per year, using pure limestone ($CaCO_3$) from the fossil reef visible behind the buildings and shale from Seebe, 7 km downstream; see pp. 59–60. Emissions of dust and fly ash in the stack gases are suppressed with electrostatic precipitators. Most of the steep rock faces visible elsewhere in the photograph are composed of dolomite, $(Ca,Mg)CO_3$, which contributes to the high calcium and magnesium hardness of the Bow River water (Chapter 12). Photograph by the author.

Contents

Contents.

Foreword

THIS BOOK is the outgrowth of a one-semester course that I have given since 1967 for third- and fourth-year undergraduate students of chemical engineering. The subject matter is also an extremely important, yet frequently neglected, component of a balanced undergraduate programme for chemistry students. This volume has been prepared largely because most current inorganic chemistry textbooks fail to give adequate systematic consideration to the immense economic, social, and environmental impact of inorganic chemistry. It is hoped that this book will find a niche in university and college chemistry programs, either as a concise text for an independent course in applied inorganic chemistry or as a complement to one of the currently fashionable academic texts in a full-year course on inorganic chemistry.

It is not the purpose of this book to provide an encyclopedia of specific technologies, which tend to be ephemeral. Rather, these (possibly obsolescent) topics are used to introduce some basic principles of applied inorganic chemistry that are of continuing relevance. Concepts or phenomena discussed in ostensibly unrelated sections of the book are extensively cross-referenced so as to emphasize the common scientific underpinnings of diverse technological practices. Conversely, some subjects of current academic interest, such as bioinorganic chemistry, which have had relatively little socioeconomic impact to date and which involve highly complex chemistry, have been given less consideration.

A background of first-year general chemistry and introductory organic chemistry is assumed. It has been my experience that modern theories of chemical bonding do not contribute to the understanding of applied inorganic chemistry in proportion to the intellectual effort they require of the student; therefore, such theories are not developed here. Simple electron-pair concepts of covalency and the coordinate bond are adequate for a working understanding of most chemical phenomena and are covered in virtually all high school or first-year university courses nowadays. Many such courses, and most introductory courses in organic chemistry, also present the rudiments of molecular orbital theory. For students with this background, I have included a few applications of molecular-orbital concepts to explain bonding in metal carbonyls and organometallics, but these are not essential to the purpose of the book.

Of much greater technological importance is the interplay of thermodynamics and kinetics in choosing strategies for putting chemical knowledge
to practical use. A synopsis of useful thermodynamic and kinetic concepts
is given with the expectation that it will be supplemented by more rigorous courses in thermodynamic principles. I have found it expedient to
take some possibly illicit shortcuts, such as skirting the issue of single-ion
activities, but have given appropriate caveats wherever such simplifications
have been made.

Space does not permit me to thank all those who have contributed to
this project in various ways over the years. I would like to record my
special gratitude to Drs. G. Bélanger (Hydro Québec), V. I. Birss (University of Calgary), G. L. Bolton (Sherritt–Gordon), S. Didzbalis (Inco),
G. Strathdee (Potash Corporation of Saskatchewan) and H. L. Youngblood
(Lafarge Canada) for technical information and advice. I thank my col-
leagues at The University of Calgary, in particular Professors P. M. Boor-
man and H. L. Yeager, for their enthusiastic encouragement and material
assistance; the staff of the University of Calgary Press, especially Linda
Cameron and John King, for their generous help and professional guidance
at every stage of the production of this book; Kim Wagstaff for the comput-
erized typesetting of the text and the diagrams; and Leslie (Green) Janzen,
Christine Schill, and Laurie Malinsky for typing the manuscript.

T. W. Swaddle
Calgary, Alberta
November, 1990

Chapter 1

Chemical Energetics

1.1 Kinetics and Thermodynamics

THERE ARE two very basic questions that a chemist or chemical engineer must ask concerning a given chemical reaction:

(a) how *far* does it go, if it is allowed to proceed to equilibrium? (Indeed, does it go in the direction of interest at all?)

(b) how *fast* does it progress?

Question (b) is a matter of chemical *kinetics* and reduces to the need to know the *rate constants* (customarily designated k) for the various steps involved in the reaction mechanism. Note that the rate equation for a particular reaction is not necessarily obtainable by inspection of the stoichiometry of the reaction, unless the mechanism is a one-step process—and this is something that usually has to be determined by experiment. Chemical reaction time scales range from fractions of a nanosecond to millions of years or more. Thus, even if the answer to question (a) is that the reaction is expected to go to essential completion, it may be so slow as to be totally impractical in engineering terms. A brief review of some basic principles of chemical kinetics is given in Section 1.4.

Question (a) is in the province of chemical *thermodynamics*, and amounts to evaluating the *equilibrium constant* (K). Unlike the rate equation, the equilibrium expression for a typical reaction, e.g.:

$$a\text{A} + b\text{B} \rightleftharpoons c\text{C} + d\text{D} \qquad (1.1)$$

can be written down by inspection of the stoichiometry:

$$K^\circ = \frac{\{\text{C}\}^c \{\text{D}\}^d}{\{\text{A}\}^a \{\text{B}\}^b} \qquad (1.2)$$

1

where the braces represent the *activities* of the chemical species A, B, C and D. In simple terms, *activity* is a thermodynamically effective concentration, and is related to the stoichiometric concentration (moles per liter of solution, *molar*, M; or moles per kg of solvent, *molal*, m) by the activity coefficient γ:

$$\{X\} = \gamma[X] \qquad (1.3)$$

where the square brackets denote stoichiometric concentration. For ideal systems (in practice, for gaseous reactions at low total pressures, or solution reactions at high dilution), γ is unity. For simplicity, we will in general assume that this is the case and will thus equate activity with concentration. For ionic solutions, this simple definition of activity coefficient fails because we can never measure the activity of, say, a cation M^{m+} without anions X^{x-} being present at the same time; instead, we usually define a mean ionic activity $a_\pm$ and coefficient $\gamma_\pm$ as:

$$a_\pm = \gamma_\pm c(m^m x^x)^{1/(m+x)} \qquad (1.4)$$

where c is the molal concentration of the electrolyte $M_x X_m$.

The superscript $^\circ$ in Eqn. 1.2 indicates a true thermodynamic equilibrium constant; we will use plain K where concentrations have been used in place of activities.

1.2   Equilibrium and Energy

A rigorous treatment of chemical thermodynamics[1] is beyond the scope of this book. However, there are several thermodynamic relationships that can provide important insights, even if we resort to a few oversimplifications of thermodynamic concepts. In an overview of inorganic chemistry and its applications, it is more important to appreciate what thermodynamics can tell us than to worry about its rigor or theoretical significance.

Perhaps the most important equation relates the *thermodynamic equilibrium constant* K° to the *standard free energy change* ΔG° of the reaction:

$$\Delta G^\circ = -RT \ln K^\circ \qquad (1.5)$$

where R is the gas constant (8.3143 J K^{-1} mol^{-1}) and T is the temperature in kelvins. The *standard Gibbs free energy* G° represents the capacity of the system for doing external work at constant pressure; thus, the thermodynamic driving force of a chemical reaction is the tendency to minimize this free energy. When G° is minimized, the reaction ceases, i.e., has reached equilibrium.

Activities, and hence K° values, are necessarily dimensionless quantities and are defined with reference to a convenient *standard state*. The standard state now universally adopted in the Système International (SI) is:

- pressure $(P) = 100$ kPa (1 bar),
- temperature $(T) = 298.15$ K ($25.00\,°$C),
- solute concentrations hypothetically 1 molal,
- ideal gas behavior (i.e., as if $P \to 0$), and
- ideal solution behavior (i.e., as if infinitely dilute).

This last restriction explains the *hypothetical* standard concentration of 1 molal—obviously, 1 mol kg^{-1} is quite concentrated, and the standard state conditions are extrapolated from high dilution. Gas concentrations are conveniently expressed as partial pressures:

$$\text{concentration} = \frac{n \text{ mol}}{\text{volume } V} = \frac{P}{RT}(\text{ideally}) \propto P, \text{at } 25\,°\text{C} \qquad (1.6)$$

so the standard state for a gaseous reactant is 1 bar partial pressure with (extrapolated) ideal behavior. Thus, if we start with our reactants and products all at their standard conditions and allow the reaction to proceed to equilibrium, an amount of energy $\Delta G°$ becomes available for external work. In the context of doing external electrical work, an oxidation-reduction reaction can generate a *standard electromotive force $E°$* given by:

$$\Delta G° = -nFE° \qquad (1.7)$$

where n is the number of moles of electrons transferred in the reaction and F is the charge of one mole of electrons (*Faraday's constant*, 96 487 coulombs). However, the change in standard heat content, or *enthalpy change ($\Delta H°$)*, associated with the reaction is not the same as $\Delta G°$, since some of the heat content can never be extracted. We speak of the *standard entropy* of a substance $S°$ as being its isothermally unavailable heat content per kelvin:

$$\Delta G° = \Delta H° - T\Delta S°. \qquad (1.8)$$

$\Delta H°$, the enthalpy change of the reaction, can be calculated by subtracting the heat contents of the reactants from those of the products (Hess's law):

$$\Delta H° = \sum H°(\text{products}) - \sum H°(\text{reactants}). \qquad (1.9)$$

However, the absolute heat contents of the individual substances are generally not available. The arbitrary assumption is therefore made that $H°$ is zero for the chemical elements in their most stable form at $25\,°$C and 1 bar. The heat contents of chemical compounds are then defined as their *standard heat of formation $\Delta H_f°$* from their elements at $25\,°$C and 1 bar. For the formation of liquid water:

$$H_2(g) + \tfrac{1}{2}O_2(g) \xrightarrow[\text{1 bar}]{25\,°\text{C}} H_2O(l) \qquad (1.10)$$

ΔH_f° is -285.830 kJ mol^{-1}; for water vapor, ΔH_f° is -241.818 kJ mol^{-1}, the difference being simply the heat of evaporation of 1 mole of water under standard conditions. For the formation of carbon monoxide:

$$C(s) + \tfrac{1}{2}O_2(g) \rightleftharpoons CO(g) \qquad (1.11)$$

ΔH_f° is -110.525 kJ mol^{-1}, relative to $\Delta H_f^\circ = 0$ for graphite. (Diamond, the other familiar form of elemental carbon, is actually less stable than graphite at 25 °C and 1 bar; $\Delta H_f^\circ = 1.895$ kJ mol^{-1}.) We can therefore calculate a standard heat of reaction for the water–gas reaction:

$$C(s) + H_2O(g) \rightleftharpoons H_2(g) + CO(g) \qquad (1.12)$$

to be $(-110.525) - (-241.818) = +131.293$ kJ mol^{-1}. Note that reaction 1.12 is *endothermic* (the plus sign means that heat is taken into the system), whereas reactions 1.10 and 1.11 are *exothermic* (the reactions give out heat to the surroundings). Many heats of formation or of reaction can be measured by calorimetry (i.e., by measuring the temperature rise of a thermally insulated apparatus of known heat capacity when the reaction of interest is carried out in it), or can be obtained from other ΔH_f° data, as shown above for the water–gas reaction 1.12. If we know ΔH° and also know the *standard entropy change* (ΔS°) for a given reaction, we can calculate its equilibrium constant (K°) from a combination of Eqns. 1.5 and 1.8:

$$\ln K^\circ = \frac{\Delta S^\circ}{R} - \frac{\Delta H^\circ}{RT}. \qquad (1.13)$$

Now S°, the standard entropy of a single substance, unlike H°, *can* be calculated absolutely if we know the *standard heat capacity* C_p° of that substance as a function of temperature from zero kelvin.

$$S^\circ = \int_0^T \frac{C_p^\circ}{T} dT = \int_0^T C_p^\circ d(\ln T) \qquad (1.14)$$

Entropies are thus determinable from statistical mechanics or from calorimetry. They are listed along with ΔH_f° for many substances in Appendix C and in various reference books.[2,3] Values of S° for $H_2(g)$, $O_2(g)$ and $H_2O(l)$ are 130.684, 205.138, and 69.91 J K^{-1} mol^{-1}, respectively.

It is important to note that the entropies of gases are larger than those of liquids or solids. This is because entropy is a function of the degree of randomness or disorder at the molecular level. Customarily, S° values are given in *joules* K^{-1} mol^{-1}, whereas ΔG° and ΔH° are given in *kilo*-joules; remember to multiply the latter by 1000 when doing calculations. The entropy of formation of liquid H_2O (reaction 1.10) is then:

$$\Delta S_{f\ H_2O(l)}^\circ = S_{H_2O(l)}^\circ - S_{H_2(g)}^\circ - \tfrac{1}{2}S_{O_2(g)}^\circ = -163.34 \text{ J K}^{-1} \text{ mol}^{-1}. \quad (1.15)$$

The free energy of formation of liquid water at 25 °C is then:

$$\Delta G_f^\circ = \Delta H_f^\circ - T\Delta S_f^\circ = -237.13 \text{ kJ mol}^{-1} \qquad (1.16)$$

and the thermodynamic equilibrium constant for reaction 1.10 is:

$$K^\circ = \frac{\{H_2O(l)\}}{\{H_2(g)\}\{O_2(g)\}^{1/2}}. \qquad (1.17)$$

The activity of the pure water phase, which is separate from the gaseous reactants, can be set to unity, and the gas activities may be approximated to their partial pressures in bars:

$$K = \frac{1}{P_{H_2}P_{O_2}^{1/2}} \text{ bar}^{-3/2} = \exp\frac{-\Delta G^\circ}{RT} = 3.5 \times 10^{41} \text{ bar}^{-3/2}. \qquad (1.18)$$

Thus, the equilibrium pressure of hydrogen created by dissociation of air-free liquid water would be only 2×10^{-28} bar, or one molecule in 200 m^3. In effect, then, reaction 1.10 goes to total completion because it is highly *exergonic*, i.e., because ΔG_f° is so strongly negative, and liquid water simply does not dissociate detectably at 25 °C (even if this were kinetically favored).

Usually, ΔH_f° is the dominant term in Eqn. 1.16. For example, the formation of nitric oxide:

$$\tfrac{1}{2}N_2(g) + \tfrac{1}{2}O_2(g) \rightleftharpoons NO(g) \qquad (1.19)$$

is endothermic ($\Delta H_f^\circ = +90.25$ kJ mol^{-1}) and also *endergonic*, i.e., ΔG_f° is positive ($+86.55$ kJ mol^{-1}) corresponding to $K^\circ \ll 1$. In other words, equilibrium 1.19 lies very far to the left at 25 °C. Thus, thermodynamics predicts that NO should decompose almost completely to nitrogen and oxygen at room temperature. In fact, it does not, because the reaction is so slow (unless catalyzed).

Note that reactions 1.10, 1.11, and 1.19 involve fractional stoichiometric coefficients on the left. This is because we wanted to define conventional *enthalpies (etc.) of formation* of *one mole* of each of the respective products. However, if we are not concerned about the conventional thermodynamic quantities of formation, we can get rid of fractional coefficients by multiplying throughout by the appropriate factor. For example, reaction 1.10 could be doubled, whereupon ΔG° becomes $2\Delta G_f^\circ$, $\Delta H^\circ = 2\Delta H_f^\circ$, and $\Delta S^\circ = 2\Delta S_f^\circ$, and the right-hand sides of Eqns. 1.17 and 1.18 must be squared so that the new equilibrium constant $K' = K^2 = 1.23 \times 10^{83}$ bar^{-3}. Thus, whenever we give a numerical value for our equilibrium constant or the associated thermodynamic quantities, we must make clear how we chose to define the equilibrium. The concentrations we calculate from it will, of course, be the same, no matter how it was defined. Sometimes, as in Eqn. 1.18, the units given for K will imply the definition, but in certain reactions (such as 1.19), K is dimensionless.

1.3   Temperature Dependence of Equilibrium

Equation 1.13 shows that, *if* $\Delta S°$ and $\Delta H°$ for a reaction are known and can be taken to be independent of temperature, then in principle we can calculate the equilibrium constant K_T at any temperature T.

$$\ln K_T = a - bT^{-1} \tag{1.20}$$

This approximation is good enough to allow extrapolations of equilibrium constants over modest temperature ranges for many commonly encountered reactions. The reason is that the molar heat capacities $C_p°$ of the reactants and the products tend to cancel, giving a standard heat capacity change of reaction $\Delta C_p°$ which is often negligible:

$$\Delta C_p° = \sum C_p°(\text{products}) - \sum C_p°(\text{reactants}). \tag{1.21}$$

Reaction 1.19 is a case in point. We have two molecules (O_2, N_2) of similar heat capacities reacting to give two others (2NO) that have thermodynamic properties intermediate between O_2 and N_2; so $\Delta C_p° \approx 0$. Consequently, application of the equations:

$$\Delta H_T = \Delta H° + \int_{298}^{T} \Delta C_p° dT \tag{1.22}$$

$$\Delta S_T = \Delta S° + \int_{298}^{T} \Delta C_p° d(\ln T) \tag{1.23}$$

tells us that ΔH_T and ΔS_T will be essentially independent of temperature. If, however, $\Delta C_p°$ is *not* negligible, we must resort to evaluation of Eqns. 1.22 and 1.23. If $\Delta C_p°$ is significant but approximately constant over the temperature range chosen, Eqn. 1.24, derived by combining 1.22 and 1.23, is better than Eqn. 1.20. $\Delta C_p°$ can be calculated from tables of standard (25 °C) thermodynamic properties.[2]

$$\ln K_T = a - bT^{-1} + c\ln T \tag{1.24}$$

1.4   Chemical Kinetics: Some Basic Principles[4]

Chemical kinetics (i.e., reaction rates) and equilibria are interrelated in that chemical equilibria are *dynamic*—that is, reaction continues in both the forward and reverse directions at equilibrium, but the *rates* in the opposing directions are exactly equal. It follows that the energetic factors considered above also influence reaction rates. Kinetic phenomena, however, are most in evidence far from equilibrium, and are more difficult to treat theoretically than are equilibria. Thus, although thermodynamics can tell us whether a particular reaction *should* proceed from reactants to specific products under given conditions of temperature, pressure, concentration, etc., it tells us nothing about how the reaction could take place at the molecular level (the reaction mechanism) or how *fast* equilibrium will be approached. The reaction may be "infinitely slow" for lack of a favorable mechanism. Reaction kinetics and mechanism are intimately related; the rate of a reaction is largely determined by (*a*) how the reacting molecules organize themselves in space, ready to set up new bonds and break old ones, (*b*) how many of the potential reactant molecules have enough energy to get over the bond making/breaking energy barrier (activation energy), and (*c*) (for reactions in solution) how the solvent molecules rearrange to facilitate this activation process.

The rate of a reaction is expressed as the increase in the amount of a particular product, or the decrease in the amount of a reactant, per second at a time *t*. For reactions in a fluid phase, it is usual to substitute concentrations or, for gases, partial pressures, for amounts. (See Section 7.5 for treatment of the kinetics of some solid-state reactions.) Note that the numerical value of the rate may depend upon how we choose to define it. For example, in the oxidation of aqueous iodide ion by arsenic acid to give triiodide ion and arsenious acid:

$$H_3AsO_4 + 3I^- + 2H^+ \rightleftharpoons H_3AsO_3 + I_3^- + H_2O \qquad (1.25)$$

the rate may be defined as $d[I_3^-]/dt$, or $-d[H_3AsO_4]/dt$, or $-d[I^-]/dt$, etc., but the stoichiometric coefficients in reaction 1.25 tell us that the last rate is numerically three times either of the other two, so we must state clearly what we mean by "reaction rate."

The *rate equation*, sometimes loosely called the "rate law," relates the rate at time *t* to the activities (less rigorously, the concentrations) of the reactants remaining at that instant. Unlike the equilibrium expression (cf. Eqn. 1.2), however, the form of the rate equation generally *cannot* be obtained merely by inspection of the stoichiometric equation unless the reaction is known to proceed to completion in a single step. It must be determined experimentally, and this includes finding out whether the reaction is or is not a single-step process. The activities (in practice, concentrations or, for gases, partial pressures) and the powers to which they are raised (reaction *orders*) in the experimentally determined rate equation tell us which molecules, or fragments thereof, are actually involved in the rate-determining step of the reaction. Any reactant particles that are not involved in this bottleneck configuration or *transition state* will not affect the observed reaction rate, but will be consumed relatively rapidly in subsequent steps. Thus, it is found that the rate of the forward reaction in Eqn. 1.25 is first-order with respect to each of arsenic acid, iodide ion, and hydrogen ion (and so *third-order overall*). We cannot determine the order with respect to water, as this is the solvent and its activity is essentially unity at all times.

$$\frac{d[\mathrm{I_3}^-]}{dt} = k_f [\mathrm{H_3AsO_4}][\mathrm{I}^-][\mathrm{H}^+] \qquad (1.26)$$

In Eqn. 1.26, k_f is the forward rate constant, and the composition of the transition state is $\{\mathrm{H_4AsO_4I}\}$, although it could contain additionally (or be short of) the elements of one or more water molecules, since we cannot determine the order with respect to the solvent. Equation 1.26 cannot be arrived at from reaction 1.25, but consideration of the concentration factors in the two equations tells us at once the rate law for the reverse reaction (Eqn. 1.28, rate constant k_r), since, according to reaction 1.25, the equilibrium expression has to be:

$$K = \frac{[\mathrm{H_3AsO_3}][\mathrm{I_3}^-]}{[\mathrm{H_3AsO_4}][\mathrm{I}^-]^3[\mathrm{H}^+]^2} \qquad (1.27)$$

and, at equilibrium, the forward and reverse rate must be equal.

$$-\frac{d[\mathrm{I_3}^-]}{dt} = \frac{k_r[\mathrm{H_3AsO_3}][\mathrm{I_3}^-]}{[\mathrm{I}^-]^2[\mathrm{H}^+]} \qquad (1.28)$$

It follows that the equilibrium constant K is given by k_f/k_r. The reverse reaction is *inverse* second-order in iodide, and inverse first-order in H^+. This means that the transition state for the reverse reaction contains the elements of arsenious acid and triiodide ion *less* two iodides and one hydrogen ion, i.e., $\{\mathrm{H_2AsO_3I}\}$. This is the same as that for the forward reaction, except for the elements of one molecule of water, the solvent, the participation of which cannot be determined experimentally. The concept of a common transition state for the forward and reverse reactions is called the *principle of microscopic reversibility.*

However, more than one reaction pathway may exist, in which case the rate equation will contain sums of terms representing the competing reaction pathways. For example, one of the oxidation reactions that convert the atmospheric pollutant sulfur dioxide to sulfuric acid (a component of "acid rain") in water droplets in clouds involves dissolved ozone, $\mathrm{O_3}$ (see Sections 2.3 and 2.5):

$$\mathrm{SO_2(g)} + \mathrm{H_2O(l)} \rightleftharpoons \mathrm{H_2SO_3(aq)} \rightleftharpoons \mathrm{HSO_3}^-(\mathrm{aq}) + \mathrm{H}^+(\mathrm{aq}) \qquad (1.29)$$

$$\mathrm{HSO_3}^- + \mathrm{O_3} \rightarrow \mathrm{H}^+ + \mathrm{SO_4}^{2-} + \mathrm{O_2} \qquad (1.30)$$

The rate equation for 1.30 turns out to be:

$$-\frac{d[\mathrm{HSO_3}^-]}{dt} = \left(k_1 + \frac{k_2}{[\mathrm{H}^+]}\right)[\mathrm{HSO_3}^-][\mathrm{O_3}] \qquad (1.31)$$

which implies two parallel pathways: a pH-independent one with a transition state composition $\{HSO_6\}^-$ and rate constant k_1, and a pH-dependent one with transition state $\{SO_6\}^{2-}$ and rate constant k_2. (Actually, pH also affects the overall aqueous ozone oxidation process through the solubility equilibrium of SO_2, reaction 1.29.)

Rate equations of considerable complexity can result from *chain reactions*, such as the reaction of bromine with hydrogen in the gas phase between 200 and 300 °C to form hydrogen bromide. These are reactions in which a chain carrier is created in an initiation step (here, a $\dot{B}r$ atom from dissociation of Br_2), and goes on to create more carriers ($\dot{B}r + H_2 \rightarrow HBr + \dot{H}$, followed by $\dot{H} + Br_2 \rightarrow HBr + \dot{B}r$, and so on) until a recombination step ends the chain. The rate equation for HBr formation has been shown to be:

$$\frac{d[HBr]}{dt} = \frac{k_1[H_2][Br_2]^{1/2}}{1 + k_2[HBr]/[Br_2]}. \tag{1.32}$$

Chain reactions are important in certain polymerizations, organic halogenation reactions, combustion processes, explosions, etc. Usually they involve a *free radical* (a molecule or atom with an odd valence electron, e.g., $\dot{B}r$), since these will create another free radical each time they react with an ordinary molecule having only paired electrons. Detailed discussions are available in standard texts.[4]

The temperature dependence of a rate constant is usually given empirically by the Arrhenius equation:

$$k = A \exp \frac{-E_a}{RT} \tag{1.33}$$

in which E_a is the Arrhenius activation energy (typically on the order of 50 to 100 kJ mol^{-1}) and the pre-exponential factor A includes the frequency of collision between the reactant molecules, the probability of their mutual orientation being favorable for reaction, etc. Thus, $\ln k$ is a linear function of the reciprocal Kelvin temperature, so that if we know k at any two temperatures, or at one temperature when E_a is known, we can use Eqn. 1.33 to calculate k at any other temperature. This does not apply for reactions proceeding by two or more parallel pathways of different E_a (unless one can evaluate Eqn. 1.33 for the various pathways separately, e.g., if one can solve Eqn. 1.31 for k_1 and k_2 at two temperatures), or for multistep reactions if a different step becomes rate-determining as the temperature changes. In either case, a curved plot of $\ln k$ against T^{-1} will result. Simple Arrhenius treatments work well in a great many cases.

An alternative to Eqn. 1.33 that is popular with modern kineticists is the *Eyring equation* (1.34), which derives from the notion that the transition state is in (very unfavorable) equilibrium with the reactants but decays with a *universal frequency* given by $k_B T/h$, where k_B is Boltzmann's constant and h is Planck's constant. Then, by analogy with Eqns. 1.5 and 1.8, we have:

$$k = \frac{k_B T}{h} \exp \frac{-\Delta G^{\ddagger}}{RT} = 2.083 \times 10^{10}\, T \exp \frac{-(\Delta H^{\ddagger} - T\Delta S^{\ddagger})}{RT} \quad (1.34)$$

where $\Delta G^{\ddagger}$, $\Delta H^{\ddagger}$, and $\Delta S^{\ddagger}$ are the free energy, enthalpy, and entropy *of activation*, respectively. Thus, $\ln(k/T)$ should be a linear function of T^{-1}. This is slightly different from the Arrhenius expectation, but in practice data usually fit either equation equally well within the inevitable experimental errors. For reactions in solution, $\Delta H^{\ddagger} = E_a - RT$. The Eyring approach has the advantage that the pseudothermodynamic activation parameters can be readily related to the true thermodynamic quantities that govern the equilibrium of the reaction; the Arrhenius equation is easier to use for simple interpolations or extrapolations of rate data.

1.5   Ionization Potential and Electron Affinity

The nth *ionization potential* (IP) is the energy required to remove an electron from an atom $M^{(n-1)+}$ in the gaseous state. Thus, for aluminum, we have:

$$Al(g) \rightleftharpoons Al^+(g) + e^- \qquad \text{First } IP: \ 577 \text{ kJ mol}^{-1}$$
$$Al^+(g) \rightleftharpoons Al^{2+}(g) + e^- \qquad \text{Second } IP: 1816$$
$$Al^{2+}(g) \rightleftharpoons Al^{3+}(g) + e^- \qquad \text{Third } IP: \underline{2744}$$
$$Al(g) \rightleftharpoons Al^{3+}(g) + 3\,e^- \qquad\qquad\quad 5137 \text{ kJ mol}^{-1}.$$

In general, ionization potentials *decrease* as we descend the Periodic Table within a given group. This is as we might expect, since the atoms increase in size. The first IP is 899 kJ mol^{-1} for beryllium, falling to 503 kJ mol^{-1} for barium, down Group 2. As we cross the Periodic Table from left to right, IPs tend to rise:

	Al	Si	P	S	Cl	Ar	
First IP:	577	786	1061	1000	1255	1521	kJ mol^{-1}.

According to Slater,[8] this is because electrons in the same quantum shell (here, the $3p$ orbitals) screen each other's "view" of the nuclear charge by only ≈ 0.35 unit. Thus, going from Al to Si, the nuclear charge increases by $+1.00$, but the added electron screens only $+0.35$ of this. Electrons in lower shells screen the nuclear charge by essentially $+1.00$ unit, as seen by the outermost electrons. This same effect explains the *lanthanide contraction*—the steady shrinking of lanthanide(III) ion radii from 106 to 85 pm as we fill the $4f$ quantum shell from La^{3+} ($4f^0$) to Lu^{3+} ($4f^{14}$).

There is a very large increase in IP as we pass from the $n(s, p)$ to the $(n-1)(s, p)$ quantum shell; thus, the first IP of Na$^+$ (removal of the lone

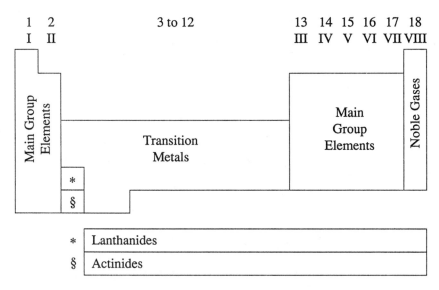

Figure 1.1 Principal features of the Periodic Table. The Arabic group numbers 1–18 are now recommended in place of the traditional Roman I–VIII (A and B). Group names include: alkali metals (1), alkaline earths (2), chalcogens (16), and halogens (17). (See inside front cover for the detailed structure of the Periodic Table.)

3s electron) is 496 kJ mol^{-1}, as against 2081 kJ mol^{-1} for Ne (removal of a 2p electron from a complete shell).

The nth *electron affinity* (EA) of an atom $X^{(n-1)-}$ was traditionally defined as the energy *released* when an electron is added to a gaseous atom. This, however, goes against the thermodynamic convention of defining energy changes as positive when energy is *gained* by the system. We will always use the thermodynamic convention instead. Confusion can arise because EAs can be either positive or negative (unlike IPs, which are always positive as normally—and thermodynamically correctly—defined). If, in a table, the first EAs of the halogens are negative, then we know that the modern thermodynamically correct convention is being used.

Trends in EA are less clear-cut than those in IP, but in general EA values are most negative for elements at the upper right of the Periodic Table, notably the halogens:

$$\text{First } EA\text{:} \quad \begin{aligned} &\text{F} \quad -322 \text{ kJ mol}^{-1} \\ &\text{Cl} \quad -349 \\ &\text{Br} \quad -325 \\ &\text{I} \quad -295 \end{aligned}$$

In general, EAs become positive for $n > 1$ and as we start a new quantum shell—e.g., on going from F (adding the last electron to the $2p$ shell) to Ne (starting the $3s$ shell). Slater's approach makes it clear why this is so.

$$\begin{array}{llll} & \text{O} & \text{F} & \text{Ne} \\ \text{First } EA: & -141 & -322 & +29 \text{ kJ mol}^{-1} \\ \text{Second } EA: & +710 & & \\ \text{O} \rightarrow \text{O}^{2-} & +569 \text{ kJ mol}^{-1} & & \end{array}$$

Taking the trends in IP and EA together, we see that the elements at the left (especially lower left) of the Periodic Table, including the transition elements and the lanthanides and actinides, have the greatest tendency to form cations, while the elements at the right (especially O, F and Cl) tend to form anions. Those in the upper middle region will tend to form compounds by sharing pairs of electrons, i.e., by covalent rather than ionic bonding. It can be anticipated that, even in covalent bonds, the electrons will often be unequally shared by the bound atoms, giving rise to a certain degree of *ionic character* or *polarity* to the bond, whenever the *electronegativities* (χ) of the atoms are different.

1.6   Electronegativity and Bond Energies

Electronegativity (χ) is defined as a measure of the power of an atom *in a molecule* to draw electrons to itself. The stressed words are necessary to distinguish electronegativity from electron affinity. An atom in a molecule may attract electrons from other atoms (a function of the negative of EA, presumably) but must also retain its own electrons against the attractions of the other atoms (a function related to IP). Accordingly, Mulliken proposed an electronegativity scale that summed $-EA$ and IP.

$$\chi(\text{Mulliken}) = \frac{IP - EA}{2} \tag{1.35}$$

The most widely used scale, however, is that of Linus Pauling. It is instructive in that it relates bond structure (ionic character) directly to thermochemically measured bond energies and hence, ultimately, to the bulk-matter thermodynamic properties discussed above. Unfortunately, there is sometimes confusion over what exactly we mean by "bond energy." We could define a "mean bond energy" E, which is the contribution of an A—B pair to the total energy of the molecule AB_n. Thus, for the gas methane (CH_4),

$$E = \frac{\text{heat of atomization}}{\text{number of bonds}} = \frac{1657}{4} = 414 \text{ kJ mol}^{-1}.$$

However, if we take this molecule apart stepwise, we get four different results which we may call the *bond dissociation energies* D (the nomenclature is arbitrary):

$$CH_4 \rightarrow CH_3 + H \quad D = 431\,kJ\;mol^{-1}$$
$$CH_3 \rightarrow CH_2 + H \qquad\quad 364$$
$$CH_2 \rightarrow CH + H \qquad\quad 523$$
$$CH \rightarrow C + H \qquad\quad\; 339$$
$$CH_4 \rightarrow C + 4H \qquad\quad \underline{1657}\,kJ\;mol^{-1}\;(=4E)$$

Clearly, one cannot use tabulated "bond energy" data uncritically. In any event, the relationship with the heat of formation of CH_4 can be illustrated by an enthalpy cycle of the following kind:

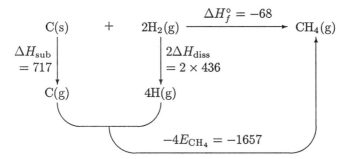

Here, ΔH_{sub} is the heat of sublimation (direct vaporization) of graphite, ΔH_{diss} is the bond energy of molecular hydrogen. The energies are quoted in kJ mol^{-1}.

$$\Delta H_f = \Delta H_{sub} + 2\Delta H_{diss} - 4E_{CH_4} = -68\;kJ\;mol^{-1}$$

Table 1.1 lists a few typical single-bond energies for homoatomic (A—A or B—B) and heteroatomic (A—B) bonds. One might naively expect the energy D_{AB} of an A—B bond to be about the average of the A—A and B—B bond energies, but heteroatomic bonds are nearly always stronger

TABLE 1.1
Typical Single-Bond Dissociation Energies (kJ mol^{-1})

Homoatomic		Heteroatomic	
H_2	436	HF	565
F_2	159	HCl	431
Cl_2	243	HBr	368
Br_2	192	HI	297
I_2	150	H_3C—H	431
C—C	347		
Si—Si	222		

TABLE 1.2
Some Pauling Electronegativities

F	4.0	Br	2.9	H	2.2	Al	1.6
O	3.5	I	2.6	P	2.1	Li	1.0
Cl	3.2	S	2.58	Li	1.0	Na	0.9
N	3.0	C	2.55	Si	1.8	Cs	0.8

than this average. Pauling maintained that this was due to an ionic contribution Δ_{AB} to D_{AB}:

$$\Delta_{AB} = D_{AB} - \frac{D_{AA} + D_{BB}}{2} \tag{1.36a}$$

or

$$\Delta_{AB} = D_{AB} - \sqrt{D_{AA} \times D_{BB}}. \tag{1.36b}$$

In Eqn. 1.36a, the arithmetic mean of D_{AA} and D_{AB} is taken, as is usual, but some authors have used the geometric mean as in Eqn. 1.36b to avoid negative Δ_{AB} values in a few cases involving very weak bonds. Either way, the Pauling electronegativity *difference* between A and B is defined by:

$$\Delta_{AB} = (\chi_A - \chi_B)^2 \text{ in electron volts} \tag{1.37}$$
$$= 96.5(\chi_A - \chi_B)^2 \text{ in kJ mol}^{-1}.$$

The conversion factor of 96.5 is Faraday's constant (divided by 1000 for *kilo*-joules). Since the electronegativity difference is squared, the ionic contribution Δ_{AB} represents a positive contribution to the A—B bond energy whether $\chi_A > \chi_B$ or vice versa. If we set the value of χ for any one element, we can, in principle, use Eqns. 1.36 and 1.38 to calculate χ for any other element from thermochemical data:

$$|\chi_A - \chi_B| = 0.1018\sqrt{\Delta_{AB}} \quad (\Delta_{AB} \text{ in kJ mol}^{-1}). \tag{1.38}$$

These *Pauling* or *thermochemical* electronegativities are obviously only semiquantitative; as we noted above, D_{AB} varies considerably depending upon what other atoms are attached to A or B, and Eqn. 1.36a gives somewhat different D_{AB} values from Eqn. 1.36b. Several more refined χ scales have been proposed,[5] but none are completely satisfactory. Pauling electronegativities (typical values of which are listed in Table 1.2) have the virtue of relating thermochemical properties directly to bond character.

1.7   Electronegativity and Chemical Properties

An obvious use of an electronegativity scale is to predict the direction of electrical polarity of a covalent bond with ionic character. Table 1.2 tells us that the C—H bond in alkanes is polar in the same sense as, although to a much lesser degree than, the O—H bonds in water:

$$R_3C^{\delta-}\!\!-\!H^{\delta+} \qquad\qquad 2\delta\!-\!O\overset{\displaystyle /H^{\delta+}}{\underset{\displaystyle \backslash H^{\delta+}}{}}$$

On the other hand, the polarity of the Si—H bond in silanes (the silicon analogues of alkanes) is reversed:

$$R_3Si^{\delta+}\!\!-\!H^{\delta-}.$$

Thus, the hydrogens in silanes (or organosilanes, if R is an organic group) behave more like hydride (H^-) ions than like protons. Whereas triphenylmethane (R = C_6H_5) is unreactive towards water, triphenylsilane hydrolyzes (reacts with water) to give triphenylsilanol [$(C_6H_5)_3Si$—OH] and hydrogen gas (cf. $H^- + H_2O \rightarrow OH^- + H_2$):

$$(C_6H_5)_3Si^{\delta+} - \vdots - H^{\delta-}$$
$$HO^{\delta-} - \vdots - H^{\delta+} \qquad\longrightarrow\qquad (C_6H_5)_3SiOH + H_2.$$

Similarly, it comes as no surprise to learn that Si—Cl bonds are much more susceptible to hydrolysis than C—Cl bonds, since Si—Cl has the greater electronegativity difference. These naive predictions can be misleading, however. The R_3C—F bond should be very polar, and hydrolysis might be expected to give R_3C—OH and HF. But, in fact, fluorocarbons are usually extremely inert to hydrolysis, even by aqueous alkali, because the *overall* energy change on replacing the C—F and HO—H bonds by C—OH and H—F is, on balance, unfavorable (see Exercise 5.5).

Essentially wholly ionic compounds, consisting of a crystalline lattice of M^{m+} and X^{x-} ions held together by simple electrostatic attraction rather than covalent bonding, can be expected for M of very low χ (e.g., most metals, but especially the heavier members of main groups 1 and 2) and X of high χ (notably F, O, and Cl). Simple electrostatic considerations, confirmed by X-ray crystallography, indicate that highly charged anions X^{x-} (and also H^-) will be large with easily polarized electron clouds. Conversely, small and/or highly charged cations M^{m+} have high electrical potentials at their surfaces and will tend to draw the electron clouds of nearby anions or molecules toward themselves. Consequently, such M^{m+} or X^{x-} tend to indulge in electron-pair *sharing* (covalency) to at least some degree, rather than to form truly ionic compounds (*Fajans' rules*). Examples:

NaF, NaCl, MgF_2, $MgCl_2$: ionic, high-melting solids

AlF_3: ionic, high-melting solid

$AlCl_3$: solid which is easily sublimed; exists as covalent Al_2Cl_6 units in vapor state

SiF_4, $SiCl_4$: covalent gas, covalent liquid

Similarly, very few true ionic hydrides, borides, nitrides, or carbides exist (see Section 10.1).

1.8   Catenation: Inorganic Macromolecules

The tendency of atoms of certain elements to form chains with themselves (homoatomic catenation) or in alternation with other atoms (heteroatomic catenation) is of extreme importance in chemistry. The immense subject of organic chemistry and, indeed, life as we know it depend on the special ability of carbon to catenate, while, from the chemical engineering standpoint, catenation and the associated ability to form molecular rings and cages provide opportunities to tailor-make materials of desired mechanical, electrical, thermal, chemical, or catalytic properties.

Which elements can be expected to homocatenate? They must have χ large enough to attract electrons for covalent bonding, but not so large as to attract them *too* strongly, i.e., they must be "willing" to share electrons. This means elements with mid-range values of χ, i.e., elements in the upper-central region of the main groups of the Periodic Table. From Table 1.2, we see that carbon and sulfur are the prime candidates. Carbon, having a normal valency of four, readily forms elaborate branched chains, rings, or networks (organic molecules). In the limiting case, this three-dimensional network gives diamond, the hardest known substance.

Sulfur, however, is usually only divalent, unless oxidized, and its catenation chemistry is therefore more limited and less familiar. Solid α-sulfur contains very stable 8-membered crown-like rings, S_8:

but many other ring-sizes and open chain lengths are possible. Thus, in addition to the familiar hydrogen sulfide H_2S (a very toxic gas with the well-known "rotten eggs" smell), there exist numerous *sulfanes* (ostensibly analogues of alkanes and silanes), $H—S_x—H$, which occur in natural gas wells and may cause problems if they decompose to deposit solid S_8 on depressurization. Much important aqueous sulfur chemistry involves oxosulfur species of the type $^-O_3S—S_x—SO_3^-$ ($x = 0$ to 20 or more), such as the tetrathionate ion ($x = 2$), which results when thiosulfate ion ($S_2O_3{}^{2-}$) is oxidized by, e.g., iodine (the basis of the classical analytical technique known as *iodometry*):

$$O{=}\overset{\overset{\textstyle O^-}{|}}{\underset{\underset{\textstyle O}{\|}}{S}}{-}S^- + I_2 + {}^-S{-}\overset{\overset{\textstyle O^-}{|}}{\underset{\underset{\textstyle O}{\|}}{S}}{=}O \longrightarrow O{=}\overset{\overset{\textstyle O^-}{|}}{\underset{\underset{\textstyle O}{\|}}{S}}{-}S{-}S{-}\overset{\overset{\textstyle O^-}{|}}{\underset{\underset{\textstyle O}{\|}}{S}}{=}O + 2I^- \quad (1.39)$$

Silicon has $\chi \approx 1.8$, so the Si—Si bond is rather weak, and, while Si is usually tetracovalent like carbon, the silanes Si_xH_{2x+2} are much less numerous, and much more reactive, than the alkanes. On the other hand, Si—O bonds have a large electronegativity difference (3.5 to 1.8), enough to contribute a large Δ_{SiO} to the bond strength while remaining covalent and thus directional. Consequently, —Si—O—Si—O— (siloxane) chains are very stable indeed. Furthermore, Si is almost always tetravalent, so that branched Si—O chains and, indeed, sheets or three-dimensional networks are common both in naturally occurring minerals and in synthetic silicates, many of which are of primary economic importance (Chapter 8).

This propensity of silicon to form branched Si—O structures does not mean that siloxane chemistry is directly analogous to carbon (organic) chemistry. Carbon readily forms multiple bonds, whether to itself, as in ethene ($H_2C{=}CH_2$, "ethylene") and ethyne ($HC{\equiv}CH$, "acetylene"), or to another atom, such as oxygen in $R_2C{=}O$ (a ketone). The multiple-bonding tendency of Si is negligible; only very recently have Si=Si double bonds been characterized, and, even so, only in compounds of the form $R_2Si{=}SiR_2$, where R is a very bulky organic group that prevents polymerization to give Si—Si chains. Thus, attempts to make ketone analogues like $R_2Si{=}O$, where R is some organic group, invariably lead to polymeric substances $[R_2SiO]_n$ called *silicones* (by analogy with *ketones*), although a more meaningful name might be organopolysiloxanes.

Silicones can be oils, greases, waxes, or waxy solids with several very useful and adjustable properties: high thermal stability, excellent electrical insulation, remarkable water-repellency, and low chemical reactivity (though aqueous OH^- or HF can break the Si—O—Si links). They are made by hydrolysis of organochlorosilanes R_nSiCl_{4-n}, which are produced from elemental silicon by the *Rochow process*:

$$RCl(g) + Si(s) \xrightarrow[\text{Cu catalyst}]{\text{heat}} R_nSiCl_{4-n} \quad (1.40)$$

$$nR_2SiCl_2 + nH_2O \rightarrow \ \overset{\displaystyle R}{\underset{\displaystyle R}{-Si-O}} \left[\overset{\displaystyle R}{\underset{\displaystyle R}{-Si-O}} \right] \overset{\displaystyle R}{\underset{\displaystyle R}{-Si-O-}} + 2nHCl. \qquad (1.41)$$

↑
repeating
element

Inclusion of $RSiCl_3$ will lead to chain-branching:

$$-\overset{|}{\underset{|}{Si}}-O-\overset{|}{\underset{|}{Si}}-O-\overset{|}{\underset{|}{Si}}-$$
$$\underset{}{\overset{|}{O}}$$
$$-\overset{|}{\underset{|}{Si}}-$$

while R_3SiCl will cause chain termination:

$$\cdots \overset{\displaystyle R}{\underset{\displaystyle R}{Si}}-O-\overset{\displaystyle R}{\underset{\displaystyle R}{Si}}-R.$$

Another group of inorganic heteroatomic chain compounds with considerable industrial potential is the *phosphazenes*, which are based on a phosphorus-nitrogen skeleton. The electronegativity difference here ($\chi_P = 2.1$, $\chi_N = 3.0$) is less than in the siloxane link, but some double-bonding is involved between N and P, which strengthens the framework. Simple chlorophosphazenes result from the reaction of phosphorus pentachloride with ammonium chloride in an appropriate medium such as chlorobenzene:

$$nPCl_5 + nNH_4Cl \rightarrow \left[-N{=}\overset{\displaystyle Cl}{\underset{\displaystyle Cl}{P}}- \right]_n + 4nHCl. \qquad (1.42)$$

The chlorine atoms can be replaced by organic groups R (e.g., $R = C_6H_5$, by reaction with phenyllithium) to give polymers of very high thermal stability, although the N—P links may tend to hydrolyze slowly. Perfluoroalkoxy side chains ($R = C_xF_{2x+1}O-$) confer high resistance to hydrolysis. Phosphazenes of this type may be used to coat fabrics as water-repellents and flame-retardants, while the low-temperature flexibility (i.e., low glass transition temperatures) and excellent solvent resistance of phosphazene plastics hold high promise for special applications such as fuel hoses for use in cold environments. The cost of such materials, however, is still prohibitively high for large-scale commercial use. Allcock[6] has reviewed the chemistry and uses of inorganic macromolecules.

1.9 Multiple Bonding and its Chemical Consequences

As noted in Section 1.8 with reference to silicon, multiple bonding is not universal across the Periodic Table, but is very important for carbon and its neighbor nitrogen. One might expect the energies of double and triple bonds to be twice and three times those of comparable single bonds, but Table 1.3 shows that this is only roughly true. The N—N bond energies have important chemical consequences. The N—N single bond is very weak, but the N≡N bond is very strong. Thus, the toxic liquid hydrazine $(H_2N—NH_2)$ is very reactive and easily oxidized to N_2. It finds application as a rocket fuel. In industry, it is used as a reducing agent for such things as corrosion control (Section 14.7), electrodeless plating of metals onto metallic or non-metallic surfaces (e.g., copper onto glass or plastics for circuit boards), or removal of toxic chromates from waste water as in:

$$4CrO_4{}^{2-} + 3N_2H_4 + 4H_2O \rightarrow 4Cr(OH)_3(s)\downarrow + 3N_2 + 8OH^-. \qquad (1.43)$$

The extraordinary strength of the triple bond in N_2 means that many reactions of nitrogen compounds (e.g., reaction 1.43) tend to produce N_2 gas, often with a large release of energy. If this occurs rapidly, we have the sudden production of a hot, expanding gas—in other words, the makings of a dangerous explosion. Many nitrogen compounds such as lead azide $(Pb(N_3)_2)$ or ammonium nitrate (NH_4NO_3) are therefore only *kinetically* stable; that is, they exist only because of substantial activation energy barriers to their violent decomposition to N_2, etc. Once such a reaction is initiated, the heat of reaction ensures that the activation barrier is easily overcome, and an explosion results. For lead or other metallic azides, slight

TABLE 1.3
Typical Multiple-Bond Dissociation Energies (kJ mol^{-1})

	Bond Order		
	1	2	3
C—C	347	611	837
C—N	293	615	891
N—N	159	418	946

shock is often enough to detonate the solid. Ammonium nitrate is stable enough in normal handling, and *gentle* heating to its melting point (170 °C) leads to the smooth evolution of nitrous oxide, or "laughing gas":

$$NH_4NO_3 \xrightarrow{\text{heat}} N_2O + 2H_2O. \tag{1.44}$$

However, the reaction is exothermic, and if the temperature rises above 250 °C or if the solid is strongly shocked, violent decomposition to N_2 rather than to N_2O (which has a positive free energy of formation) may result:

$$NH_4NO_3 \rightarrow N_2 + 2H_2O + \tfrac{1}{2}O_2. \tag{1.45}$$

Ammonium nitrate is produced in large volume by the synthetic fertilizer industry, and plant operators must be aware of the potential hazard posed by reaction 1.45. In 1921, an attempt was made at a plant in Oppau, Germany to break up a stockpile of caked NH_4NO_3 by dynamiting it. Over 1000 people died in the resulting explosion. In 1947, a fire broke out on a freighter carrying 1400 tonnes of NH_4NO_3 and led to an explosion that killed some 600 people and destroyed several ships, refineries and a large rubber plant in Texas City, Texas. Other oxidizing agents besides nitrate ion can react explosively with NH_4^+. For example, on May 4, 1988, near Henderson, Nevada, a plant which produced ammonium perchlorate as a rocket fuel oxidant was destroyed by a series of massive explosions involving 4000 tonnes of NH_4ClO_4. Two senior company executives were killed, while 350 local residents were injured and thousands of buildings, some over 30 km away, were damaged.

All commercial explosives[7] are nitrogen-based. Airport security units may soon make use of this fact in neutron beam scanning of luggage; N atoms are highly susceptible to activation by neutrons, and the resulting characteristic 10.8 MeV gamma radiation will promptly reveal the presence of explosives. Examples of explosives include NH_4NO_3–fuel oil mixtures, trinitrotoluene (TNT), $[CH_2N(NO_2)]_3$ (RDX), nitrocellulose, and nitroglycerine (glyceryl trinitrate). The last is a very shock- and temperature-sensitive explosive oil. It is normally stabilized by absorption on the diatomaceous earth kieselguhr (Alfred Nobel's "dynamite") or on a combustible solid phase such as nitrocellulose (gelignite and cordite) or a mixture of wood pulp and sodium nitrate (modern dynamite). Even the old-fashioned "black powder" (75% KNO_3, 15% carbon, and 10% sulfur) relies largely on the violent oxidation of C and S by nitrate ion to give nitrogen and other hot gases. An approximate equation is:

$$2KNO_3 + 2C + S \rightarrow N_2 + CO + CO_2 + K_2SO_3. \tag{1.46}$$

The other side of the coin is that the reverse process, the conversion of inert atmospheric N_2 into needed fertilizers (NH_3 or ammonium salts, or nitrates) is very difficult. This problem is discussed in more detail in Chapter 3. We must not, however, suppose that *all* multiple bonds are necessarily inert. In fact, the data of Table 1.3 imply that this is a special feature of *triply* bonded dinitrogen, since the N≡N bond is rather weak. Acetylene, for example, reacts readily with hydrogen gas (especially if catalyzed) to form ethane because the energy required to reduce the triple C≡C bond to single C—C and to break two H—H bonds is more than compensated for by the formation of four new *heteroatomic* bonds:

$$
\begin{array}{lll}
\text{H—C} \equiv \text{C—H} & \text{put in:} & 837 - 347 = \quad 490 \\
\quad + & & \\
\quad 2\text{H}_2 & \text{put in:} & 2 \times 435 = \quad 870 \\
\quad \downarrow\ {\scriptstyle \Delta H} & & \\
\text{H}_3\text{C} \;-\; \text{CH}_3 & \text{get out:} & 4E_{\text{CH}} = \underline{-1657} \\
& & \Delta H = \ -297\,\text{kJ mol}^{-1}.
\end{array}
$$

Further information on the issues raised in Sections 1.5 to 1.9 may be obtained from the standard texts.[8]

References

1. (*a*) P. A. Rock, "Chemical Thermodynamics," University Science Books: Mill Valley, California, 1984. (*b*) G. N. Lewis and M. Randall, "Thermodynamics," McGraw-Hill: Toronto, 1961 (revised by K. S. Pitzer and L. Brewer). (*c*) R. H. Parker, "An Introduction to Chemical Metallurgy," 2nd edn., Pergamon Press: Oxford, 1978, Chapters 1–4, presents a condensed account of thermodynamic and kinetic concepts relevant to applied inorganic chemistry.

2. D. D. Wagman et al. "The NBS Tables of Chemical Thermodynamic Properties," American Chemical Society: Washington, D.C., 1982 (also published as *Journal of Physical and Chemical Reference Data*, 1982, *11*, Supplement No. 2).

3. R. C. Weast and M. J. Astle (eds.), "CRC Handbook of Chemistry and Physics," CRC Press: Boca Raton, Florida, 1990 (revised annually).

4. (*a*) A. A. Frost and R. G. Pearson, "Kinetics and Mechanism," John Wiley & Sons: New York, 1953. (*b*) K. J. Laidler, "Reaction Kinetics" (2 vols.), Pergamon Press: Oxford, 1963.

5. L. C. Allen, "Electronegativity in the average one-electron energy of the valence-shell electrons in ground-state free atoms," *Journal of the American Chemical Society*, 1989, *111*, 9003–9014.

6. H. R. Allcock, "Inorganic macromolecules," *Chemical and Engineering News*, March 18, 1985, 22–36.

7. J. E. Dolan, "Molecular energy—the development of explosives," *Chemistry in Britain*, 1985, *21*, 732–737.

8. (*a*) J. E. Huheey, "Inorganic Chemistry," 3rd edn., Harper and Row: New York, 1983. (*b*) C. F. Bell and K. A. K. Lott, "Modern Approach to Inorganic Chemistry," Butterworths: London, 1972. (*c*) F. A. Cotton and G. Wilkinson, "Advanced Inorganic Chemistry," 5th edn., Wiley-Interscience: New York, 1988. (*d*) N. N. Greenwood and A. Earnshaw, "Chemistry of the Elements," Pergamon Press: Oxford, 1984.

Exercises

1.1 At any given temperature T, the thermodynamic equilibrium constant K° for a given reaction is related to the standard enthalpy of reaction ΔH_T° and the entropy of reaction ΔS_T° by:

$$- RT \ln K_T^\circ = \Delta G_T^\circ = \Delta H_T^\circ - T\Delta S_T^\circ \tag{1}$$

which, at the standard reference temperature of 298.15 K, becomes

$$\ln K_{298}^\circ = \frac{\Delta S_{298}^\circ}{R} - \frac{\Delta H_{298}^\circ}{298.15R}. \tag{2}$$

If the standard heat capacity change of reaction is ΔC_p°, we can write:

$$\Delta H_T^\circ = \Delta H_{298}^\circ + \int_{298}^{T} \Delta C_p^\circ dT \tag{3}$$

$$\Delta S_T^\circ = \Delta S_{298}^\circ + \int_{298}^{T} \frac{\Delta C_p^\circ}{T} dT. \tag{4}$$

(*a*) Assuming that ΔC_p° is independent of temperature (an acceptable simplification, if not strictly correct), show that:

$$\ln K_T^\circ = \ln K_{298} + \left(\frac{\Delta H_{298}^\circ}{298.15} - \Delta C_p^\circ\right)\left(\frac{T - 298.15}{RT}\right)$$
$$+ \frac{\Delta C_p^\circ}{R} \ln \frac{T}{298.15} \tag{5}$$

which has the general form:

$$\ln K_T = a - \frac{b}{T} + c\ln T \tag{6}$$

where a, b, and c are constants.

(b) For the self-ionization of water:

$$H_2O(l) \rightleftharpoons H^+(aq) + OH^-(aq) \qquad (7)$$

Olofsson and Hepler report $K^\circ_{298} = 1.002 \times 10^{-14}$, $\Delta H^\circ_{298} = 55.815$ kJ mol^{-1}, and $\Delta C^\circ_{p298} = -224.6$ J K^{-1} mol^{-1}. Calculate K° for $0\,^\circ$C, on the basis of a constant ΔC°_p (Eqn. 5). We customarily associate the neutral point of water with pH 7.00; this is valid at $25\,^\circ$C, but what is the corresponding pH at $0\,^\circ$C?

(c) Repeat calculation (b) with the assumption that ΔC°_p is negligible (i.e., that ΔH° and ΔS° can be regarded as constants).

(d) Calculations (b) and (c) show that heat capacity effects can be neglected for approximate calculations of K° over modest temperature ranges. Show, however, that the data given in (b) predict that K° for the ionization of water will pass through a *maximum* at about $274\,^\circ$C, if ΔC°_p and the pressure remain constant. What would the neutral pH be at this maximum? What factors complicate these high-temperature predictions?

[*Answers:* (b) 1.15×10^{-15}, pH 7.47; (d) pH 5.67.]

1.2 (a) Suppose a chemical reaction has two simultaneous (parallel) pathways of comparable importance in the middle of a particular temperature range, but different activation energies. Sketch the plot of $\ln k$ against T^{-1} over that range.

(b) Repeat exercise (a) for a single-path, multi-step reaction, to show how the identity of the rate-determining step may change as the temperature is increased.

(c) Catalysts are often said to work by providing a reaction path of lower activation energy. If so, would you expect catalysts to be proportionally more effective at high temperatures or low?

1.3 Given the following mean bond energies in kJ mol^{-1}:

H—H	435	Cl—Cl	243	F—F	159	C—C	347
H—Cl	431	C—H	414	C—Cl	331		

and assuming the electronegativity χ_H of hydrogen to be 2.0, calculate values for the electronegativities of carbon and chlorine. If the electronegativity of fluorine is 4.0, estimate the C—F bond energy. (The measured value is 490 kJ mol^{-1}.) Note that some minor inconsistencies may be noted when these calculations are carried out by different routes; remember that electronegativity is only a semi-quantitative concept. Furthermore, if one calculates, say, $|\chi_{Cl} - \chi_H|$, it is not clear whether χ_{Cl} is larger or smaller than χ_H; remember, however, that Cl tends to form Cl$^-$ and H prefers to form H$^+$.

1.4 Phosphazenes of the type $(NPCl_2)_n$ have high thermal stability, but hydrolyze slowly on exposure to water or moist air. What would you expect the hydrolysis products to be? (The Pauling electronegativities of N, P and Cl are 3.0, 2.2, and 3.2, respectively.)

1.5 Compare the estimated enthalpy change for the reaction of acetylene with hydrogen to give ethane:

$$H-C{\equiv}C-H + 2H_2 \rightarrow H_3C-CH_3$$

with that of nitrogen to give hydrazine:

$$N{\equiv}N + 2H_2 \rightarrow H_2N-NH_2$$

given the mean bond energies $E_{C-H} = 414$, $E_{C-C} = 347$, $E_{C{\equiv}C} = 837$, $E_{H-H} = 435$, $E_{N{\equiv}N} = 946$, $E_{N-N} = 159$, and $E_{N-H} = 389$ kJ mol^{-1} (all negative, in the thermodynamic sense). Note that high triple bond strength does not necessarily imply lack of reactivity.

1.6 At room temperature, lead azide is an explosive solid that can be detonated by shock (or by heating to 350 °C), while sodium azide exists as stable white crystals that cannot be detonated (unless allowed to react with extraneous material). Rationalize these observations with the aid of Appendix C. (*Hint:* Are there temperatures above or below which ΔG_f° is negative, if heat capacity effects are negligible?) Why are heavy metal azides so explosive?

Chapter 2

The Atmosphere and Atmospheric Pollution

THE IMPORTANCE of main group inorganic chemicals may be gauged in terms of production tonnage, as they make up most of the top dozen or so industrial chemicals (Table 2.1). Indeed, sulfuric acid consumption is often used as an informal measure of a nation's economic activity, since H_2SO_4 is used in a host of chemical, metallurgical, and general manufacturing operations. Prominent amongst these "heavy" inorganic chemicals are products obtained from the atmosphere: elemental oxygen and nitrogen (isolated by fractionation of liquid air), and the nitrogen-derived chemicals ammonia, nitric acid, ammonium nitrate, and urea.[1–3] Conversely, many of the more serious atmospheric pollutants (CO, CO_2, O_3, NO_x, SO_2) are also main group inorganic substances.[4–7] We will consider this latter aspect first.

2.1 Carbon Dioxide

Ordinary air contains a variable amount of water vapor, but its composition on a dry basis by *volume* is 78.08% N_2, 20.95% O_2, 0.934% Ar, and, presently, only 0.034% CO_2, plus minor amounts of Ne, He, CH_4, Kr, H_2, N_2O, and Xe. Nevertheless, carbon dioxide has a major influence on climate through the "*greenhouse effect*": sunlight penetrates the atmosphere, is mostly absorbed by the Earth, and is re-emitted as infrared radiation, but an important fraction of this is trapped by water vapor, CO_2, and certain other gases such as methane in the atmosphere, leading to a significant warming of the environment. An extreme example of this is found on the planet Venus, which has an atmosphere of CO_2 and N_2 at 100 bars total pressure—and a ground temperature of $\approx 460\,°C$. However, Venus is much closer to the Sun.

TABLE 2.1
Inorganic Chemicals in the Top Fifty,
in Terms of Tonnage, Produced in the United States in 1989[a]

Rank	Product	Mt[b]	Rank	Product	Mt
1	sulfuric acid	39.4	14	ammonium nitrate	6.9
2	nitrogen	24.4	17	carbon dioxide	4.9
3	oxygen	17.1	29	hydrochloric acid	2.4
5	ammonia	15.3	31	ammonium sulfate	2.1
6	lime	15.0	35	potash	1.52
7	phosphoric acid	10.5	38	carbon black	1.32
8	chlorine	10.1	43	aluminum sulfate	1.07
9	sodium hydroxide	10.0	44	titanium dioxide	1.01
11	sodium carbonate	9.0	45	calcium chloride	0.87
12	nitric acid	7.2	46	sodium silicate	0.79
13	urea	7.0	48	sodium sulfate	0.73

[a] Source: *Chemical and Engineering News*
[b] t = metric tonne = 1000 kg

Normally, the mean CO_2 concentration is buffered by the oceans, which hold CO_2 in solution at the surface and also precipitate it in the depths as limestone ($CaCO_3$), and by green plants (biomass), which convert it to carbohydrates. With the spread of industrialization and increases in human population, however, the CO_2 level has increased markedly (currently, by 0.6 parts per million per annum) through fossil fuel (coal, oil, natural gas) burning and loss of the forests. At present, an average of 1.1 tonnes of CO_2 is released annually for each person on Earth; for North Americans, this figure is 5 tonnes. This is expected to lead to a doubling of the CO_2 concentration and a warming of all climates by several kelvins over the next century. The consequences are hard to predict, but a rise in sea level (through expansion of seawater and partial melting of the polar ice caps) and the spreading of deserts seem likely. In addition to CO_2, chlorofluorocarbons (see Sections 2.3 and 5.5) and methane contribute to the greenhouse effect, and their concentrations in the atmosphere are also increasing markedly. Much of this additional methane is thought to come from the digestive processes of cattle, the world population of which is increasing rapidly.

The greenhouse effect apart, CO_2 is an innocuous substance, although when evolved in caves or other enclosures it can displace air by virtue of its higher density and thus present a risk of asphyxiation. An extreme example of mass asphyxiation by CO_2 occurred in August, 1986, in Cameroon when

Lake Nyos "turned over," so decompressing a large volume of CO_2-laden water from its depths, whereupon about a cubic kilometer of CO_2 gas spread over the surrounding land, killing 1700 people and several thousand head of livestock.

Carbon dioxide has a conveniently low critical point (31 °C, 7.39 MPa), and the supercritical fluid is used as a non-toxic extractant of excess fats from foodstuffs and in decaffeinating coffee. The chief future use of super-critical CO_2, however, is likely to be the enhancement of recovery of oil that cannot be extracted from wells by conventional techniques.

2.2 Carbon Monoxide

The bonding in carbon monoxide is best explained by pointing out that it is isoelectronic with (has the same number of electrons as) N_2, and the bond is effectively triple. The difference in nuclei, however, results in much higher reactivity for CO than for N_2 and in the propensity of CO for forming co-ordinate bonds to many transition metal centers, as in $Fe(CO)_5$. Molecular orbital theory explains this in terms of the ability of a low-lying empty antibonding orbital concentrated on the C end to accept electrons from a filled d orbital of the metal, while the resulting build-up of negative charge on C allows its unused electron pair to be donated back to the metal:

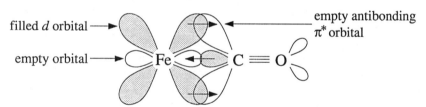

The toxicity of CO (and, in a sense, of the isoelectronic cyanide ion, CN^-) is a direct consequence of this, as CO will compete very effectively with O_2 for the iron centers of the hemoglobin ("Hb") in blood:

$$HbO_2 + CO \underset{}{\overset{K \gg 1}{\rightleftharpoons}} HbCO + O_2 \qquad (2.1)$$

$$HbO_2 =$$

Quite small concentrations of the odorless, colorless CO in air can therefore lead to headache and extreme drowsiness, and eventually to death by preventing O_2 from being delivered to tissues where it is required for metabolism. Fortunately, reaction 2.1 is readily reversible, so first aid for CO poisoning consists in moving the victim to the fresh air and making the person breathe vigorously. The victim will recover completely if this is done soon enough.

Nevertheless, CO poisoning is a serious hazard associated with incomplete combustion of carbon or its compounds (particularly at high temperatures, where it is the favored oxide of C if the O_2 supply is limited—Section 15.6), e.g., in automobile exhausts, tobacco smoke, domestic wood or coal stoves, or industrial operations. Natural processes, such as forest fires, photolysis of CO_2 over lakes, and decay of vegetation, account for the production of about six times as much CO as is produced through human activity. However, the *concentrations* are rarely high enough to be toxic, whereas accumulations of CO in city streets, parking structures and buildings can produce toxic symptoms. Nature also *removes* CO from the air by the action of soil bacteria, so *dilution* of CO is sufficient to protect health and the environment. The high concentrations of CO in automobile exhausts may be reduced by use of a *catalytic converter* (see Section 2.4) which accelerates the air-oxidation of CO to the relatively benign CO_2.

2.3   Ozone

Ozone (O_3), an *allotrope* (or modification) of elemental oxygen, is a very endothermic colorless gas with a characteristic sharp smell. It is formed by the action of ultraviolet light or electrical discharges on ordinary oxygen gas.

$$\tfrac{3}{2}O_2 \xrightarrow[\text{discharge}]{\text{UV or electrical}} O_3 \quad \Delta H = +142 \text{ kJ mol}^{-1} \tag{2.2}$$

Consequently, substantial concentrations of ozone can accumulate around electrical equipment and photocopying machines. Rooms containing these must be well ventilated, since ozone is toxic and can cause headaches and irritation of the mucous membranes, even in low concentrations. It has been suggested that ozone may be the cause of the excess death rate from leukemia (blood cancer) in electrical workers. Ozone is also extremely damaging to plant life—much more so than acid rain (Section 2.5). It also attacks rubber and may cause rubber insulation on electrical equipment to perish rapidly.

Nevertheless, ozone is beneficial in at least three ways. It is an excellent disinfectant and is being used increasingly to sterilize municipal water supplies (Section 12.7). It can also be used as an oxidant in the chemical process industries. Finally, O_3 is an important constituent of the *ozone layer* in the stratosphere. (The *stratosphere* is that part of the atmosphere between 12 and 47 km in altitude. The region below that level is called the *troposphere*.) This ozone layer absorbs short-wavelength ultraviolet light that would otherwise be very damaging to living things on the Earth's surface. For example, it causes skin cancer and eye damage in humans. There has been much concern that the ozone layer could be destroyed by introduction of excessive amounts of nitrogen oxides from supersonic commercial aircraft operating in the stratosphere:

$$NO + O_3 \rightarrow NO_2 + O_2. \tag{2.3}$$

Reaction with photolyzed chlorofluorocarbons, such as CF_2Cl_2 and $CFCl_3$, which are used as supposedly *inert* refrigerants and aerosol spray propellants, may also damage it. Since 1987, it has been clear that a large "hole" in the ozone layer develops over Antarctica during local Spring. Breakaway sections of this hole have swept as far north as Australia and New Zealand. A similar phenomenon has now been recognized in the Arctic. Furthermore, the chlorine monoxide molecule ClO has been shown to be present in these holes and to correlate inversely with the surviving O_3 level, as expected from the following simplified reaction scheme:[8, 9]

$$CF_2Cl_2 \text{ or } CFCl_3 \xrightarrow[\text{stratosphere}]{\text{photolysis in}} \dot{C}l \tag{2.4}$$

$$\dot{C}l + O_3 \longrightarrow ClO + O_2 \tag{2.5}$$

$$O_3 \xrightarrow{\text{UV}} O(^1D) + O_2(^1\Delta) \tag{2.6}$$

$$O + ClO \longrightarrow \dot{C}l + O_2 \tag{2.7}$$

$$O(^1D) + CF_2Cl_2 \text{ or } CFCl_3 \longrightarrow ClO, \text{ etc.} \tag{2.8}$$

Note that since the atomic chlorine consumed in reaction 2.5 is regenerated in 2.7, once formed, it acts as a *catalyst* for ozone decomposition. Reaction 2.6 represents the decomposition of ozone by ultraviolet light, so that reactions 2.6 and 2.2 form a dynamic system that accounts for the absorption of ultraviolet light by the ozone layer. Reaction 2.6 produces atomic oxygen in an excited (^1D) electronic state, *singlet oxygen*. This is important in the troposphere too, as $O(^1D)$ reacts with water vapor to produce chemically reactive hydroxyl radicals:

$$O(^1D) + H_2O \rightarrow 2\dot{O}H. \tag{2.9}$$

An excited state of *molecular* oxygen, $O_2(^1\Delta)$, is also produced in reaction 2.6 and is also called *singlet oxygen*. *Singlet* means that all electron spins are paired; both atomic and molecular oxygen have two unpaired electrons in their ground states, which are called *triplet* states.

Singlet oxygen and $\dot{O}H$ radicals are probably responsible for much of the biological hazards posed by ozone, which is finally receiving recognition as one of the most objectionable of the common air pollutants. Ozone is produced in substantial concentrations by industrial activity and, indirectly, from automobiles. The action of light on atmospheric NO_2, which is itself formed photochemically from unburned gasoline and NO in car exhausts (see reactions 2.16 and 2.17), gives monatomic oxygen, which then reacts with O_2 to give O_3. It is at least partially responsible for the widespread damage to forests that is occurring in northern Europe and northeastern North America.

2.4   Nitrogen Oxides

Nitrous oxide (N_2O, see Section 1.9) is a colorless, odorless gas with mildly anaesthetic properties ("laughing gas") and is widely used as the propellant in whipped cream spray cans. It would seem an unlikely pollutant, but it is formed in Nature by bacterial reduction of nitrates and, on diffusing to the stratosphere, becomes involved in the ozone cycle (reactions 2.2, 2.3, and 2.6) following conversion to nitric oxide (NO):

$$N_2O + O(^1D) \rightarrow 2NO. \tag{2.10}$$

There is therefore concern that the ever-increasing use of synthetic nitrate fertilizers may result in further depletion of the ozone layer. Eventually, stratospheric NO is returned to the earth as nitric acid (see below), but the overall dynamics of the very complex atmospheric chemistry are not fully understood.

The oxides NO, NO_2 (nitrogen dioxide) and N_2O_4 (dinitrogen tetroxide), commonly lumped together as "NO_x", pose much more pressing pollution problems. The colorless gas nitric oxide can form by direct combination of atmospheric N_2 and O_2, but the reaction is very endothermic (therefore favored by high temperatures) and both the equilibrium and the rate constants for NO formation at 298 K are extremely unfavorable:

$$N_2 + O_2 \rightleftharpoons 2NO \quad \Delta H = +180.6 \text{ kJ mol}^{-1}. \tag{2.11}$$

Even at 3000 K, the yield of NO from air is still only a few percent, but the equilibrium is *rapidly* established (see Exercises 2.2 and 3.1). Consequently, whenever air is subjected to very high temperatures and then rapidly quenched to ≤ 1000 K, the small but significant high-temperature NO yield is "frozen in," as happens in internal combustion engines, and in high-temperature furnaces when vented. Pollution by NO_x is therefore usually blamed upon automobile traffic, but heavy industry is a major culprit too. Nitric oxide reacts quite rapidly with the oxygen of the air to form the red-brown gas NO_2:

$$2NO + O_2 \rightarrow 2NO_2 \tag{2.12a}$$

$$NO + O_3 \rightarrow NO_2 + O_2. \tag{2.12b}$$

This in turn dimerizes to yellow N_2O_4, which is normally encountered as a gas (the liquid boils at $21\,^\circ C$ at 1 bar):

$$2NO_2 \rightleftharpoons N_2O_4. \tag{2.13}$$

This dimerization is not unexpected, since NO_2 has one unused valence electron left on the N. Reaction 2.13 is exothermic, so that the NO_2/N_2O_4 mixture is browner at high temperature and at high dilutions. In high concentrations near room temperature, it is more yellow. The brown smog layers often seen over cities are due primarily to dilute NO_2, with some contribution from particulate matter.

In high concentrations, the pungent-smelling NO_2 is toxic and can cause septic pneumonia. The effects on humans of chronic exposure to low NO_2 concentrations are not well understood at present, but the known carcinogenic activity of nitrosated secondary amines $R_2N\!-\!N\!\!=\!\!O$, which might form by interaction of certain protein-related material with NO_x, gives food for thought. At present, concern centers upon two sequences of reaction that NO_x undergoes in the atmosphere.

(a) NO_2 in the troposphere reacts with hydroxyl radicals (from reaction 2.9, via reactions 2.2, 2.6, and 2.10) to form nitric acid:

$$NO_2 + \dot{O}H \rightarrow HNO_3. \tag{2.14}$$

This is eventually washed into the soil, some distance from the source of NO_x, by rain. Thus, NO_x, along with oxides of sulfur (Section 2.5), is connected with the phenomenon of "acid rain," which is at least partly responsible for the destruction of forests (Section 2.3). It also contributes to the acidification of lakes, leading to fish-kills through disruption of the food-chain or dissolution of aluminum from normally insoluble soil minerals (nominally $Al(OH)_3$); aluminum ions are toxic to most fish:

$$Al(OH)_3 + 3H^+ \rightleftharpoons Al^{3+}(aq) + 3H_2O. \tag{2.15}$$

Mobilized aluminum has also been linked to forest damage, since, in sufficient concentration, it is directly toxic to the roots of spruce trees and many other plants.

On the other hand, nitrate ion (NO_3^-) is a valuable fertilizer, and some HNO_3 precipitation may be beneficial to vegetation so long as the excess acidity is neutralized (e.g., by a natural $CaCO_3$ content of the soil or by deliberate liming). Besides, it seems that SO_2 (Section 2.5) and/or ozone (Section 2.3) are the chief causes of airborne environmental damage.

(b) Automotive emissions of NO_x and unburned hydrocarbons can react photochemically under strong sunlight to form *photochemical smog*, a white aerosol that is intensely irritating to the eyes and mucous membranes. This problem is serious in urban areas like Los Angeles, California, where a combination of local weather and topography keeps automobile-contaminated air trapped under intense sunlight for extended periods. The chemistry of smog formation is complex, involving over 50 free-radical gas reactions (see, e.g., Fergusson,[5] pp. 235–36), but can be crudely summarized thus:

$$\underset{\text{unburnt fuel}}{R\!-\!H} \xrightarrow{\text{light}} \dot{H} + \dot{R} \xrightarrow{O_2} R\dot{O}\dot{O} \qquad (2.16)$$

$$\left.\begin{array}{l} NO + O_3 \text{ or} \\ NO + R\dot{O}\dot{O} \end{array}\right\} \longrightarrow NO_2 \qquad (2.17)$$

$$R\dot{O}\dot{O} + NO_2 \longrightarrow \underset{\text{peroxynitrates}}{ROONO_2} \qquad (2.18)$$

$$R\dot{O}\dot{O} + NO \rightarrow R\dot{O} \xrightarrow{O_2} \underset{\text{aldehydes}}{R'CHO} \qquad (2.19)$$

$$R'CHO + \dot{O}H \xrightarrow{O_2} R'CO\!-\!O\dot{O} \xrightarrow{NO_2} \underset{\text{``PAN''}}{R'CO\!-\!OONO_2.} \quad (2.20)$$

The worst irritants in photochemical smogs are peroxyacyl nitrates, "PAN." It seems that they produce singlet oxygen when they hydrolyze, and this may account for their biological action. Other objectionable components of photochemical smogs include aldehydes, organic hydroperoxides (ROOH), and peroxynitrates.

Clearly, the best single approach to these problems is to reduce NO_x emissions in automobile emissions. Three basic strategies are: (*i*) to reduce the combustion temperature (since reaction 2.11 is endothermic), e.g., by reducing the compression ratio of the engine to $\approx 8.5{:}1$ from the more usual $10.5{:}1$, with consequent loss of engine efficiency; (*ii*) to make the concentration of O_2 after combustion as low as possible by using fuel-rich mixtures (e.g., 11:1 fuel-to-air by weight, rather than the efficient 15:1 which is near the theoretical value)—a wasteful strategy that leaves much unburnt fuel and CO; (*iii*) to remove NO and unburnt fuel from the exhaust gases with a catalytic converter.

The last option has been implemented in North America. Catalytic converters are necessarily two-stage, since they must first *reduce* NO to N_2 (using unburnt fuel or CO as the reductant), and then *oxidize* the remaining CO and hydrocarbons to CO_2 with injected air

($\approx$400 °C). The catalysts usually employed are platinum or platinum-rhodium alloys, which also catalyze the oxidation of sulfur in the fuel ($\approx$0.04%) to SO_3 (Section 4.2), so that some sulfuric acid (H_2SO_4) is produced, but this appears not to be a serious problem.

2.5 Sulfur Dioxide and Trioxide

Sulfur dioxide is a pungent, easily liquefiable (bp -10 °C) gas which is produced in large amounts through combustion of fossil fuels (notably coal), the roasting of sulfide ores, pulp and paper mill discharges, and in Nature through volcanic activity. Its direct toxicity is moderate, but prolonged exposure can be fatal, especially to victims who already have bronchial problems. Some 4000 deaths were attributed to the SO_2 content of the disastrous London smog of 1953. Nowadays, coal-burning in British cities is closely regulated, and coal-smoke smogs have virtually disappeared.

Sulfur dioxide emissions continue to cause problems, however, in the context of acid rain. Sulfur dioxide dissolves readily in water to form sulfurous acid, H_2SO_3, which is a fairly strong acid itself, with $pK_a(1) = 1.91$ and $pK_a(2) = 7.18$, at 25 °C and infinite dilution ($pK_a = \log_{10} K_a$).

$$H_2SO_3 \xrightleftharpoons{K_a(1)} H^+ + \underset{\text{bisulfite}}{HSO_3^-} \qquad (2.21a)$$

$$2HSO_3^- \rightleftharpoons H_2O + \underset{\text{metabisulfite}}{S_2O_5^{2-}} \qquad (2.21b)$$

$$HSO_3^- \xrightleftharpoons{K_a(2)} H^+ + \underset{\text{sulfite}}{SO_3^{2-}} \qquad (2.21c)$$

The sulfur-derived acidity in acid rain, however, derives from H_2SO_4 rather than from H_2SO_3. The reaction of gaseous SO_2 with oxygen to give SO_3 (a colorless substance, mp 17 °C, bp 45 °C), and hence H_2SO_4 by hydrolysis, is extremely slow. It is likely that oxidation is effected through the action of hydrogen peroxide or ozone in clouds, through the photochemical effect of ultraviolet light, or through catalysis by dust particles:[6, 9]

$$H_2O + SO_2 + \tfrac{1}{2}O_2 \text{ (or } \tfrac{1}{3}O_3, \text{ or } H_2O_2) \rightarrow H_2SO_4(l)(\text{aerosol}). \qquad (2.22)$$

In the dry state, i.e., in the absence of clouds, the chief oxidant is probably $\dot{O}H$ (cf. reaction 2.14):

$$SO_2 + \dot{O}H \rightarrow HS\dot{O}_3 \xrightarrow{\dot{O}H} H_2SO_4(\text{aerosol}). \qquad (2.23)$$

The sulfuric acid in turn may react with traces of ammonia in the air to give particulate NH_4HSO_4 and $(NH_4)_2SO_4$. Sulfuric acid is a stronger acid than sulfurous ($pK_a(1) < 0$, $pK_a(2) = 1.99$ at 25 °C and infinite dilution), and rain as acidic as pH 2.1 has been recorded at a research station in Hubbard Brook, New Hampshire. Acid rain destroys building materials (especially marble), kills fish and vegetation, accelerates metallic corrosion (Sections 14.5 and 14.7), and can be directly harmful to humans (e.g., the "alligator skin" condition reported in Cubatão, Brazil). Sulfate rain is not completely without redeeming features, as many soils (e.g., in southern Alberta, Canada) are sulfur-deficient, but on balance its acidity is unacceptable, and sulfur oxide emissions must be controlled at source. Several control measures are possible:

(a) *Desulfurization of fuels.* This is difficult and expensive. Of several processes that have been developed for desulfurizing coals, the one proposed by Edgel P. Stambaugh at the Battelle Memorial Institute (Columbus, Ohio) is interesting in that it involves hydrothermal conversion of sulfur to sulfide ion and its recovery as the easily handled element through the Claus process (see also Section 4.1).

$$\text{S in coal} + \text{NaOH(aq)} \xrightarrow{220-350\,^\circ\text{C}} \text{S}^{2-}\text{(aq)}$$
$$\downarrow \text{CO}_2(\text{g}) \qquad\qquad (2.24)$$
$$\text{S}_8 \xleftarrow{\text{Claus}} \text{H}_2\text{S(g)} + \text{Na}_2\text{CO}_3(\text{aq})$$

(b) *Scrubbing of stack gases.* Sulfur-containing fuels are burnt as received, and the SO_2 is removed by passing the flue gases up a column of wet limestone:

$$\text{CaCO}_3(\text{s}) + \text{SO}_2(\text{g}) \xrightarrow{\text{H}_2\text{O}} \text{CaSO}_3(\text{s}) + \text{CO}_2(\text{g}). \qquad (2.25)$$

This is easier than pretreatment of the fuel, but the waste is calcium *sulfite* which, unlike the more familiar sulfate (gypsum, Plaster of Paris), is quite toxic to plants and is therefore difficult to dispose of safely.

(c) *Recovery of* SO_2 *as* H_2SO_4. Direct conversion of SO_2 in flue gases to useful sulfuric acid (as in Section 4.2) is economically attractive, but usually the SO_2 will contain many impurities that may poison the catalyst necessary to oxidize SO_2 to SO_3.

(d) *Fluidized bed combustion.* In the temperature range 820 to 870°C, limestone reacts with SO_2 and air to give calcium sulfate rather than sulfite, and this product can be readily disposed of or utilized, e.g., as roadbed cement. Fennelly[10] describes a two-stage fluidized bed furnace (Fig. 2.1) in which coal fluidized in an air-stream is burned

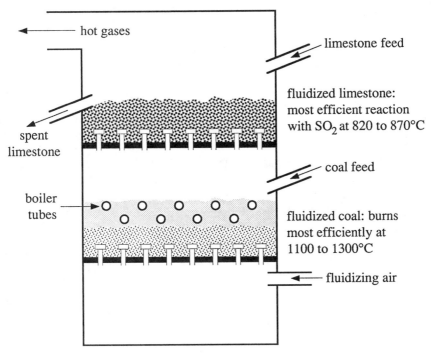

Figure 2.1 A two-stage fluidized-bed coal-fired boiler that minimizes sulfur dioxide emissions (adapted from P. F. Fennelly, *American Scientist*, 1984, *72*, 254).

at 1100 to 1300°C (the temperature at which combustion is most efficient) around the boiler tubes, giving highly efficient heat transfer without formation of "hot spots." The gases then pass to the fluidized limestone bed where over 90% of the SO_2 is removed:

$$CaCO_3 \rightarrow CaO + CO_2 \qquad (2.26)$$

$$CaO + SO_2 + \tfrac{1}{2}O_2 \rightarrow CaSO_4. \qquad (2.27)$$

Reaction 2.26 is endothermic and 2.27 exothermic, and the combination is almost thermoneutral ($\Delta H \approx 0$). Since the combustion temperature is lower than in traditional boiler designs, less NO_x is emitted, and the ash does not melt and so cannot foul the heat transfer surfaces.

(e) *Use of low-sulfur fuels.* Sulfur dioxide emissions from coal-burning utilities could be substantially reduced by the simple expedient of using low-sulfur coals. Transportation costs and political considerations must, however, be taken into account.

References

1. W. Büchner, R. Schliebs, G. Winter, and K. H. Büchel, "Industrial Inorganic Chemistry," VCH Publishers: New York, 1989 (trans. D. R. Terrell).

2. R. Thompson (ed.), "The Modern Inorganic Chemicals Industry," Special Publication No. 31, Chemical Society: London, 1977.

3. "Chemistry in the Economy," The American Chemical Society: Washington, D.C., 1973.

4. R. M. Harrison (ed.), "Pollution: Causes, Effects and Control," Special Publication No. 44, Royal Society of Chemistry: London, 1983.

5. J. E. Fergusson, "Inorganic Chemistry and the Earth," Pergamon Press: Oxford, 1982.

6. A. Cocks and T. Kalland, "The chemistry of atmospheric pollution," *Chemistry in Britain*, September 1988, *24*, 884–888.

7. M. B. Hocking, "Modern Chemical Technology and Emission Control," Springer-Verlag: New York, 1984.

8. (*a*) F. S. Rowland, "Chlorofluorocarbons and the depletion of stratospheric ozone," *American Scientist*, 1989, *77*, 36–45. (*b*) P. S. Zurer, "Studies on ozone destruction expand beyond Antarctic," *Chemical and Engineering News*, May 30, 1988, 16–25. (*c*) T.-L. Tso, L. T. Molina, and F. C.-Y. Wang, "Antarctic stratospheric chemistry of chlorine nitrate, hydrogen chloride and ice: release of active chlorine," *Science*, 1987, *238*, 1253–1260. It appears that HCl and $ClONO_2$ (chlorine nitrate) serve as chlorine reservoirs in stratospheric polar clouds.

9. (*a*) J. H. Seinfeld, "Urban air pollution: state of the science," *Science*, 1989, *243*, 745–752. (*b*) S. E. Schwarz, "Acid deposition: unraveling a regional phenomenon," *Science*, 1989, *243*, 753–763. (*c*) M. B. McElroy and R. J. D. Salawitch, "Changing composition of the global stratosphere," *Science*, 1989, *243*, 763–770. (*d*) S. H. Schneider, "The greenhouse effect: science and policy," *Science*, 1989, *243*, 771–781.

10. P. F. Fennelly, "Fluidized bed combustion," *American Scientist*, 1984, *72*, 254.

Exercises

2.1 Wet limestone scrubbing of stack gases to remove SO_2 gives hydrated calcium sulfite

$$CaCO_3 + SO_2 + \tfrac{1}{2}H_2O \rightarrow CaSO_3 \cdot \tfrac{1}{2}H_2O + CO_2 \qquad (1)$$

rather than the comparable hydrated sulfate

$$CaCO_3 + SO_2 + \tfrac{1}{2}H_2O + \tfrac{1}{2}O_2 \rightarrow CaSO_4 \cdot \tfrac{1}{2}H_2O + CO_2 \qquad (2)$$

Assume the gases are at 298 K and that other standard conditions prevail, and determine whether the prevalence of reaction 1 at near-ambient conditions is a result of thermodynamic or kinetic constraints.

	$CaSO_3 \cdot \tfrac{1}{2}H_2O$	$CaSO_4 \cdot \tfrac{1}{2}H_2O$	O_2	
ΔH_f°	-1311.7	-1574.65	0	kJ mol^{-1}
S°	121.3	134.8	205.138	J K^{-1} mol^{-1}

2.2 Given that the standard heat of formation of nitric oxide is 90.25 kJ mol^{-1} and that the standard entropies of N_2, O_2 and NO are 191.61, 205.138, and 210.761 J K^{-1} mol^{-1}, respectively, derive the equilibrium constant for the formation of one mole of NO at 298 K and 1 bar, assuming ideal gas behavior.

With the composition of air given in Section 2.1, calculate the equilibrium pressure of NO that would form in air at 298 K and 100 kPa if the reaction were kinetically feasible.

[*Answers:* 6.8×10^{-16}; 2.8×10^{-11} Pa.]

2.3 From the standpoint of abatement of the contribution of CO_2 to the greenhouse effect, natural gas (methane) is preferred as a fuel over gasoline and, particularly, coal. Explain why this is so.

Chapter 3

Agricultural Chemicals[1-6]

THE FOOD we eat is derived ultimately from plants, algae, and micro-organisms that convert inorganic matter such as carbon dioxide, water, nitrogen, sulfates, and phosphates into carbohydrates, proteins, fats, vitamins, etc. The complex biochemical processes involved are usually driven by sunlight (through photosynthesis). Agriculture therefore depends upon our supplying these inorganic nutrients to plants. Cereals, vegetables, fruit-bearers or animal fodder require nutrients in forms that they can use. In particular, we need to convert (or "fix") the inert N_2 of the atmosphere into soluble, reactive compounds such as nitrates, ammonia, and ammonium salts. Other major fertilizer components are sulfate, potassium, and phosphate ions. It may also be necessary to provide trace nutrients such as cobalt compounds, or to remove excess soil acidity by treatment with lime (CaO).

3.1 Natural Sources of Fixed Nitrogen

Apart from minor sources such as thunderstorms (Section 3.2), atmospheric nitrogen is fixed in Nature by certain soil bacteria, blue-green algae, and the root nodules of legumes. This is accomplished either by oxidation to nitrate:

$$\tfrac{1}{2}N_2(g) + \tfrac{1}{2}H_2O(l) + \tfrac{5}{4}O_2(g) \rightarrow HNO_3(aq) \quad \Delta H = -65 \text{ kJ mol}^{-1} \quad (3.1)$$

or by reduction to ammonia or ammonium salts:

$$\tfrac{1}{2}N_2(g) + \tfrac{3}{2}H_2O(l) \rightarrow NH_3(aq) + \tfrac{3}{4}O_2(g) \quad \Delta H = +348 \text{ kJ mol}^{-1}. \quad (3.2)$$

Reaction 3.1 might seem to be thermodynamically favored, but in fact no kinetically easy route from triply bonded N_2 to $NO_3{}^-$ exists, since the endothermic intermediate NO (Section 2.4) is likely to be involved. As written, reaction 3.2 has prohibitive energetics, but in practice the process is more complex than this. For example, the fact that free O_2 is not formed but is in effect consumed in other biochemical reactions, makes for a favorable energy balance. The limiting factor is again kinetics, as plausible intermediates such as hydrazine (H_2N—NH_2) are endergonic compounds.

Natural nitrogen-fixing systems overcome these kinetic barriers with certain *enzymes* known as nitrogenases which contain iron and molybdenum atoms. Enzymes are biological catalysts, usually having complicated protein-based structures, often with complexed metal ions as their active sites. The exact structure and mode of action of the various nitrogenases are the subjects of intensive current research with important industrial consequences. The prospect of mimicking their ability to fix N_2 in mild aqueous conditions is very attractive in view of the high energy and capital requirements of current NH_3 or HNO_3 manufacture, as described below.

3.2   Direct Combination of Nitrogen and Oxygen

As noted in Section 2.4, atmospheric N_2 and O_2 combine endothermically in small but significant yield and at a sufficiently rapid rate above about 2000 K, but the gases must be quenched rapidly if the high-temperature yield of NO is to be recovered for subsequent conversion to nitric acid:

$$\tfrac{1}{2}N_2 + \tfrac{1}{2}O_2 \rightleftharpoons NO \quad \Delta H^\circ = 90.3 \text{ kJ mol}^{-1}. \tag{3.3}$$

Lightning strikes may contribute some fixed N to the biosphere by reaction 3.3 in regions where thunderstorms are frequent. Electric arcs have been used in this way in industry. In the 1950s, the *Wisconsin process,*[7] was developed for NO production. In this process, air is blasted over hot refractory pebbles (MgO, preheated to $\approx$2200 °C by burning fuel-oil or natural gas) and is then quenched rapidly. The great increase in energy costs since that time, however, has made the Wisconsin and arc processes increasingly uneconomical when compared to the catalytic oxidation of ammonia in nitric acid production (Section 3.4).

3.3   Ammonia Synthesis

The direct gas-phase synthesis of ammonia from nitrogen and hydrogen (the *Haber process*, 1908) is presently the cornerstone of the extremely important fertilizer industry:

$$N_2 + 3H_2 \rightleftharpoons 2NH_3 \quad \Delta H^\circ = -92.2 \text{ kJ mol}^{-1}. \tag{3.4}$$

1 vols.   3 vols.   2 vols.

TABLE 3.1

Equilibrium Yields of Ammonia According to Equation 3.4

Temperature	at 25 bars	at 400 bars
100 °C	91.7%	99.4%
400 °C	8.7%	55.4%
500 °C	2.9%	31.9%

It is implicit in reaction 3.4 that the equilibrium yield of ammonia is favored by high pressures and low temperatures (see Table 3.1). However, compromises must be made with the capital cost of high-pressure equipment and with the slow rate of reaction at low temperatures, even when a catalyst is used. In practice, Haber plants are usually operated at 80 to 350 bars and at 400 to 540 °C, and several passes are made through the converter. The catalyst (Section 9.3) is typically finely divided iron (supplied as magnetite, Fe_3O_4 which is reduced by the H_2) with a KOH promoter on a support of refractory metallic oxide. The upper temperature limit is set by the tendency of the catalyst to sinter above 540 °C. To increase the yield, the gases may be cooled as they approach equilibrium.

The nitrogen required is obtained by fractional distillation of liquid air. The hydrogen used to be obtained by electrolysis of liquid water. (If cheap surplus electrical capacity becomes available in the future, this method may well be reintroduced. Catalytic photolysis of water using sunlight is another possible future source of H_2.) The *Haber–Bosch process* of 1916 used *water-gas*, which is a mixture of H_2, CO and CO_2 made by alternating blasts of steam and air over coke at red heat.

$$H_2O(g) + C(s) \xrightarrow{\approx 800\,°C} H_2(g) + CO(g) \quad \Delta H = +175 \text{ kJ mol}^{-1} \quad (3.5)$$
$$2H_2O(g) + C(s) \rightarrow 2H_2(g) + CO_2(g) \quad \Delta H = +178 \text{ kJ mol}^{-1} \quad (3.6)$$

The periodic air blasts to reheat the coke are necessary because of the marked endothermicity of these reactions. The CO content is used to generate further H_2 in the *water–gas shift reaction*:

$$H_2O(g) + CO(g) \rightleftharpoons CO_2(g) + H_2(g) \quad \Delta H = -41 \text{ kJ mol}^{-1}. \quad (3.7)$$

Here, the exothermicity dictates a low reaction temperature (450 °C or less) and the use of a catalyst (Fe_2O_3/Al_2O_3, usually) if good yields are to be obtained in a reasonable time. Usually, the reaction is carried out in two steps: first temperatures of 450 to 500 °C are used and a substantial degree of conversion quickly results. Then the temperature is dropped to about 200 °C to optimize the yield. The CO_2 is removed by scrubbing the gas with water, in which CO_2 is much more soluble than H_2.

At present, H_2 is usually made from natural gas (methane, CH_4) or naphtha by *steam reforming*, first introduced in the 1930s:

$$CH_4(g) + H_2O(g) \rightleftharpoons CO(g) + 3H_2(g) \quad \Delta H = +206 \text{ kJ mol}^{-1}. \quad (3.8)$$

A temperature of 700 to 800 °C is optimal, with a nickel catalyst. Alternatively, methane can be partially oxidized with pure oxygen:

$$CH_4 + \tfrac{1}{2}O_2 \rightleftharpoons CO + 2H_2 \quad \Delta H = -36 \text{ kJ mol}^{-1}. \quad (3.9)$$

At first sight, steam reforming seems to give more H_2 per CH_4 than reaction 3.9, but it is endothermic, and additional methane must be burned to provide the heat input. In practice, reaction 3.8 is still preferred, inasmuch as it is more economical to provide the needed heat than to recover the waste heat of reaction 3.9. In either case, the CO may be used to make more H_2 by reaction 3.7.

Ammonia is readily liquefied under pressure or on cooling (bp -33 °C, critical point 132.5 °C and 114.0 bars). The liquid ("*anhydrous ammonia*") was formerly used as a refrigerant and is currently transported and even injected into soils for fertilizer. However, since both liquid and gaseous ammonia are corrosive to the flesh, ammonia is commonly converted in aqueous solution to either ammonium sulfate or ammonium nitrate, which are then recovered by evaporation as crystalline solids for safe shipment, storage, and use:

$$2NH_3(g) + H_2SO_4(aq) \rightarrow (NH_4)_2SO_4(aq). \quad (3.10)$$

Recently, the use of urea has gained favor as an agricultural source of nitrogen. It is made by reaction of ammonia with the CO_2 by-product of the water–gas shift reaction (3.7):

$$CO_2 + 2NH_3 \rightarrow \underset{\text{ammonium carbamate}}{NH_4[CO_2NH_2]} \xrightarrow[200\,°C]{150 \text{ bars}} \underset{\text{urea}}{(H_2N)_2CO + H_2O.} \quad (3.11)$$

3.4   Nitric Acid and Ammonium Nitrate

The *catalytic* oxidation of ammonia by air over platinum gauze at $\approx$900 °C gives nitric oxide (reaction 3.12), which is then oxidized to nitric acid by air and *liquid* water in a "nitrous gas absorber" (reactions 3.13 and 3.14):

$$NH_3 + \tfrac{5}{4}O_2 \rightarrow NO + \tfrac{3}{2}H_2O(g) \quad \Delta H° = -226.3 \text{ kJ mol}^{-1} \quad (3.12)$$
$$NO + \tfrac{1}{2}O_2 \rightarrow NO_2 \quad \Delta H° = -57.1 \text{ kJ mol}^{-1} \quad (3.13)$$

$$3NO_2 + H_2O(l) \rightarrow 2HNO_3(l) + NO \quad \Delta H° = 71.7 \text{ kJ mol}^{-1}. \quad (3.14)$$

The net reaction is thus:

$$NH_3 + 2O_2 \rightarrow HNO_3 + H_2O(l) \tag{3.15}$$

with $\Delta H° = -447.1$ kJ mol^{-1} if the product is aqueous nitric acid (hypothetical 1.0 molal) or -413.8 kJ mol^{-1} for 100% liquid HNO$_3$. *Specific catalysis of reaction 3.12 is* essential; otherwise the thermodynamically favored oxidation of ammonia to N$_2$ will occur instead. (Recall that NO is an *endergonic* compound; Sections 2.4 and 3.2.)

$$NH_3 + \tfrac{3}{4}O_2 \rightarrow \tfrac{1}{2}N_2 + \tfrac{3}{2}H_2O(g) \quad \Delta H° = -316.6 \text{ kJ mol}^{-1} \tag{3.16}$$

For the same reason, the gas leaving the catalyst should be substantially free of NH$_3$; otherwise reaction 3.17 will occur while the gas is still hot:

$$6NO + 4NH_3 \rightarrow 5N_2 + 6H_2O. \tag{3.17}$$

Note that the most important function of the catalyst, here and in many other instances, is not so much its overall catalytic activity as its *selectivity* in promoting the one reaction (3.12) over its competitors. In biological systems, enzymes have evolved to extraordinary degrees of selectivity; often, they will catalyze one specific biochemical reaction and no others.

The *nitrous gas absorption* step (reactions 3.13 and 3.14) is slow, especially if concentrated HNO$_3$ is required, since cooling to 2°C is then necessary. Consequently, large countercurrent towers of stainless steel are needed, with associated high capital cost. The recovery of the heat of reaction of this step is inefficient because of the low temperature of the source gases that must be maintained. It has been suggested that the energy of reaction 3.12 could be more effectively recovered if it is run in a fuel cell (see Exercise 13.8).

Some nitric acid is used for the manufacture of explosives, etc., but much is converted on-site to the potentially explosive high-nitrogen fertilizer, ammonium nitrate (Section 1.9). Ammonia gas from the Haber plant is absorbed in aqueous HNO$_3$, and the NH$_4$NO$_3$ solution is evaporated to a liquid melt ($< 8\%$ H$_2$O) for crystallization, but care must be taken to keep the pH above about 4.5 and to exclude any material (chlorides, organic compounds, metals) that might catalyze the explosive decomposition of NH$_4$NO$_3$. Obviously, it is also wise to keep the melt mass low and to vent it to avoid pressure build-up. Solid NH$_4$NO$_3$ is very *hygroscopic* (i.e., it picks up water from the air). Non-oxidizable drying agents such as clays are usually added to suppress this effect and the consequent caking.

3.5   Sulfuric Acid

The manufacture of sulfuric acid is discussed in Chapter 4. We note here simply that 90% of the sulfur produced industrially is made into H$_2$SO$_4$.

Of this, two-thirds is consumed in fertilizer manufacture, either directly in making ammonium sulfate fertilizer, or indirectly in producing superphosphate (Section 3.6), potassium sulfate (Section 3.7), and other products. Sulfur is an important plant nutrient, and some soils are sulfur-deficient and accordingly require additions of sulfate or elemental sulfur. However, sulfur is less frequently the limiting factor in plant growth than is fixed N or soluble phosphorus. The chief role of $SO_4{}^{2-}$ in $(NH_4)_2SO_4$ fertilizer is as a benign vehicle for the ammonium ion (cf. K_2SO_4, Section 3.7).

3.6   Phosphates

Phosphorus is essential for plant growth and is often the limiting nutrient in aquatic ecosystems. Consequently, if abundant supplies of phosphate (e.g., from detergents in sewage plant discharges) are introduced into lakes, explosive growth of algae usually results (eutrophication of the lake). As the algae consume most of the oxygen in the lake, other aquatic life is virtually wiped out. Similarly, weed growth in the Bow River downstream of the city of Calgary, Alberta, thrives on phosphates remaining in discharged treated sewage, resulting in this case in a world-class trout fishery, but weed growth interferes with the water intakes of downstream water users. The simplest means of removing phosphates (present mainly as monohydrogen phosphate ion, $HPO_4{}^{2-}$) from waste water is to add lime, since *basic* (i.e., OH^--containing) calcium phosphates are insoluble in water:

$$5Ca(OH)_2 + 3HPO_4{}^{2-}(aq) \rightarrow Ca_5(PO_4)_3OH(s)_\downarrow + 6OH^- + 3H_2O. \quad (3.18)$$

Conversely, the usual mineral source of phosphorus is insoluble rock phosphate $[Ca_3(PO_4)_2]$, and this must be converted to *acidic* calcium phosphates (i.e., to Ca salts of $HPO_4{}^{2-}$ and $H_2PO_4{}^-$), in order to make a fertilizer soluble enough to be readily utilized by plants. This is done by treating rock phosphates with concentrated H_2SO_4. The product *superphosphate* that results is typically 32% $CaHPO_4$ plus $Ca(H_2PO_4)_2 \cdot H_2O$, 3% absorbed H_3PO_4, and 50% $CaSO_4$. Sometimes, the P content is expressed as a percentage of phosphorus pentoxide (here, 20% P_2O_5).

Alternatively, we may first make phosphoric acid (H_3PO_4) itself (a syrupy liquid much like concentrated H_2SO_4) and use this to convert further rock phosphate to *concentrated superphosphate* (83% $CaHPO_4$ or $Ca(H_2PO_4)_2$, 3% H_3PO_4, and only 5% $CaSO_4$; in effect 48% P_2O_5). The key principle in either case is that H_2SO_4 is a much stronger acid than H_3PO_4 (for example, *in water*, their first pK_as are < 0 and 2.12, respectively) and so phosphate ions are protonated at the expense of H_2SO_4:

$$Ca_3(PO_4)_2(s) + 3H_2SO_4(l) \rightarrow 2H_3PO_4(l) + 3CaSO_4(s). \quad (3.19)$$

The $CaSO_4$ may form as gypsum ($CaSO_4 \cdot 2H_2O$) or anhydrite ($CaSO_4$), depending on reaction conditions. The important engineering point is to obtain an easily filterable solid. Another possibility is to neutralize the H_3PO_4 with ammonia to obtain $(NH_4)_3PO_4$, a readily soluble fertilizer with very high available N and P contents.

3.7 Potash

Fertilizers are usually designated with three numbers, indicating respectively their nitrogen, phosphate, and potash* contents. Thus, a 20–3–4 fertilizer contains 20% N, 3% P expressed as P_2O_5, and 4% potassium expressed as K_2O. About 95% of all potash produced goes into fertilizers. In Saskatchewan, it is encountered mainly as deposits of the chloride, KCl (*sylvite*—or *sylvinite*, if mixed with NaCl), surrounded by a mineralization zone of carnallite, $KCl \cdot MgCl_2 \cdot 6H_2O$, and is recovered either as the solid or by solution mining, although the latter involves more expenditure of energy. The solid KCl (density 1.98 g cm^{-3}) can be freed of the denser NaCl (2.16 g cm^{-3}) by flotation in a fluid of intermediate density or by crystallization from solution. The solubility of KCl in 100 mL water rises from 34.7 g at 20 °C to 56.7 g at 100 °C, whereas that of NaCl is little affected by temperature. Thus, a solution saturated with KCl and NaCl at 20 °C will dissolve substantially more KCl from sylvinite on heating to 100 °C, and relatively pure KCl will crystallize out on cooling. When chlorides are undesirable (e.g., where excess Cl$^-$ tends to accumulate in the soil as in the Netherlands or in arid, irrigated regions, or where a crop such as grapes is sensitive to Cl$^-$) KCl is converted to K_2SO_4:

(*a*) In the *Mannheim process* the solid is heated at relatively low temperatures with concentrated H_2SO_4 to get solid $KHSO_4$, which is then heated strongly with more KCl to yield K_2SO_4:

$$KCl + H_2SO_4 \rightarrow KHSO_4 + HCl \quad \text{(exothermic)} \quad (3.20)$$
$$KHSO_4 + KCl \rightarrow K_2SO_4 + HCl \quad \text{(endothermic)}. \quad (3.21)$$

This illustrates the general principle that an involatile powerful protonator (H_2SO_4, bp 290–317 °C) will displace a volatile weaker acid (hydrogen chloride, bp -85 °C) on heating.

(*b*) The *Hargreaves process* follows a similar principle:

$$4KCl(s) + 2SO_2 + O_2 + 2H_2O(g) \rightarrow 2K_2SO_4 + 4HCl. \quad (3.22)$$

---

*The term potash in its strict sense refers to potassium carbonate (K_2CO_3), which was formerly made by leaching wood *ash* with water in iron *pots*. In common usage, it refers to any source of potassium.

(*c*) A third method uses complex precipitation processes involving aqueous solutions of (usually) magnesium sulfate minerals, e.g., kainite, $KCl \cdot MgSO_4 \cdot 3H_2O$:

$$KCl + KCl \cdot MgSO_4 \cdot 3H_2O \rightarrow K_2SO_4 + MgCl_2 + 3H_2O. \qquad (3.23)$$

Alternatively, KCl can be converted to KNO_3, which obviously has high fertilizer value:

$$3KCl + 4HNO_3(65\%) \xrightarrow{75\,°C} 3KNO_3 + Cl_2 + NOCl + 2H_2O. \qquad (3.24)$$

The nitrosyl chloride is oxidized separately with air to NO_2 and thence to HNO_3, so that the net reaction becomes:

$$2KCl + 2HNO_3 + \tfrac{1}{2}O_2 \rightarrow 2KNO_3 + Cl_2 + H_2O. \qquad (3.25)$$

References

1. R. Thompson (ed.), "The Modern Inorganic Chemicals Industry," Special Publication No. 31, Chemical Society: London, 1977.

2. J. E. Fergusson, "Inorganic Chemistry and the Earth," Pergamon Press: Oxford, 1982.

3. J. A. Kent (ed.), "Riegel's Handbook of Industrial Chemistry," 8th edn., Van Nostrand-Reinhold: New York, 1983.

4. G. T. Austin, "Shreve's Chemical Process Industries," 5th edn., McGraw-Hill: New York, 1985.

5. "Fertilizer Manual," International Fertilizer Development Center, United Nations Industrial Development Organization.

6. W. Büchner, R. Schliebs, G. Winter, and K. H. Büchel, "Industrial Inorganic Chemistry," VCH Publishers: New York, 1989 (trans. D. R. Terrell).

7. E. D. Ermenc, "Wisconsin process pebble furnace fixes atmospheric nitrogen," *Chemical Engineering Progress*, 1956, *52(4)*, 149–153.

Exercises

3.1 The rate of reaction of air over a hot MgO pebble bed to form nitric oxide is given by:

$$\frac{dP_{NO}}{dt} = k_f P_{N_2} P_{O_2} - k_r P_{NO}^2$$

where the rate constants in $bar^{-1}s^{-1}$ are:

Temperature	k_f	k_r
2000 °C	0.6	370
2200 °C	7	2000

Assume the partial pressure of NO to be negligible relative to the partial pressures of N_2 and O_2, which can be taken to be 0.78 and 0.21 bar, respectively. On this basis, what is the equilibrium partial pressure of NO (a) at 2000 °C, and (b) at 2200 °C? (c) What is the heat of formation of (one mole of) NO in this temperature regime? What will be the initial rate of loss of NO if air at equilibrium is shock-cooled (d) from 2200 to 2000 °C, (e) from 2000 to 1800 °C, and (f) from 2000 to 1500 °C? What do your results imply for the operation of a Wisconsin process plant?
[*Answers:* (a) 0.0163 bar; (b) 0.0239 bar; (c) +90 kJ mol^{-1}; (d) −0.113 bar s^{-1}; (e) −8.0 × 10^{-3} bar s^{-1}; (f) −2.6 × 10^{-4} bar s^{-1}.]

3.2 Use the enthalpy, entropy, and heat capacity data of Appendix C together with Eqns. 2 and 5 from Exercise 1.1 to calculate the equilibrium constant K for the formation of ammonia from hydrogen and nitrogen (a) at 25 °C and (b) at 500 °C. Use these results to calculate the equilibrium percentage yield of ammonia (c) at 25 °C and (d) at 500 °C, if the total pressure is 100 bars at all times and the hydrogen and nitrogen partial pressures are in the ratio 3:1. Assume ideal gas behavior and that ΔC_p° is independent of temperature, and note that, if the partial pressure of nitrogen is x, those of hydrogen and ammonia will be $3x$ and $(100 - 4x)$ bars, respectively. Problem (c) can be solved using an approximation, but (d) is best solved iteratively.
[*Answers:* (a) 5.9 × 10^5 bar^{-2}; (b) 1.098 × 10^{-5} bar^{-2}; (c) 99.4%; (d) 8.9%.]

3.3 From the thermodynamic data of Appendix C, show that the product of the reaction of ammonia gas with oxygen would be nitrogen, rather than nitric oxide, under standard conditions and in the absence of kinetic control by (e.g.) specific catalysis of NO formation by platinum. (Assume the other product to be water *vapor*).
[*Answer:* The free energy changes per mole of NH_3 are −326.4 and −239.8 kJ, respectively, so that the first reaction is strongly favored over the second, but *both* are feasible.]

Chapter 4

Sulfur
and its Compounds

4.1 Elemental Sulfur[1-4]

ELEMENTAL SULFUR occurs naturally in association with volcanic vents and, in Texas and Louisiana, as underground deposits. The latter are mined by injecting air and superheated water, which melts the sulfur and carries it to the surface in the return flow (the *Frasch process*). Most of the world's sulfur, however, now comes as a by-product of the desulfurization of fossil fuels. Thus, Albertan "sour" natural gas, which often contains 30% or more (90%, in some cases) hydrogen sulfide (H_2S), as well as hydrocarbons (mainly methane) and small amounts of CO_2, carbonyl sulfide (COS), and water, is "sweetened" by scrubbing out the H_2S and then converting it to elemental S in the *Claus process*. The Claus process is applicable in any industrial operation that produces H_2S (cf. Section 2.5); it converts this highly toxic gas to the non-toxic, relatively unreactive, and easily transportable solid sulfur.

Hydrogen sulfide is weakly acidic, and so may be removed from natural gas by absorption in a counter-current of an appropriate basic solution, usually aqueous monoethanolamine ("MEA," $HO-CH_2-CH_2-NH_2$) or diethanolamine ("DEA," $HO-CH_2-CH_2-NH-CH_2-CH_2-OH$):

$$HOCH_2CH_2NH_2(aq) + H_2S(g) \rightleftharpoons HOCH_2CH_2NH_3^+(aq) + HS^-(aq). \quad (4.1)$$

The H_2S can be regenerated by stripping the MEA solution with steam, since reaction 4.1 is readily reversible. (This would not be feasible if the absorbant were a strong base such as NaOH.) There is a problem in that CO_2 and COS react slowly with MEA to give oxazolidone and thiooxazolidone, respectively:

$$
\begin{array}{ccc}
\text{H}_2\text{C}\!-\!\text{NH} & \qquad & \text{H}_2\text{C}\!-\!\text{NH} \\
\diagup \qquad \backslash & & \diagup \qquad \backslash \\
\text{H}_2\text{C} \qquad\quad \text{C}\!=\!\text{O} & & \text{H}_2\text{C} \qquad\quad \text{C}\!=\!\text{S} \\
\backslash \quad \diagup & & \backslash \quad \diagup \\
\text{O} & & \text{O}
\end{array}
$$

These must be reconverted to MEA periodically with NaOH. DEA has the advantage of giving an easily regenerable compound with COS. The recovered H_2S is burned partially to SO_2 in a limited air supply in the *front-end furnace*:

$$2H_2S + 3O_2 \rightarrow 2SO_2 + 2H_2O \tag{4.2}$$

and the SO_2 and unchanged H_2S are then made to react:

$$2H_2S + SO_2 \rightarrow 3S + 2H_2O. \tag{4.3}$$

Reaction 4.3 is exothermic, so low temperatures give more complete conversion, but it is then slow. A compromise is to work at a somewhat elevated temperature ($450\,^\circ$C) with an appropriate catalyst (Fe_2O_3 or γ-Al_2O_3). The net reaction is:

$$2H_2S + O_2 \rightarrow 2H_2O + 2S. \tag{4.4}$$

In practice, conversions of up to 98% are achieved; residual H_2S is then burned to SO_2 and (environmental considerations permitting) is usually disposed of up the stack.

Elemental sulfur exhibits complicated allotropy, i.e., it exists in many modifications.[4] The stable, prismatic crystal form at room temperature, α- or orthorhombic sulfur, is built up of stacks of S_8 rings (Section 1.8). If it is heated quickly, it melts at $112.8\,^\circ$C. However, if it is heated *slowly*, it changes to needle-like crystals of β- or monoclinic sulfur, which is the stable form above $95.5\,^\circ$C and which melts at $119\,^\circ$C. Both β-S and the yellow mobile melt (below $160\,^\circ$C) are composed exclusively of S_8 rings. Solids containing S_7, S_9, S_{10}, S_{12} and other rings are known, but all slowly revert to S_8 below $160\,^\circ$C.

Above $160\,^\circ$C, however, molten S becomes increasingly brown and extremely viscous, with maximum viscosity at about $200\,^\circ$C. This is caused by the opening of the S_8 rings and the formation of chains of up to $100\,000$ S atoms. Beyond $200\,^\circ$C, the average chain length decreases and the sulfur becomes somewhat more fluid again. The deep red color above $250\,^\circ$C is due largely to S_3 and S_4 fragments, while above the normal boiling point of $444.60\,^\circ$C the vapor contains mainly S_3, S_4, S_5 and S_7 units. If molten sulfur is poured into ice-water, much of the long-chain structure is "frozen in," and the white, plastic product is known as catena- or λ-sulfur. However, it eventually reverts to α-sulfur. Obviously, these facts have important consequences for the industrial handling of elemental sulfur. It must also be remembered that sulfur can be set on fire in air.

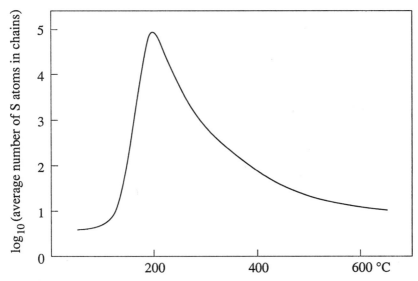

Figure 4.1 Dependence of molecular chain length, and hence viscosity, of molten sulfur on temperature.

Most sulfur (90%) is converted to sulfuric acid. When the price of sulfur is low, as was the case through much of the 1970s, it is often uneconomic to transport it to distant markets. Many large-scale uses for surplus sulfur have been suggested, including "thermopave" (S-containing asphalt), "sulfurcrete" (a concrete made from pebbles and molten sulfur), and foamed sulfur for road-bed insulation in very cold climates.[5]

4.2   Sulfuric Acid[1-3, 6, 7]

Most sulfuric acid is currently produced by burning elemental sulfur from the Claus or Frasch processes in air to obtain sulfur dioxide, catalytically oxidizing this with air to sulfur trioxide, and then hydrolyzing the SO_3 to H_2SO_4. Alternative sources of SO_2 include the roasting of sulfide ores (e.g., pyrite, FeS_2) and the burning of high-S fossil fuels (Section 2.5).

As noted in Section 2.5, there are two important complications in the conversion of SO_2 to liquid H_2SO_4: the oxidation of SO_2 is generally slow and must be catalyzed, while the direct reaction of SO_3 with water tends to produce intractable aerosols (mists) of H_2SO_4. The *lead chamber process*, which dates back to 1746, employs nitrogen oxides as the catalyst; the intermediate $HO-SO_2-O-NO$, or "nitrosylsulfuric acid," is formed and is easily hydrolyzed to liquid (or aqueous) H_2SO_4 and nitrogen oxides, which are recycled. The details of the chemistry remain obscure even today, and plant operation is said to be more of an art than a science.[2]

At present, most sulfuric acid is made by the *contact process*, which has been in use since 1831. In this, the exothermic air-oxidation of SO_2 is catalyzed by vanadium pentoxide (V_2O_5) or platinum, the yield of SO_3 being limited on the first pass to some 60% because the temperature rises to 600 °C or more. Usually, three more passes over the catalyst are made, and the yield can be increased to 98%. The SO_3 vapor is then absorbed into 100% H_2SO_4, and water is added to the resulting mixture of disulfuric ($H_2S_2O_7$) and sulfuric acids (known as *oleum*) until the $H_2S_2O_7$ is all hydrolyzed to H_2SO_4. This obviates the aerosol problem.

$$2SO_2 + O_2 \underset{V_2O_5}{\overset{Pt \ or}{\rightleftharpoons}} 2SO_3 \qquad (4.5)$$

$$SO_3 + O{=}\overset{\displaystyle OH}{\underset{\displaystyle O}{\overset{|}{\underset{\|}{S}}}}{-}OH \rightarrow O{=}\overset{\displaystyle OH}{\underset{\displaystyle O}{\overset{|}{\underset{\|}{S}}}}{-}O{-}\overset{\displaystyle OH}{\underset{\displaystyle O}{\overset{|}{\underset{\|}{S}}}}{=}O \qquad (4.6)$$

$$H_2S_2O_7 + H_2O \rightarrow 2H_2SO_4 \qquad (4.7)$$

Concentrated sulfuric acid is a powerful protonating agent, and oleum is even more so. Oleum is used to catalyze some important petrochemical reactions that proceed by formation of a carbocation intermediate (Section 8.3). This protonating power of concentrated H_2SO_4 results in it being a powerful desiccant; it absorbs water avidly from the air, and chars paper or sugar by extracting the *elements* of water from it (paper and sugar are carbohydrates, i.e., have the *empirical* formula $C_x(H_2O)_y$).

4.3   Sulfur Chemicals in the Pulp and Paper Industry[1, 2, 8, 9]

Although the finest quality paper usually has a substantial rag content, most paper is made from wood. Wood consists of three main components: (*a*) cellulose fibers, which are polysaccharides, i.e., polymerized sugars that hydrolyze in aqueous acid to simple sugars like glucose; (*b*) lignin, which binds the fibers together and is of variable composition, but contains many linkages of the type:

TABLE 4.1
The Principal Chemical Wood Pulping Processes

Kraft Process	*Sulfite Process*
• alkaline: carbon-steel vessels can be used, though stainless steel is better	• acidic: must use corrosion-resistant (stainless steel) vessels
• gives long fibers, therefore strong paper (kraft = strength in Swedish)	• shorter fibers because of acid hydrolysis; paper is weaker, and residual acid may cause long-term embrittlement
• any kind of wood may be used	• spruce or fir chips only
• product is brown unless bleached	• product is white
• typical use: packaging, grocery bags	• typical use: general white paper

and (*c*) oils, fats, waxes, proteins, etc., inside the cells, which break down to give terpenes, fatty acids, tannins, etc.

Thus, the objective of wood pulping is to break down the lignin structure to give soluble anions such as:

without excessive hydrolysis of cellulose. Although low-grade paper such as newsprint can be made by pulping wood mechanically (*groundwood pulp*), better quality material is produced by chemical pulping, since lignin remaining in paper hastens embrowning and decay. The two chief methods are the *kraft* (or *sulfate*) *process* and the *sulfite process*, the salient features of which are summarized in Table 4.1. The manner in which the chemicals are recycled and the by-products recovered (as much for environmental protection as for economic reasons) provides an instructive example of chemical engineering practice.

The kraft process. Wood chips are digested at about 180 °C and 8 bars in aqueous $NaOH/Na_2S/Na_2CO_3$ for 2 to 5 hours. The lignin is attacked by OH^- and S^{2-} (actually mostly HS^-) to give soluble alcohols, acid anions, and some mercaptans (RSH) and organic sulfides (R_2S), which often cause an exceedingly unpleasant stench in the neighborhood of the kraft mill. However, some 1.5 to 3.2 kg dimethylsulfide per tonne cellulose can usually be recovered from the exhaust gases and may be oxidized in the gas phase with air, using nitric oxide as catalyst, to give the versatile solvent dimethylsulfoxide ("DMSO," $(CH_3)_2S{=}O$):

$$(CH_3)_2S + NO_2 \rightarrow (CH_3)_2SO + NO \qquad (4.8)$$
$$+\tfrac{1}{2}O_2$$

Turpentine may also be recovered from the exhaust gas condensate. The spent aqueous solution, known as *black liquor*, is centrifuged to give *tall oil*, a source of oleic and linoleic acids, which are used in making soaps and greases. The aqueous phase is then evaporated to dryness and the residue is ignited so that the organic content is charred to elemental carbon. To this molten residue, which contains NaOH, *salt cake* (Na_2SO_4) is added. Salt cake is obtainable from natural brines in Texas or Searle's Lake, California, or can be made[2, 8] by heating ordinary salt with concentrated sulfuric acid (cf. K_2SO_4 manufacture, Section 3.7):

$$2NaCl(s) + H_2SO_4(l) \rightarrow 2HCl(g) + Na_2SO_4. \qquad (4.9)$$

The role of the carbon is to *reduce* the sulfate ion of the salt cake to sulfide (cf. pyrometallurgy, Chapter 15). The carbon goes to carbonate rather than to CO_2 because of the alkaline medium:

$$Na_2SO_4 + 2C + 4NaOH(l) \rightarrow 2Na_2CO_3 + Na_2S + 2H_2O. \qquad (4.10)$$

The product is cooled and dissolved in water to give *green liquor*—the green color being due to complexing of traces of iron by sulfide. (This same green material develops around the yolks of overcooked hard-boiled eggs, where slow decomposition of the egg-white proteins gives H_2S.) Addition of hydrated lime ($Ca(OH)_2$, Chapter 5) regenerates NaOH:

$$Ca(OH)_2 + Na_2CO_3(aq) \rightarrow CaCO_3(s)_\downarrow + 2NaOH(aq) \qquad (4.11)$$

by virtue of the poor solubility of $CaCO_3$, which precipitates and is returned to the lime kiln. The solution of NaOH and Na_2S plus some remaining Na_2CO_3 is called *white liquor*—the iron sulfides having been removed by the liming. This is sent back to the digester for the next batch operation.

The brownish pulp is acceptable for making packaging material but requires bleaching for use as writing or printing paper. Chlorine has traditionally been used as a pulp bleach and is readily available on-site in pulp mills that have their own NaOH plants (Section 5.3). However, it tends to chlorinate as well as to oxidize organic compounds in the pulp, producing small but potentially dangerous quantities of *dioxins*. Laboratory tests

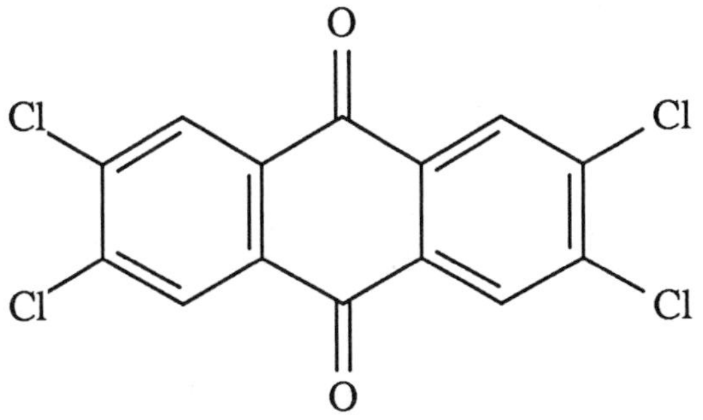

Figure 4.2 Structure of 2,3,7,8-tetrachlorodibenzo-p-dioxin (TCDD).

on animals have demonstrated that 2,3,7,8-tetrachlorodibenzo-p-dioxin (or TCDD), in particular, is an extremely potent toxin and carcinogen. The modern use of chlorine dioxide gas (ClO_2, Section 5.4) followed by further treatment with aqueous "chloride of lime" (crude $Ca(ClO)_2$, made by the reaction of hydrated lime with chlorine) gives lower dioxin levels as well as a better, acid-free paper. An even safer alternative, however, is to use aqueous hydrogen peroxide (H_2O_2) as the bleaching agent. (The pure liquid is too unstable for industrial use.) Using H_2O_2 has obvious advantages since no TCDD can be formed and since the H_2O_2 can itself be disposed of (as necessary) simply by catalyzing its decomposition to water and O_2 (Section 11.5). Hydrogen peroxide is readily made by electrolyzing aqueous H_2SO_4 or by applying the anthraquinone process.[3] No doubt, the hydrogen peroxide method will soon largely replace the others. The paper is then "filled" with a white opaque solid (usually titanium dioxide, Section 15.8) to give a durable white paper.

The sulfite process. In countries where supplies of soft white conifer woods are abundant (e.g., Canada), the sulfite process is widely used, whereas elsewhere, the kraft method predominates. The chemistry involves the attack of sulfurous acid (H_2SO_3) on $C{=}C$ double bonds and ketone groups in lignin to form soluble sulfonates. Since the acidic reducing solutions cause severe corrosion of ordinary steels, stainless steel fittings and digesters lined with acid-resistant brick are needed. The solutions also cause some undesirable hydrolysis of the cellulose fibers. It is therefore usual to use a buffer solution of HSO_3^- and H_2SO_3 in 3:1 ratio rather than straight aqueous H_2SO_3. This keeps the hydrogen ion concentration down (see Exercise 4.1) and minimizes the volume of solution by increasing the solubility of SO_2.

The sequence is as follows: (*a*) Sulfur dioxide from burning sulfur or roasting pyrite (FeS_2) is passed over wet limestone ($CaCO_3$) until a total of 7% SO_2 has dissolved:

$$SO_2(g) + H_2O(l) \rightarrow H_2SO_3(aq) \xrightarrow{CaCO_3} Ca(HSO_3)_2(aq) + CO_2(g). \quad (4.12)$$

(*b*) The resulting liquor is used to digest the wood chips. Live steam may be injected at 7 bars to give a temperature of 150 °C over a 6 to 12-hour period. Alternatively, the solution may be heated indirectly to ≈130 °C with steam coils for about 24 hours, giving a somewhat stronger pulp. (*c*) Recycling the chemicals involves recovery of SO_2 from the relief gases and evaporation of the liquor to a syrupy consistency before addition of slaked lime to precipitate $CaSO_3$, which is returned to the digester. The dissolved calcium lignin-sulfonates can be worked up to give useful products such as vanillin, but usually the chief concern is environmental protection. Dimethylsulfide can also be recovered from the spent liquor and converted to DMSO as in reaction 4.8.

Sulfite paper has a relatively short life span, since residual acid will continue to hydrolyze the cellulose and cause embrittlement. Further sources of acid include aluminum sulfate (which is added together with resin to suppress the bleeding or feathering of ink into the paper) and SO_2 and NO_x from the atmosphere. Much of the world's library collections and archives will soon be lost as the paper crumbles. Various deacidification treatments (e.g., with ammonia, morpholine, cyclohexylamine carbamate, or diethylzinc) have been proposed and tried, but at best they can only halt the process of embrittlement—they cannot reverse it.[10, 11]

References

1. G. T. Austin, "Shreve's Chemical Process Industries," 5th edn., McGraw-Hill: New York, 1985.

2. J. A. Kent (ed.), "Riegel's Handbook of Industrial Chemistry," 8th edn., Van Nostrand-Reinhold: New York, 1983.

3. W. Büchner, R. Schliebs, G. Winter, and K. H. Büchel, "Industrial Inorganic Chemistry," VCH Publishers: New York, 1989 (trans. D. R. Terrell).

4. F. A. Cotton and G. Wilkinson, "Advanced Inorganic Chemistry," 5th edn., Wiley-Interscience: New York, 1988.

5. T. W. Swaddle, "Sulfur utilization—a challenge to Canadians," *Chemistry in Canada*, 1974, *26*, 22–24.

6. R. Thompson (ed.), "The Modern Inorganic Chemicals Industry," Special Publication No. 31, The Chemical Society: London, 1977.

7. J. A. Fergusson, "Inorganic Chemistry and the Earth," Pergamon Press: Oxford, 1982, p. 180.

8. F. A. Lowenheim and M. K. Moran, "Faith, Keyes and Clark's Industrial Chemicals," 4th edn., Wiley-Interscience: New York, 1975.

9. K. Goel, "Chemicals from spent sulfite liquor," *Canadian Chemical News*, April, 1987, 9–11.

10. (*a*) C. J. Shanhani and W. K. Wilson, "Preservation of libraries and archives," *American Scientist*, 1987, *75*, 240–251. (*b*) M. Sun, "The big problem of brittle books," *Science*, 1988, *240*, 598–600.

Exercises

4.1 In the sulfite pulp process, the digestion fluid is typically made by dissolving sulfur dioxide in water in the presence of limestone until the solution contains 7.0% by weight SO_2, three-quarters of which is in the form of bisulfite ion. Why is this done? In support of your answer to this question, calculate the pH of (*a*) the digestion solution prepared as above, (*b*) a hypothetical 7% solution of SO_2 alone, and (*c*) an actual saturated solution of SO_2 alone (2.9% SO_2), if the first acid dissociation constant of sulfurous acid is 1.7×10^{-2}, at ambient temperature and pressure.

[*Answers:* (*a*) 2.25; (*b*) 0.88; (*c*) 1.09.]

4.2 On the basis of the following information, estimate the equilibrium constant for the formation of sulfur trioxide from the dioxide and oxygen in the gas phase (*a*) at 25 °C, and (*b*) at 800 °C. (*c*) What assumption(s) have you had to make in answering parts (*a*) and (*b*)?

	ΔH_f (kJ mol^{-1})	$S°$ (J K^{-1} mol^{-1})
$SO_2(g)$	−296.830	248.22
$SO_3(g)$	−395.72	256.76
$O_2(g)$	0	205.138

[*Answers:* (*a*) 2.59×10^{12} bar$^{-1/2}$; (*b*) 0.80 bar$^{-1/2}$.]

Chapter 5

Alkalis and Halogens

INDUSTRIAL alkalis (excluding the weak alkali ammonia, Section 3.2) and the halogens may be considered together since the manufacture of caustic soda (sodium hydroxide) from salt inevitably gives chlorine, too.[1-6] Furthermore, the chemistry of chlorine and the other halogens in alkaline media is also of practical importance. However, for many industrial purposes, the cheapest source of alkali is lime (calcium oxide), which is used in steel-making and other metallurgical operations ($\approx$45% of U.S. production of lime), in pollution control (Chapter 2), in water treatment (Section 3.6; Chapter 12), in pulp and paper production (Section 4.3), in the reduction of soil acidity, in cement and concrete manufacture, and in many chemical processes.

5.1 Lime Burning

Like chalk, limestone is largely calcium carbonate and dissociates reversibly to lime and carbon dioxide at sufficiently high temperatures (*calcination*):

$$CaCO_3(s) \rightleftharpoons CaO(s) + CO_2(g). \tag{5.1}$$

This reaction is endothermic, but the temperature is not usually allowed to exceed 1100 K, despite incompletion of reaction, since an unreactive modification of lime would be formed. (This is a common occurrence when inorganic solids are strongly heated.) Fortunately, removal of the gaseous product helps drive the reaction to the right. The carbon dioxide may be recovered along with the lime for on-site use in (e.g.) the *Solvay process* for Na_2CO_3 (Section 5.2), but usually it is sent up the stack. In any event, the stack gases should be freed of particulate matter by electrostatic precipitation or scrubbing with water. A pure limestone source is important. The ubiquitous silica (SiO_2) or silicate impurities form molten calcium silicate ($CaSiO_3$) in the kiln, and this may prevent free flow of the lime.

There is a tendency for magnesium ion to substitute for Ca^{2+} in minerals, and magnesium calcium carbonate containing equal numbers of Mg and Ca ions is a readily recognizable distinct mineral (*dolomite*, $(Ca,Mg)CO_3$). The value of lime, however, lies largely in the moderate solubility (giving ≈ 0.05 mol OH^- per liter) of $Ca(OH)_2$, known as *hydrated lime* in the U.S.A. and *slaked lime* in the U.K.:

$$CaO(s) + H_2O(l) \rightarrow Ca(OH)_2(s) \tag{5.2}$$

$$Ca(OH)_2(s) \rightleftharpoons Ca^{2+}(aq) + 2OH^-(aq). \tag{5.3}$$

Unlike $Ca(OH)_2$, $Mg(OH)_2$ is poorly soluble (1.6×10^{-4} mol L^{-1}). However, this can be turned to advantage if $Mg(OH)_2$ is required for reconversion by heating to magnesium oxide (*magnesia*), which is used as a refractory (i.e., high temperature resistant) structural material (mp 2800 °C). In particular, magnesium may be recovered from seawater (0.13% Mg^{2+}), if it is treated with calcined dolomite, $(Ca,Mg)O$. The $Mg(OH)_2$ that forms from the $(Ca,Mg)O$ remains undissolved, but the $Ca(OH)_2$ so produced dissolves and causes precipitation of the $Mg^{2+}(aq)$ as further $Mg(OH)_2$. Not all commercial magnesia is made by this route, however; if the mineral magnesite ($MgCO_3$) is available in sufficient purity, it can readily be calcined to give MgO and CO_2.

Reaction 5.2 is strongly exothermic; water poured onto fresh lime (*quicklime*) may boil. Lime is therefore often used as a water-removing agent (desiccant) for gases or organic solvents.

Lime kilns are frequently associated with cement-making.[7] The lime-sand cements in use since Roman times gain their mechanical strength from the slow reaction of $Ca(OH)_2$ with the CO_2 of the air to form interlocking crystals of $CaCO_3$. The sand acts primarily as a matrix around which this process occurs.

Portland cement, however, sets from within. It was discovered in 1824, and is made by firing limestone ($\approx 75\%$) with clays at ≈ 1400 °C to form a clinker, which is then powdered. This material consists of various anhydrous calcium silicates (Ca_3SiO_5, β-Ca_2SiO_4) and aluminates ($Ca_3Al_2O_6$), which react exothermically with water by a complicated mechanism[7b, c] to form a hard mass of hydrated compounds. It is usually necessary to add $CaSO_4$ (gypsum or anhydrite) to the powder to prevent it from setting too quickly when moistened. The tensile strength of such cements is limited by the presence of relatively large pores (0.1–1.0 mm). Modern research has produced cements that are substantially free of these (*macro-defect-free* or *MDF* cements) and have very promising mechanical properties, e.g., they can be formed into cement springs.[7b, c]

5.2 Soda Ash

Sodium carbonate is a widely used source of mild alkali, either in its hydrated form of big, glassy $Na_2CO_3 \cdot 10H_2O$ crystals ("washing soda") or as the powdery anhydrous solid Na_2CO_3 ("soda ash"). Most North American soda ash is now derived from natural alkaline brines (California) or from underground deposits of "trona" (Wyoming):

$$2(Na_2CO_3 \cdot NaHCO_3 \cdot 2H_2O) \xrightarrow{\text{heat}} 3Na_2CO_3 + CO_2 + 3H_2O. \qquad (5.4)$$

Formerly, most sodium carbonate was made by the famous *Solvay process*, which has been in use since 1869 but is no longer competitive with trona. Nevertheless, it is worth consideration as a classic example of chemical engineering practice. The *net* Solvay reaction is:

$$2NaCl + CaCO_3 \xrightarrow{NH_3 (aq)} Na_2CO_3 + CaCl_2 \qquad (5.5)$$

but it obviously does not proceed directly according to reaction 5.5. Several steps are involved in which ammonia (which was an expensive by-product of coal distillation before the advent of the Haber process, Section 3.2) is used and recycled, as follows:

(*a*) *Ammonia absorption.* NH_3 in the recycled gases is dissolved in brine until a 6 mol NH_3 L^{-1} solution is obtained. Since this is an exothermic process, cooling is necessary.

(*b*) *Carbonation.* CO_2 from the on-site lime kiln (Section 5.1) is absorbed in the ammoniated brine:

$$NaCl + NH_3 + CO_2 + H_2O \xrightarrow{aq} NaHCO_{3\downarrow} + NH_4Cl. \qquad (5.6)$$

Reaction 5.6 is exothermic as well, so cooling to $20\,°C$ is necessary. The precipitated sodium bicarbonate is filtered. (The reaction is stopped at about 75% completion; otherwise, NH_4HCO_3 is also precipitated.)

(*c*) *Calcination.* The $NaHCO_3$ product is heated at $\approx150\,°C$ to form soda ash:

$$2NaHCO_3(s) \xrightarrow{150\,°C} Na_2CO_3(s) + H_2O(g) + CO_2(g). \qquad (5.7)$$

(*d*) *Ammonia recovery.* The solution phase ("mother liquor") is heated to expel excess CO_2, and then lime from the kiln is added to regenerate the ammonia for recycling:

$$2NH_4Cl(aq) + Ca(OH)_2 \rightarrow CaCl_2(aq) + 2NH_3(g) + 2H_2O(l). \quad (5.8)$$

Thus, the process consumes only brine (aqueous NaCl) and limestone (which are cheap) and energy. The only waste product is calcium chloride, which can be used for de-icing roads. Inevitably, there are complications; in particular, the plant design must provide for the elimination of solid contaminants such as clay minerals (from the limestone) and $CaSO_4$ (from the sulfate ion usually present in brines).

5.3   Caustic Soda: the Chloralkali Industry[1-6]

The annual production of caustic soda (sodium hydroxide) in the U.S.A. is about 10 million tonnes, of which 50% is consumed by the chemical industry and a further 20% by pulp and paper plants. Sodium hydroxide is made by electrolyzing strong brine (i.e., decomposing it with an electric current). Saturated brine contains about 360 g or 6.2 mol NaCl per kg water. The other products (both useful, but usually sold barely above cost) are gaseous chlorine and hydrogen. The essential feature of any *chloralkali cell* is the separation of the anode reaction (where chloride ion is oxidized to chlorine) from the cathode reaction (in which OH^- and H_2 are the end products):

$$Cl^-(aq) \longrightarrow \tfrac{1}{2}Cl_2(g) + e^- \tag{5.9}$$

$$Na^+(aq) + e^- [\rightarrow Na] \xrightarrow{H_2O} Na^+(aq) + OH^-(aq) + \tfrac{1}{2}H_2(g). \tag{5.10}$$

The principal types of chloralkali cell currently in use are the *diaphragm* (or *membrane) cell* and the *mercury cell.*

(a) *The diaphragm cell.* The anode and cathode compartments are separated by a porous diaphragm. Formerly, the diaphragm was made of asbestos, but now, other materials, specially created for chloralkali electrolysis, have been introduced.

In the obsolescent asbestos diaphragm cell, the product on the cathode side is typically 11% in NaOH and 16% NaCl (i.e., about 2.7 mol of each per kg of solution). Evaporation to about 50% NaOH causes most of the NaCl to crystallize out, leaving about 1% NaCl in solution; this is pure enough for many industrial uses. Be this as it may, the Si—O links in the asbestos are attacked by the alkali (Section 1.8) and the diaphragm soon deteriorates.

The new diaphragm materials are mostly membranes made of highly alkali-resistant fluorocarbon polymers (Section 5.5) in which cation-exchanging functional groups such as sulfonate have been incorporated.[8] One such material is "Nafion" (Du Pont tradename):

$$\cdots - CF - CF_2 - \cdots \quad \} \text{ inert fluorocarbon "backbone"}$$

$$\begin{array}{c} | \\ O \\ | \\ CF_2 \\ | \\ CF - CF_3 \\ | \\ O \\ | \\ CF_2 - CF_2 - SO_3^- \quad Na^+ \text{ (exchangeable)} \end{array}$$

Because the cations are readily exchangeable for others in such materials, these membranes allow rather free passage of Na^+ from anode to cathode compartments to match current flow in the external circuit. However, since OH^- or Cl^- penetration is negligible, substantially pure NaOH solution can be made in a *membrane cell.*

(b) *The mercury cell.* These devices depend upon the ability of mercury metal (which is liquid at ambient temperatures) to dissolve sodium metal to form an *amalgam* (sodium amalgam, Na/Hg), which is much less reactive towards water than is metallic sodium. Thus, the elemental sodium, which is arguably an intermediate in reaction 5.10, can be trapped as Na/Hg, transported out of the electrolysis cell (the dilute amalgam is also free-flowing), and made to react with pure,

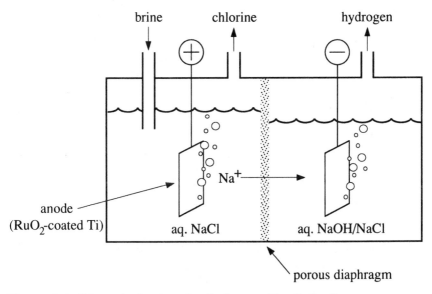

Figure 5.1 Schematic drawing of a diaphragm chloralkali cell.

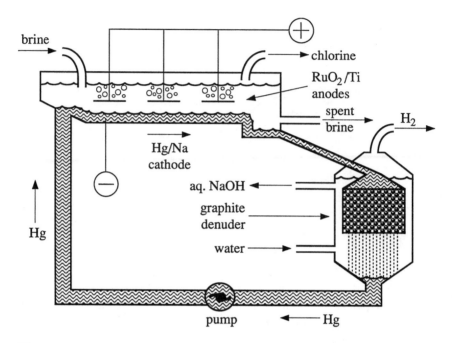

Figure 5.2 Mercury chloroalkali cell.

chloride-free water in a separate vessel to give pure aqueous NaOH and hydrogen gas. A graphite *denuder* is used to accelerate this last process:

At the anodes:

$$Cl^- \rightarrow \tfrac{1}{2}Cl_2 + e^- \tag{5.11}$$
$$\text{(some } O_2 \text{ is also liberated)}$$

At the cathode:

$$Na^+(aq) + Hg(l) + e^- \rightarrow Na/Hg(l) \tag{5.12}$$

At the denuder:

$$Na/Hg(l) + H_2O(l) \rightarrow NaOH(aq) + Hg(l) + \tfrac{1}{2}H_2(g). \tag{5.13}$$

The mercury cell thus gives very pure NaOH and, in terms of energy consumption, is also more economical than the diaphragm cell. However, the inevitable leakage of some mercury into local rivers or lakes can have (and has had) serious consequences because of the bacterial conversion of Hg to methylmercury ion, CH_3Hg^+, which then becomes concentrated in successive steps of the food chain:

$$Hg \xrightarrow[\text{sediments}]{\text{aquatic}} CH_3Hg^+ \rightarrow fish \rightarrow humans.$$

Some oceanic fish, notably swordfish and tuna, tend to accumulate significant mercury levels from *natural* sources in this way, but there is no evidence that moderate consumption of these fish poses any toxicological hazard to humans. However, mercury contamination from *industrial* sources can reach levels high enough to pose a risk to health. In 1953, forty-three people in Minamata, Japan died, and many more suffered permanent disability, after consuming fish contaminated with methylmercury of industrial origin. Methylmercury poisoning is now commonly called *Minamata disease*. In Canada, the White Dog and Grassy Narrows Indian bands have been deprived of their traditional fish supply because of mercury contamination of the English and Wabigon Rivers system by pulp mills. It will be recalled (Section 4.3) that pulp mills consume NaOH and often use chlorine-derived products for bleaching.

Even without bacterial conversion to soluble CH_3Hg^+, elemental mercury itself is very toxic. Since it is a cumulative poison, repeated exposure to the vapor of metallic mercury (low though the vapor pressure is) can have serious, potentially fatal, long-term consequences. Mental derangement is among the symptoms of metallic mercury poisoning.* Metallic mercury poisoning and Minamata disease are extremely serious problems, and the industrial use of mercury must be tightly controlled or, where alternative technology exists, eliminated.

5.4 Chlorine and its Oxoacids

Two-thirds of the chlorine produced in North America is consumed by the organic chemicals industry (25% goes into ethylene dichloride production alone). Pulp and paper mills account for another 15%, while 5% of the total is used in water treatment. Almost all elemental chlorine is made as a by-product of caustic soda production, although the obsolete *Deacon process* has been revived (with improvements) for recycling Cl_2 on-site in plants where chlorination of hydrocarbons produces HCl:

$$2HCl(g) + \tfrac{1}{2}O_2(g) \rightleftharpoons Cl_2(g) + H_2O(g). \qquad (5.14)$$

The reaction is exothermic (see Exercise 5.4), but, since it is very slow, a catalyst is necessary. Nitric oxide, once again, serves as an oxygen carrier, as in the lead chamber process (Section 4.2) and in reaction 4.8. Even so,

---

*The expression "mad as a hatter" may have had its origins in the practice of using mercury to finish the once-fashionable beaver hats. Perhaps the "mad scientist" stereotype has similar origins, since mercury was indispensable in alchemy and in early chemistry and physics. Even Sir Isaac Newton, in the middle of his career in 1692–93, suffered a bout of psychosis which has been attributed to mercury poisoning from his alchemical work. (He had a predilection for tasting the products of his experiments.)[9]

at the elevated temperatures required, the reaction needs to be forced to completion by absorption of the steam in concentrated sulfuric acid or some other desiccant.

Chlorine dissolves in water to undergo rapid *disproportionation* (i.e., it oxidizes and reduces itself simultaneously) to hypochlorous acid and hydrochloric acid with an equilibrium constant of 5×10^{-4} mol^2 L^{-2}:

$$Cl_2(aq) + H_2O(l) \rightleftharpoons HOCl(aq) + H^+ + Cl^-. \tag{5.15}$$
$$Cl(0) \qquad\qquad\qquad Cl(I) \qquad\qquad Cl(-I)$$

Hypochlorous acid is a weak acid ($pK_a = -\log_{10} K_a = 7.4$) and decomposes slowly to give oxygen, even in dilute solution. It cannot be isolated in the free state:

$$HOCl \rightarrow \tfrac{1}{2}O_2 + H^+ + Cl^-. \tag{5.16}$$

Since hypochlorous acid is a powerful disinfectant, chlorine is widely used to render municipal water safe for drinking. If chlorine is dissolved in *cold, dilute* aqueous NaOH, relatively stable solutions of the hypochlorite ion are obtained:

$$Cl_2 + 2OH^- \rightarrow OCl^- + Cl^-. \tag{5.17}$$

These solutions are marketed as a household bleach and disinfectant ("Javex" and "Clorox," for example). However, if chlorine is passed into *hot, concentrated* alkali, the *chlorate* ion $ClO_3{}^-$ is formed:

$$3Cl_2 + 6OH^- \rightarrow 5Cl^- + ClO_3{}^- + 3H_2O. \tag{5.18}$$
$$6Cl(0) \qquad\qquad 5Cl(-I) \quad Cl(V)$$

Most chlorate salts are made in industry by direct electrolysis of acidified, concentrated brines at 40 to 45 °C in cells designed to give good mixing of the anode and cathode reaction products (contrast Section 5.3). Chlorate ion is the conjugate base of chloric acid, a moderately strong acid which cannot be isolated from aqueous solution; attempts to make it directly (e.g., by treating solid KClO$_3$ with concentrated H$_2$SO$_4$) produce gaseous chlorine oxides and may result in a violent explosion. Since chlorates are strong oxidizing agents, they are widely used in fireworks[10] and as unselective herbicides. However, the main (and rapidly expanding) use of NaClO$_3$ is as a source of chlorine dioxide for bleaching paper pulp (Section 4.3) or for sterilizing water (Section 12.7):

$$2NaClO_3 + SO_2 + H_2SO_4(aq) \rightarrow 2ClO_2(g) + 2NaHSO_4(aq). \tag{5.19}$$

This is an example of the use of SO$_2$ as a mild reductant; more powerful reductants would presumably give Cl$^-$. The orange-yellow gas ClO$_2$ is explosive in high concentrations. It should be made only where and when required and should be diluted with N$_2$ or CO$_2$ for safe handling.

Further electrolysis of chlorate solutions gives the perchlorate anion ClO_4^-, which is the anion of perchloric acid, the strongest common acid. The anion also has a negligible tendency to form complexes with metal ions in solution and so is a great favorite of physical-inorganic chemists who may want to adjust the acidity or ionic strength (Section 11.4) of a solution without introducing fresh complications. Solid perchlorates such as NH_4ClO_4 are used as oxidizers in solid rocket propellants. (The oxidized fuel is typically powdered aluminum.) However, they present a serious explosion hazard, as was demonstrated dramatically by the NH_4ClO_4 explosion in Henderson, Nevada (Section 1.9). All perchlorates, solid or liquid, are potentially dangerous oxidants. This is especially true of *concentrated* (72%) and, worse still, of 100% $HClO_4$. It is essential to keep perchlorates and organic or other oxidizable matter apart, unless one knows exactly what one is doing. For example, a drop of concentrated $HClO_4$ produces a resounding explosion with the commonly used solvent DMSO (Section 4.3).

5.5   Fluorine and its Compounds

Fluorine is less familiar than chlorine; yet, it has the greater crustal abundance, occurring as fluorite or fluorspar (CaF_2), cryolite (Na_3AlF_6), etc. It tends to replace the isoelectronic ion OH^- in solids. For example, apatite, $Ca_5(PO_4)_3OH$, the chief constituent of tooth enamel, reacts slowly with aqueous fluoride to form fluoroapatite, $Ca_5(PO_4)_3(F,OH)$, which is harder and more resistant to tooth decay. The dental profession generally considers about 1 ppm natural or added F^- to be desirable in drinking water, but too much (> 3 ppm) F^- results in mottled teeth and bone sclerosis. Concentrations in the parts-per-thousand range can cause acute toxic symptoms such as nausea—proof that one can indeed have too much of a good thing. Controversy over the benefits of artificial fluoridation of municipal water supplies has been extraordinarily emotional and shows no sign of abating.[11]

The chemistry of fluorine is dominated by its electronegativity, which is the highest of all elements. The colorless gas F_2 has an estimated standard electrode potential (Chapter 13) of +2.85 V for reduction to F^- (cf. +1.36 V for Cl_2 to Cl^-), and thus immediately oxidizes water to oxygen ($E^\circ = +1.23$ V), and 2% aqueous NaOH to the gas F_2O. Obviously, F_2 cannot be made by electrolysis of aqueous NaF, and the usual preparation involves electrolysis of HF—KF melts in a Monel (Cu—Ni alloy) or copper apparatus.

Hydrocarbons inflame spontaneously in F_2, giving fluorocarbons and HF. Controlled fluorination of organic compounds is possible by electrolyzing them in liquid HF (bp 20 °C) at a nickel anode, at voltages just below that needed to evolve F_2:

$$CH_3COOH \xrightarrow[\text{electrolyze}]{HF(l)} CF_3COOF \qquad (5.20)$$

$$C_8H_{18} \longrightarrow C_8F_{18}. \qquad (5.21)$$

Perfluoroalkanes (i.e., alkanes in which all H atoms are replaced by F) have high thermal stability and resistance to oxidation and hydrolysis. This is because F_2 is itself a better oxidant than O_2, while the high strength of the C—F bond makes it hard to hydrolyze, despite its polarity (see Exercise 5.5).

Fluorocarbon oils are expensive but are invaluable where severe conditions demand them. The now familiar solid *polytetrafluoroethylene* ("PTFE," Du Pont's "Teflon"; $F(CF_2CF_2)_nF$) has an excellent thermal range, is inert to almost all chemical reagents, and has little tendency to adhere to other materials. In the following synthetic sequence, the catalyst is typically antimony trifluoride, SbF_3:

$$HF(g) + CHCl_3(g) \xrightarrow{\text{catalyst}} CHClF_2 \xrightarrow[1000\,°C]{500\text{ to}} C_2F_4 \qquad (5.22)$$

$$C_2F_4 \xrightarrow[\text{or RO}\dot{O}]{\text{thermal,}} \cdots -CF_2-CF_2-\cdots \qquad (5.23)$$

The intermediate $CHClF_2$ is one of a group of low-boiling chlorofluorocarbons ("CFC"; "Freon" is the Du Pont tradename)[12] which are important as refrigerant fluids, as blowing agents for plastic foams, and in cleaning soldering flux residue from electronic components. They have also been used as propellants in spray-cans. The inertness of these fluorocarbons would seem to make them innocuous as atmospheric pollutants, but their very longevity allows them to accumulate to the point where they may threaten the ozone layer through photochemical release of atomic chlorine (Section 2.3, reactions 2.4 through 2.7). Some governments have therefore banned the use of chlorofluorocarbon spray propellants, and following the Montreal Protocol of 1988 there has been an international effort to phase out CFCs entirely.[12] The lesson to be learned is that *inert* does not necessarily mean *harmless;* no product or by-product of the chemical industry can be released into the environment without careful consideration of the consequences.

Fluorocarbons containing hydrogen but no chlorine, notably CF_3-CH_2F (HFC-134a),[†] pose no threat to the ozone layer and are possible

---

[†] "HFC" stands for hydrofluorocarbon (no Cl content). Sometimes, CFCs that contain hydrogen are termed "HCFCs".

The number code for CFCs and HCFCs works as follows: *First digit:* the number of C atoms *minus one* (blank if zero). *Second digit:* the number of H atoms *plus one.* *Third digit:* the number of F atoms. (The number of Cl atoms is not explicitly stated, but replacements for the $CFCl_3$ (CFC-11) and CF_2Cl_2 (CFC-12), which are now widely used as refrigerants. Unfortunately, whereas CFC-12 (the most commonly used refrigerant) is easily made from CCl_4 and HF, HFC-134a requires a costly multistep synthesis.[12] Furthermore, if a change to HFC-134a is made, existing refrigeration units will have to be modified to operate at higher pressures and will require different lubricants. The increased costs will meet with resistance, especially in many under-developed countries.

An interim solution may be the use of CFCs containing some hydrogen (HCFCs), as these compounds break down much more rapidly in the troposphere than do H-free CFCs and so deliver less chlorine to the stratosphere. Examples are $CHCl_2CF_3$ (HCFC-123) and CH_3CCl_2F (HCFC-141b). Nevertheless, *some* ozone layer damage would result from the use of HCFCs, and *all* these fluorocarbons (HFCs included) will contribute to a different atmospheric problem—the greenhouse effect (Section 2.1).

Fluorine is used in the nuclear industries of many countries to make *uranium hexafluoride* for enrichment of uranium in the fissile ^{235}U isotope:

$$UO_2(s) + 4HF(aq) \rightarrow UF_4(s) + 2H_2O \qquad (5.24)$$

$$UF_4(s) + F_2(g) \rightarrow UF_6. \qquad (5.25)$$

UF_6 is a solid (mp 64 °C) but has a high vapor pressure (15.3 kPa at 25 °C). Since ^{19}F is the only stable isotope of fluorine, the only molecular species by mass in UF_6 are $^{235}UF_6$ and $^{238}UF_6$. Repeated diffusion of $UF_6(g)$ through porous plugs (or centrifugation of the vapor) concentrates $^{235}UF_6$ relative to $^{238}UF_6$, since the speed of diffusion varies inversely as the square root of the molecular mass. This enrichment is not needed for fuelling power reactors with good neutron economy, such as the Canadian CANDU system. Gaseous diffusion plants for UF_6 are expensive, not least because UF_6 is itself an aggressive fluorinating agent and thus must be contained in special fluorine-resistant materials.

5.6 Bromine and Iodine

Bromine is a dense, red, volatile, corrosive liquid (bp 59 °C) that is best made by oxidizing the small amount of Br^- in seawater with chlorine. The resulting Br_2 vapor is then carried off in an air stream:

$$Cl_2(g) + 2Br^-(\text{seawater}) \xrightarrow{pH\,3.5} 2Cl^-(aq) + Br_2(l). \qquad (5.26)$$

This illustrates the move toward lower electronegativity (less oxidizing power) as one descends a periodic group. The bromine vapor is trapped in

---

is taken to equal the number of remaining valences.) Thus, $C_2H_2Cl_2F_2$ is CFC-132 (or Freon 132), and all single-carbon CFCs or HFCs are represented by two-digit numbers. Isomers are designated *a*, *b*, etc.

aqueous Na_2CO_3 (in effect, a mild source of alkali) as bromide and bromate ions:

$$3Br_2 + 6OH^-(aq) \rightarrow 5Br^- + BrO_3^- + 3H_2O \qquad (5.27)$$

and is recovered on acidification:

$$BrO_3^- + 5Br^- + 6H^+ \rightarrow 3Br_2 + 3H_2O. \qquad (5.28)$$

A qualitative similarity to the aqueous chemistry of chlorine will be evident. Bromine finds use as a volatile, moderate oxidant.

Solid iodine is purple, as is its vapor, but it is often brown in solution, e.g., in oxygen-containing solvents such as ethanol (*tincture of iodine* antiseptic), or in water, in which its solubility is increased by formation of a complex ion I_3^- with iodide ion. It is obtained by oxidizing the ash of dried seaweeds (kelp), or as sodium iodate, which is present in the $NaNO_3$ deposits of the Atacama desert in Chile. Most of the U.S. production, however, comes from chlorination of natural I^--bearing brines in Michigan (cf. reaction 5.26).

References

1. J. J. Leddy, "The chlor-alkali industry," *Journal of Chemical Education,* 1980, *57,* 640–641.

2. R. Thompson (ed.), "The Modern Inorganic Chemicals Industry," Special Publication No. 31, The Chemical Society: London, 1977.

3. J. A. Kent (ed.), "Riegel's Handbook of Industrial Chemistry," 8th edn., Van Nostrand-Reinhold: New York, 1983.

4. J. E. Fergusson, "Inorganic Chemistry and the Earth," Pergamon Press: Oxford, 1982.

5. G. T. Austin, "Shreve's Chemical Process Industries," 5th edn., McGraw-Hill: New York, 1985.

6. W. Büchner, R. Schliebs, G. Winter, and K. H. Büchel, "Industrial Inorganic Chemistry," VCH Publishers: New York, 1989 (trans. D. R. Terrell).

7. (*a*) C. Hall, "On the history of Portland cement after 150 years," *Journal of Chemical Education,* 1976, *53,* 222–223. (*b*) J. D. Birchall, A. J. Howard, and K. Kendall, "New cements—inorganic plastics of the future," *Chemistry in Britain,* 1982, *18,* 860–863. (*c*) D. M. Roy, "New strong cement materials: chemically bonded ceramics," *Science,* 1987, *235,* 651–658.

8. (*a*) A. Eisenberg and H. L. Yeager (eds.), "Perfluorinated Ionomer Membranes," A.C.S. Symposium Series No. 180, American Chemical Society: Washington, D.C., 1982. (*b*) S. C. Stinson, "Electrolytic cell membrane development surges," *Chemical and Engineering News*, March 15, 1982, 22–25.

9. W. J. Broad, "Sir Isaac Newton: mad as a hatter," *Science*, 1981, *213*, 1341–1344. However, see the letters to the editor: L. J. Goldwater et al., *Science*, 1981, *214*, 742; M. R. Laker et al., *Science*, 1982, *215*, 1185.

10. John A. Conkling, "Chemistry of fireworks," *Chemical and Engineering News*, June 29, 1981, 24.

11. (*a*) B. Hileman, "Fluoridation of water," *Chemical and Engineering News*, August 1, 1988, 26–42. (*b*) "Fluoridation of water," (correspondence) *Chemical and Engineering News*, October 10, 1988, 2–4.

12. (*a*) R. Pool, "The elusive replacements for CFCs," *Science*, 1988, *242*, 666–668. (*b*) P. S. Zurer, "Producers, users grapple with realities of CFC phaseout," *Chemical and Engineering News*, July 24, 1989, 7–13. (*c*) P. S. Zurer, "CFC substitutes: candidates pass early toxicity tests," *Chemical and Engineering News*, October 9, 1989, 4–6. (*d*) S. C. Stinson, "Polyurethane industry tackling CFC reductions," *Chemical and Engineering News*, October 24, 1989, 23–27.

Exercises

5.1 Here are some standard thermochemical data relevant to lime burning (limestone can be taken to be pure, solid $CaCO_3$ in its calcite form):

	Calcite	$CO_2(g)$	CaO(s)	
ΔH_f°	−1206.92	−393.509	−635.09	kJ mol^{-1}
S°	92.9	213.74	39.75	J K^{-1} mol^{-1}
C_p°	81.88	37.11	42.80	J K^{-1} mol^{-1}

(*a*) Write down the algebraic expression for the equilibrium constant K for the formation of solid lime from solid limestone.

(*b*) From the above data, evaluate K for standard conditions (1 bar, pure solid phases).

(*c*) Assume the heat capacity of reaction (ΔC_p°) to be negligible and calculate the approximate temperature at which the CO_2 pressure will reach 1 bar (i.e., the standard pressure) at equilibrium.

(*d*) Evaluate ΔC_p°. In view of the temperature range you have now estimated, was it reasonable to neglect it, for the purposes of (*c*)?

[*Answers:* (*b*) 1.43 × 10^{-23} bar; (*c*) 1110 K, or 837 °C.]

5.2 (a) From the data of Appendix C, show that dolomite is thermodynamically more stable than an equimolar mixture of calcite ($CaCO_3$)
and magnesite ($MgCO_3$) at ambient temperature.

(b) Write balanced equations to show how magnesia may be made
from dolomite and seawater.

5.3 A mercury chloralkali cell was found to require a current of 33.0 kA
in order to produce 1 tonne of chlorine per day.

(a) What is the current efficiency of the cell?

(b) Why is the efficiency less than 100%?

[*Answer:* (a) 95.4%.]

5.4 From the data in Appendix C, calculate (a) the enthalpy of reaction
and (b) the equilibrium constant for reaction 5.14 (the Deacon process) at standard conditions. (c) Above what temperature does the
reaction cease to be thermodynamically favored (i.e., ΔG° becomes
positive), if ΔC_p° can be ignored?

[*Answers:* (a) -57.20 kJ mol^{-1}; (b) 4.49×10^6 bar$^{-1/2}$; (c) 614°C.]

5.5 What do the following mean single bond energies tell us about the
susceptibility of fluorocarbons to hydrolysis?

$$
\begin{array}{ll}
\text{C—F} & 490 \text{ kJ mol}^{-1} \\
\text{O—H} & 459 \\
\text{H—F} & 565 \\
\text{C—O} & 358
\end{array}
$$

Note: HF(aq) is a "weak" acid.

Chapter 6

Ionic Solids

WHEN A metal M of low electronegativity (χ) combines with a non-metal X of high χ, the product is likely to be a solid consisting of ions M^{m+} and X^{x-}, held together in a regular pattern (the crystal lattice) by electrostatic forces, as distinct from electron-sharing bonding (covalency). The energy of these electrostatic interactions—called the *lattice energy*, U—makes the formation of the ionic solid possible by compensating for the energy inputs, such as ionization potential needed to form the ions, and is clearly dependent to some degree upon the structure of the crystal at the atomic or molecular level.[1-3]

6.1   Determination of Crystal Structure[2]

Just as the rulings on a diffraction grating create colored interference patterns in the reflected light, so layers of atoms in a crystal give rise to diffraction patterns in incident radiation of the appropriate wavelength—in this case, X-rays or beams of electrons or neutrons (which also have wave-like properties). When X-rays of wavelength λ are reflected from parallel planes of atoms of spacing d, they will reinforce each other if rays from successive planes arrive at the detector a distance λ (or $n\lambda$, where n is a positive integer) apart; otherwise, they will tend to cancel. As Fig. 6.1 shows, rays reflected from successive planes at an angle Θ will each travel $2d \sin \Theta$ further than their immediate predecessors to reach the detector. Thus, when reinforced X-rays are recorded at the detector, Eqn. 6.1 (the *Bragg equation*) must hold, and, knowing λ and measuring Θ, we can obtain d:

$$n\lambda = 2d \sin \Theta. \qquad (6.1)$$

If a *single crystal* is rotated in a monochromatic X-ray beam, a pattern of spots of reinforced X-rays can be recorded on, say, a photographic

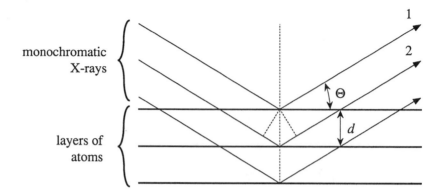

Figure 6.1 Diffraction of X-rays by layers of atoms. The path of ray 2 to the detector is longer than the path of 1 by $2(d \sin \Theta)$.

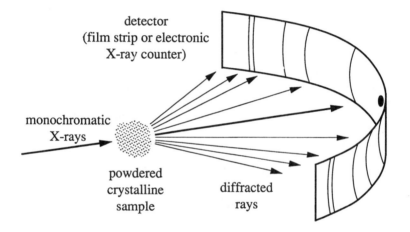

Figure 6.2 Characterization of a powdered solid by its X-ray diffraction pattern.

film placed behind the crystal perpendicular to the primary beam. (These are called *Laue photographs.* Nowadays, X-ray diffractometers use electronic photon counters as detectors.) From the spacing of these spots, the structure of the crystal can be worked out in terms of the planes of atoms present. In the case of crystals containing molecular units, these molecular structures will also show up in the analysis of the diffraction pattern. With the advent of powerful digital computers, such determinations of structure have become almost routine in modern research in synthetic chemistry.

Alternatively, when a *powdered* crystalline solid diffracts monochromatic X-radiation, the diffraction pattern will be a series of concentric rings, rather than spots, because of the random orientation of the crystals in the sample (Fig. 6.2). The structural information in this pattern is limited, but, because even those solid compounds that have the same structure but different composition will almost inevitably have different d values, each individual solid chemical compound will have its own characteristic *powder diffraction pattern*. These patterns are catalogued in the JCPDS data file,[4] and can be used to identify crystalline solids, either as pure phases or as mixtures. For example, a solid deposit accumulating in a heat exchanger can be quickly identified from its X-ray powder diffraction pattern, and its source or mechanism of formation may be deduced—is it a corrosion product (if so, what is it, and where does it come from), or a contaminant introduced with the feedwater?

6.2 Some Common Unit Cell Structures

The *unit cell* is the smallest *complete* repeating unit of a crystal structure. Six simple types of unit cell structures are commonly encountered, although many more exist:

1:1 (MX) types: halite (NaCl), cesium chloride (CsCl), zinc blende (ZnS), wurtzite (also ZnS);

2:1 (MX$_2$) types: fluorite (CaF$_2$), rutile (TiO$_2$).

Sodium chloride (halite) structure (Fig. 6.3). Sodium and chloride ions alternate in three directions at right angles. As Fig. 6.3 is drawn, the shaded ions are Na$^+$ and obviously form a *face-centered cubic* (fcc) array. However, the unshaded ions (nominally Cl$^-$) also form an fcc array, as can be seen by stacking another unit cell on top of the one shown. Thus, the Na$^+$

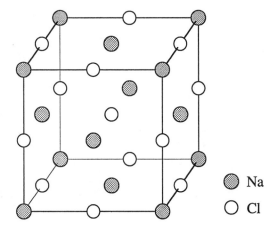

Figure 6.3 Sodium chloride unit cell.

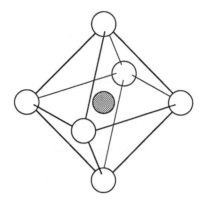

Figure 6.4 Octahedral coordination.

and Cl^- sub-lattices are interchangeable, and we could just as well have
specified the shaded ion to be chloride. The structure could be described
as interpenetrating fcc arrays of anions and cations of equal charge. The
coordination number (number of nearest neighbors) is six for both Na^+
and Cl^-. The centers of these six nearest neighbors trace out a *regular
octahedron* (Fig. 6.4), and we can therefore also describe the structure as
consisting of an fcc array of Cl^- (or Na^+) with a Na^+ (or Cl^-) ion in
all octahedral "holes" (interstices) in that array. (The usefulness of this
approach will become clearer in Section 6.4.)

The contents of the unit cell shown in Fig. 6.3 are:

$$
\left.
\begin{array}{l}
8 \times \text{one-eighth } Na^+ \text{ at each corner} \\
6 \times \text{one-half } Na^+ \text{ on each face}
\end{array}
\right\} \ 4 \text{ whole } Na^+
$$

$$
\left.
\begin{array}{l}
12 \times \text{one-quarter } Cl^- \text{ on each edge} \\
1 \times \text{a whole } Cl^- \text{ at the body center of the cube.}
\end{array}
\right\} \ 4 \text{ whole } Cl^-
$$

Each unit has a total of four NaCl. This is because, with the exception of
the Cl^- at the body center, parts of each ion lie in adjacent unit cells.

The NaCl structure is very common; examples include the halides of Li,
Na, K and Rb; AgCl and AgBr; NiO; MgO; CaO; and PbS.

Cesium chloride (CsCl) *structure (Fig. 6.5).* This can be described as
interpenetrating *simple cubic* arrays of Cs^+ and Cl^-. Again, the Cs^+ and
Cl^- positions are fully interchangeable. The structure is sometimes wrongly
called *body-centered cubic* (bcc). The terminology is appropriate only when
the shaded and unshaded atoms of Fig. 6.5 are identical, as in the room-
temperature form of metallic iron (α-Fe). In any case, the coordination
number is eight for any atom. The unit cell of CsCl contains one net CsCl
unit.

Zinc blende structure (Fig. 6.6). Zinc blende, also called *sphalerite*
(ZnS), has a cubic structure in which we again have interchangeable, inter-

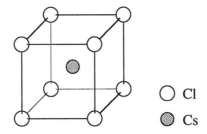

Figure 6.5 Cesium chloride unit cell.

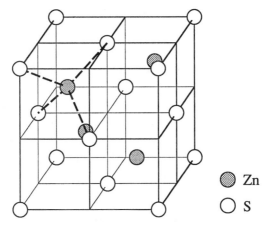

Figure 6.6 Zinc blende unit cell.

penetrating fcc arrays of Zn^{2+} and S^{2-}. If, as in Fig. 6.6, we call the shaded ions zinc and divide the unit cell into eight sub-cubes, we see that the zinc ions occupy the body centers of every *alternate* sub-cube. Furthermore, the coordination number of each ion is now four, and the nuclei of the four nearest neighbors trace out a *regular tetrahedron;* we say that the Zn^{2+} (or the S^{2-}) ions are *tetrahedrally coordinated*. (Compare this with tetrahedral bonding in organic compounds such as CH_4. Indeed, if we make all the shaded and unshaded atoms the same in Fig. 6.6, we have the diamond structure.) Thus, the Zn^{2+} ions occupy one-half of the *tetrahedral holes* in an fcc array of S^{2-} ions, and vice versa. Note that there are *two* tetrahedral (T) and *one* octahedral (O) holes per fcc sub-lattice atom; in zinc blende, the O-holes are all empty.

Wurtzite structure. Zinc sulfide can also crystallize in a hexagonal form called *wurtzite*, which is slightly less exothermically formed than the cubic zinc blende (sphalerite) modification ($\Delta H_f = -192.6$ and -206.0 kJ mol^{-1}, respectively) and hence is a high-temperature *polymorph* of ZnS. The relationship between the two structures is best described in terms of hexagonal

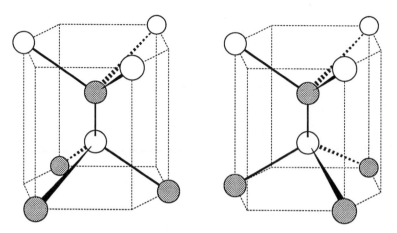

Figure 6.7 Illustrating the difference between zinc blende (left) and wurtzite (right) structures.

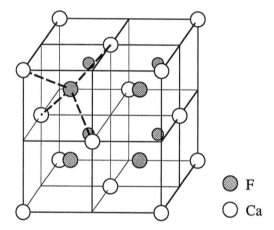

⬤ F

◯ Ca

Figure 6.8 Fluorite unit cell.

versus cubic close packing (Section 6.4) and is illustrated in Fig. 6.7. The actual unit cell of wurtzite is not shown here.

Fluorite (CaF_2) *structure (Fig. 6.8)*. Comparison with Fig. 6.6 shows that fluorite can be described as an fcc array of Ca^{2+} with F^- ions in *all* the tetrahedral holes (forming a simple cubic sub-lattice of fluorides). In this case, the Ca^{2+} and F^- sites are *not* interchangeable. This is to be expected, since we have twice as many F^- as Ca^{2+}; as noted above, there are indeed twice as many T-holes as lattice atoms. The coordination numbers of Ca^{2+} and F^- are eight and four, respectively. Other solids with this structure include the nuclear fuel UO_2.

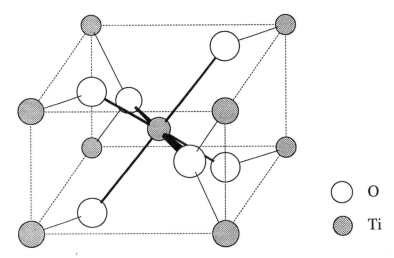

Figure 6.9 Rutile unit cell (tetragonal).

If we make the shaded spheres in Fig. 6.8 the *cations*, and the unshaded the *anions*, we have the *antifluorite* structure, which is typified by lithium oxide (an fcc array of O^{2-} with Li^+ in every T-hole).

Rutile structure (Fig. 6.9). Titanium dioxide occurs naturally as anatase, brookite, and rutile, all of which contain octahedral $TiO_6{}^{8-}$ units. The coordination number of the central Ti^{4+} is very obviously six, and a little thought confirms that the same is true of the Ti^{4+} ions at the corners. That the coordination number of the O^{2-} ions is three is seen from the nearest oxygen at the right. The anatase structure is like NaCl with every other cation missing, and the anions accordingly displaced somewhat. The heats of formation of anatase, brookite, and rutile are -939.7, -941.8, and -944.7 kJ mol^{-1}, respectively, and ΔG_f° follows this trend, so that rutile is the stable low-temperature form. The rutile structure is shared by MnO_2, SnO_2, and most divalent transition metal fluorides.

6.3 Radius Ratio Rules

Consider a set of nearest-neighbor anions surrounding a cation (which will almost inevitably be smaller) in a crystal. The maximum electrostatic attractions will result when we have as many anions as possible surrounding the cation, and in contact with it but not *quite* touching each other (since like charges repel one another). The sketch at the left of Fig. 6.10 shows a case in which anion-anion repulsions would be maximal and anion-cation attractions low. This system will seek to reduce its coordination number. Conversely, in the central sketch in Fig. 6.10, more anions could be accom-

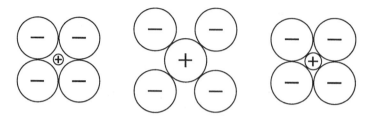

Figure 6.10 Effects of relative sizes of anions and cations in octahedral (six) coordination. Two additional anions, positioned above and below the plane, have been omitted for clarity.

modated to advantage, so the coordination number tends to increase. The limiting case, with all seven ions just in contact, is represented at the right of Fig. 6.10. (There are anions above and below the plane of the paper, making six in all, but these have been omitted for clarity.) This limiting case occurs when the sum $(r_+ + r_-)$ of the radii of the cation and anion is equal to $r_- \sqrt{2}$ (from Pythagoras' theorem), i.e., when the radius ratio $r_+/r_- = (\sqrt{2} - 1) = 0.414$. For a radius ratio less than this, we expect the coordination number to be reduced to four. Proceeding in the same way, we can calculate the ion radius ratios that set the limits between eight and six coordination, and (less importantly) four and three coordination, and so make it possible to anticipate, from tables of ionic radii,[5, 6] which of the basic structures described in Section 6.2 a given binary (i.e., two-element) ionic compound is likely to adopt (Table 6.1).

For example, the ionic radius of Mg^{2+} is usually quoted as 72 pm, and for magnesium oxide, sulfide, and telluride we have:

X^{2-}:	O^{2-}	Te^{2-}	S^{2-}
r_-:	138 pm	221 pm	184 pm
r_+/r_-:	0.47	0.33	0.39
Predicted structure:	NaCl	ZnS	ZnS
Observed structure:	NaCl	ZnS	NaCl.

This example was chosen to show that the radius-ratio structure predictions work quite well *except* near the limiting radius ratio values. Here, r_+/r_- is borderline for MgS, and we predicted the wrong structure. This is hardly surprising, since the concept of fixed ionic radii is not well founded.

First of all, we cannot measure r_+ or r_- directly. As explained in Section 6.1, X-ray diffraction measurements give us internuclear spacings corresponding to $(r_+ + r_-)$ and some non-contact separations, rather than r_+ or r_- themselves. If we can guess any one r_+ or r_-, on the basis of some theory or other, then we can estimate all the others.

Furthermore, ions are *not* hard, billiard-ball-like spheres. Since the wave functions that describe the electronic distribution in an atom or ion do not

TABLE 6.1
Predicted Dependence of Structure Type
Upon Cation/Anion Radius Ratio
for Binary Ionic Compounds

Coordination Number of M^{m+}	r_+/r_- (minimum)	Structure Type MX	Structure Type MX_2
8	0.73	CsCl	CaF_2
6	0.41	NaCl	TiO_2
4	0.22	ZnS	
3	0.15		

suddenly drop to zero amplitude at some particular radius, we must consider the surfaces of our supposedly spherical ions to be somewhat "fuzzy."

A more subtle complication is that the apparent radius of an ion increases (typically by some 6 pm for each increment) whenever the coordination number increases. Shannon[6] has compiled a comprehensive set of ionic radii that take this into account.

Finally, high pressures can induce increases in coordination number. Thus, in the Earth's mantle, four-coordination of silicon by oxygens in enstatite ($Mg_{0.9}Fe_{0.1}SiO_3$) gives way to six-coordination (a change from pyroxene-like to perovskite-type structures; see Sections 6.4 and 8.1) when the overburden of rock exceeds 670 km, at which depth the pressure is 175 kbar. Sodium chloride itself goes over to the CsCl structure at sufficiently high pressures.

6.4   Crystal Structure and the Close Packing of Spheres

Figure 6.11 shows the manner of close packing of spheres of equal radii in a single plane, A. If a second layer B of the same spheres is close packed on top of layer A (Fig. 6.12), it will be seen that each B sphere rests on three A spheres that are in mutual contact, so enclosing a void. The centers of these four spheres describe a regular tetrahedron about the void, which is therefore called a *tetrahedral interstice* or *T-hole*. A second kind of interstice, bounded by six atoms (three from each layer), is also generated, and these are called *octahedral interstices* or *O-holes* (Fig. 6.12). These O- and T-holes are entirely analogous to those described in Section 6.2 (and illustrated in Figs. 6.6 and 6.8). A different (smaller) kind of sphere could be accommodated in the T-holes between the A and B layers, much as the Zn^{2+} are accommodated in the fcc S^{2-} sub-lattice of zinc blende. In this

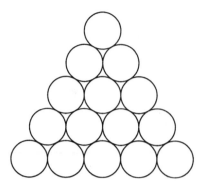

Figure 6.11 Close packing of spheres in a single layer.

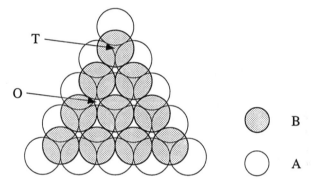

Figure 6.12 Placement of a layer B of close-packed spheres on top of a layer A, so generating octahedral (O) and tetrahedral (T) interstices between the layers.

way, crystal structures can often be represented in terms of the systematic filling of O- and/or T-holes in a close-packed array of ions X by smaller ions (usually cations) M. The X sub-lattice may be somewhat expanded from closest-packed to accommodate M (cf. the radius ratio rules in Section 6.3), but the point is that the essential geometric features of closest packing are frequently present in ionic crystal structures.

When we come to add a third layer C of atoms on top of the two layers, we find there are two close packed possibilities; this can be tested with small discs on Fig. 6.12. Each atom of layer C must rest on three of layer B. One possibility is to place atoms of layer C directly above atoms of layer A. Thus, we create a new layer just like A, and further layers are added to give a sequence ABABAB.... This is known as *hexagonal close packing* (hcp), and is exemplified by the S^{2-} or Zn^{2+} arrays in wurtzite (Fig. 6.7) or the O^{2-} array in rutile. If, however, we place the C atoms in positions directly above the octahedral holes that exist between A and B (such as

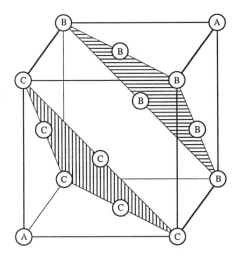

Figure 6.13 Cubic close-packed layers A, B, and C within a face-centered-cubic unit cell.

the one marked "O" in Fig. 6.12), we have a new arrangement. The fourth layer, however, would go above the A atoms (if the same packing sequence is adhered to), so the layer order would be ABCABC.... This is called *cubic close packing* (ccp), and Fig. 6.13 shows that it is identical with the face-centered cubic (fcc) array occurring in the S^{2-} or Zn^{2+} sub-lattices of zinc blende.

Many common ionic crystal structures can therefore be simply described as a ccp or hcp array of one kind of ion with one or more other kinds occupying some or all of the T- or O-holes, usually in a regular way. For example, the α-Al_2O_3 (corundum) structure is an hcp array of O^{2-} with Al^{3+} ions occupying two-thirds of the O-holes, i.e., as we move from O-hole to O-hole in a given direction, every third hole is vacant; α-Fe_2O_3 (hematite), the familiar red oxide of iron, has a similar structure. Table 6.2 summarizes some standard examples. The close-packing approach affords a simple way to describe some important *ternary* (three-element) structures. Thus, the titanium ore ilmenite, $FeTiO_3$, may be regarded as an hcp array of O^{2-} with Fe^{2+} in one-third of the O-holes, Ti^{4+} in a second third, and the remaining O-holes vacant. (Ilmenite can therefore be described as hematite in which half the Fe^{3+} are replaced by Ti^{4+} and the rest by Fe^{2+}.)

The ilmenite structure is common amongst ternary oxides of the general form ABO_3 when the ions A and B are of roughly similar radii. However, if one ion—for instance, B—is much smaller than A, and A is not too much smaller than O^{2-}, the *perovskite structure* (Fig. 6.14) is adopted. This comprises a ccp array of A^{n+} and O^{2-}, while $B^{(6-n)+}$ occupies one-fourth of the

TABLE 6.2

Binary Crystal Structures in Terms of Close-Packing

Structure Type	Close-packed Array	Filling of Interstices
zinc blende	ccp Zn^{2+}	S^{2-} in half of the T-holes
wurtzite	hcp Zn^{2+}	S^{2-} in half of the T-holes
fluorite	ccp Ca^{2+}	F^- in every T-hole
halite	ccp Cl^-	Na^+ in every O-hole
nickel arsenide	hcp As^{3-}	Ni^{3+} in every O-hole
rutile	hcp O^{2-}	Ti^{4+} in half of the O-holes
anatase	ccp[a] O^{2-}	Ti^{4+} in half of the O-holes
corundum	hcp O^{2-}	Al^{3+} in two-thirds of the O-holes

[a] distorted

O-holes; the rest remain vacant. The perovskite structure can be expected to be associated with elements B that tend to high oxidation states, such as Ti^{IV} in the mineral perovskite itself ($CaTiO_3$) or niobium(V) in $NaNbO_3$. It is also common amongst compounds of the type ABX_3, where X is F or (less commonly) Cl, Br, or I, e.g., $KNiF_3$. We noted in Section 6.3 that the normal pyroxene structure of enstatite ($MgSiO_3$) gives way to a perovskite arrangement at around 175 kbar pressure. Several perovskite-type metal oxides, such as $BaTiO_3$ and $Pb(Zr,Ti)O_3$ (lead zirconate titanate or *PZT*), have *ferroelectric* properties, which in practical terms means that a high external electrical field can cause *poling* (i.e., induction of a permanent dipole moment) in polycrystalline ceramics made from them. Field gradients of several thousand volts per cm are required to produce this poling, but the fabrication of such materials as thin films permits the recording of low-voltage electrical signals. These films have important applications in the electronics industry.[7] Finally, the non-stoichiometric (Section 7.3) compounds related to $YBa_2Cu_3O_{7-x}$, which Müller and Bednorz showed in 1986 to be superconducting at unprecedentedly high temperatures for certain ranges of x, have a complicated structure that is nevertheless derived from the perovskite type.[8, 9]

A large group of ternary oxides AB_2O_4 have structures related to that of *spinel*, $MgAl_2O_4$: a ccp array of O^{2-}, with A in one-eighth of the T-holes and B in one-half of the O-holes. If we recall that there are two T-holes and one O-hole per close-packed atom (O^{2-}), we see that this does indeed correspond to AB_2O_4. An example of an economic mineral that has this *normal* spinel structure is the chromium ore *chromite*, $FeCr_2O_4$, where A is Fe^{2+} and B is Cr^{3+}. In Nature, however, minerals rarely occur as pure

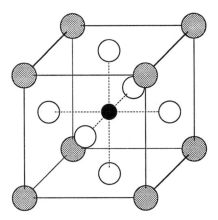

Figure 6.14 The perovskite structure. Open spheres, anion; shaded spheres, larger cation; filled sphere, smaller cation.

phases because ions of similar size can usually substitute for each other. In particular, Mg^{2+} frequently replaces Fe^{2+}, and Fe^{3+} or Al^{3+} replace Cr^{3+}. Consequently, natural chromite is more realistically formulated as $(Mg,\mathbf{Fe^{II}})(Al,Fe^{III},\mathbf{Cr})_2O_4$; $FeCr_2O_4$ is its idealized or *end-member* composition. The substituent ions replace Fe^{2+} and Cr^{3+} in a random way, much as solute molecules displace solvent molecules in a liquid solution. Consequently, chromite samples could be described as *solid solutions* of $MgAl_2O_4$ and $MgFe_2^{III}O_4$ in $Fe^{II}Cr_2O_4$.

A closely related group of AB_2O_4 oxides has the *inverse* spinel structure. Here, again, there is a ccp array of O^{2-}, but the B atoms are equally divided between T- and O-sites, and all of the A ions appear in O-, not T-, holes. Thus, we have $B(AB)O_4$ with B in one-eighth of the T-holes, A in one-quarter of the O-holes, and B in one-quarter of the O-holes.

An important example of an oxide with the inverse spinel structure is *magnetite*, a black, ferromagnetic oxide of iron (red hematite is effectively non-magnetic) containing both iron(II) and iron(III): Fe_3O_4 or $Fe^{II}Fe_2^{III}O_4$, better written as $Fe^{III}(Fe^{II}Fe^{III})O_4$. This is the usual product of the corrosion of iron at elevated temperatures with a limited oxygen supply and is also a valuable though uncommon mineral of iron, as it has a high percentage Fe content and is readily located and concentrated by virtue of its ferromagnetism. The brown ferromagnetic oxide used in Type I recording tape is *maghemite*, which contains no Fe^{2+} but has some H^+ as well as Fe^{3+} randomly distributed through the O- and T-holes of a ccp array of O^{2-}. It is often formulated as "γ-Fe_2O_3," but in reality its H^+ content is variable and the composition ranges up to HFe_5O_8, sometimes written as $5Fe_2O_3\cdot H_2O$. Since iron(II) oxide (wüstite, nominally FeO) also has a ccp O^{2-} array with Fe^{2+} in the O-holes (NaCl structure), the oxidation sequence at elevated temperatures

$$iron \rightarrow w\ddot{u}stite \rightarrow magnetite (\rightarrow maghemite) \rightarrow hematite$$

involves movement of Fe^{2+} and Fe^{3+} ions through a common cubic oxide lattice structure. The final conversion to hematite, however, requires a change-over from ccp to hcp O^{2-} (γ to α structure type).

As a final example of the use of the close-packing idea, we may note the *layer structures* typified by cadmium chloride and iodide. The iodide ions in CdI_2 form an hcp array (ABABAB...), and the Cd^{2+} occupy *alternate* layers of O-holes completely, thus giving a 1:2 stoichiometry but leaving every other layer of O-holes empty. The structure may be regarded as a stack of single-decker sandwiches with iodide "bread" (layers A and B) and a cadmium filling, but nothing between sandwiches. (The food analogy should not be pressed too far, as Cd is very toxic!) The crystal will obviously have planes of easy cleavage parallel to the layers. This structure is exhibited by many 1:2 ionic solids, including $M(OH)_2$ (if the hydrogens are ignored), where M can be Ca, Mg, Fe, Ni, Cd, etc. The $CdCl_2$ structure is also of this sandwich type, except that the anions form a ccp rather than an hcp array, with Cd^{2+} in alternating layers of O-holes.

Molybdenum sulfide (MoS_2) has an unusual layer structure in which sulfide atoms in a given upper layer sandwiching the Mo atoms are located directly above the S atoms in the lower layer, but alternate MoS_2 layers are offset as in hcp. Thus, the structure is A(Mo)A, B(Mo)B, A(Mo)A, B(Mo)B.... There is very little cohesion between successive MoS_2 sheets, and this results in a greasy consistency. Molybdenum sulfide is therefore widely used as a solid lubricant, particularly for high-temperature applications (cf. graphite).

6.5   Energetics of Ionic Compounds

To calculate the lattice energy U of an ionic crystal, consider first the potential energy E of a particular ion, say, a sodium ion in NaCl (Fig. 6.15), in terms of the coulombic attractions (of Cl^-) and repulsions (of other Na^+) of the other ions in the lattice. In NaCl, the nearest neighbors of Na^+ are: 6 Cl^- at a distance r (i.e., $r\sqrt{1}$), 12 Na^+ at a distance $r\sqrt{2}$, 8 Cl^- at a distance $r\sqrt{3}$, 6 Na^+ at a distance $2r$ (i.e., $r\sqrt{4}$), and so on. (For reasons of clarity, Fig. 6.15 shows only a few of these ions.) Then, for ions of charge $z+$ and $z-$ in a lattice with the NaCl structure ($z = 1$ for NaCl itself, 2 for MgO, etc.) where the unit cell side is $2r$, we have:

$$E = -\frac{z^2 e^2}{4\pi\varepsilon_0 r}\left(\frac{6}{\sqrt{1}} - \frac{12}{\sqrt{2}} + \frac{8}{\sqrt{3}} - \frac{6}{\sqrt{4}} + \frac{24}{\sqrt{5}} - \cdots\right). \tag{6.2}$$

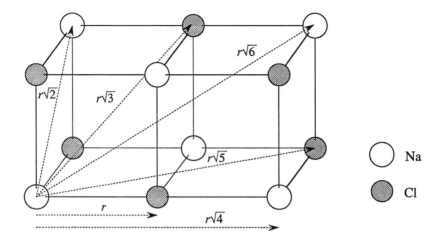

○ Na

◉ Cl

Figure 6.15 Internuclear distances in a crystal of NaCl.

Here (as usual in the SI system), the electronic charge e is 1.6021×10^{-19} A s and the permittivity of a vacuum ε_0 is 8.854×10^{-12} A^2 s^4 kg^{-1} m^{-3}. The infinite series in the brackets is convergent, and its sum (1.748) is called the *Madelung constant*, M, for the NaCl structure. Values of M are characteristic of the lattice structure but actually are not greatly different from one structure to another (e.g., M for the CsCl lattice is 1.763).

The net coulombic attractions of the ions for each other are counter-balanced at their equilibrium separation r_0 by the short-range repulsions due to interpenetration of the electronic clouds of ions "in contact." This short-range repulsion falls off sharply with increasing r. Max Born suggested that it could be represented by B/r^n, where B is a constant and n can be estimated from the compressibility of the crystal. As a rule of thumb, for ions having the electronic configurations of Ne, Ar, Kr, and Xe, $n \approx 7, 9, 10,$ and 12, respectively:

$$E = \frac{1}{4\pi\varepsilon_0}\left(-\frac{Mz^2e^2}{r} + \frac{B}{r^n}\right). \qquad (6.3)$$

At the equilibrium separation, $r = r_0$ and $dE/dr = 0$, so that

$$4\pi\varepsilon_0\frac{dE}{dr} = \frac{Mz^2e^2}{r_0^{\,2}} - \frac{nB}{r_0^{\,n+1}} = 0 \qquad (6.4)$$

whence

$$B = \frac{Mz^2e^2r_0^{\,n-1}}{n} \qquad (6.5)$$

and B can be eliminated from Eqn. 6.3:

$$E = \frac{1}{4\pi\varepsilon_0} \left(-\frac{M z^2 e^2}{r_0} + \frac{M z^2 e^2}{n r_0} \right). \tag{6.6}$$

The lattice energy U is defined as the energy *released* (U therefore being negative in the thermodynamic convention) when a mole of the requisite free gaseous ions comes together from infinite interionic separation to make up the crystal. If N is Avogadro's number (6.0225×10^{23}), we have the *Born–Landé formula*:

$$U = -NE = \frac{N M z^2 e^2}{4\pi\varepsilon_0 r_0} \left(1 - \frac{1}{n} \right). \tag{6.7}$$

Equation 6.7 can clearly be improved upon by taking a more sophisticated approach to ionic interactions, but it gives satisfactory values for U. Lattice energies can be related to the heats of formation ΔH_f° of ionic solids through the *Born–Haber cycle*, which is the counterpart of the thermochemical cycle for covalent compounds given in Section 1.6:

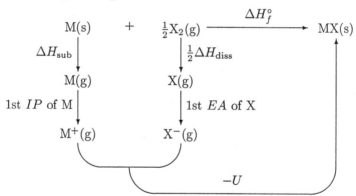

We assume that we have a solid metal M which reacts with a diatomic, gaseous non-metal X_2 (Cl_2, F_2, O_2, etc). Similar cycles can be written for solid sulfur, etc., as the non-metal. In either case, before we can connect U with ΔH_f° we must form *gaseous ions* of M and X. We need, not only the relevant ionization potentials (IP) and electron affinities (EA), but also the heats of atomization of solid M and gaseous X_2. These atomization energies are traditionally referred to as heats of sublimation ΔH_{sub} of M(s)* and of dissociation ΔH_{diss} of X_2. For NaCl itself, we have (in kJ mol^{-1}): $\Delta H_f = -411$; $IP = 502$; $\Delta H_{sub} = 108$; $EA = -354$; and $\Delta H_{diss} = 242$. So, bearing in mind that U represents energy *given out* from the system,

$$U = -(-411) + 502 + (-354) + 108 + \tfrac{1}{2}(242) = 788 \text{ kJ mol}^{-1}. \tag{6.8}$$

---

*Sublimation is an ancient alchemical term meaning direct passage from solid to vapor.

Clearly, U is the biggest number in the cycle and is the main driving force for the formation of ionic compounds. Nevertheless, the other factors can tip the balance one way or another. For example, ΔH_{sub} is particularly large for the transition metals niobium, tantalum, molybdenum, tungsten, and rhenium, with the result that, in their lower oxidation states, they do not form simple ionic compounds such as $ReCl_3$ but rather form compounds that contain *clusters* of bonded metal atoms. (In this example, Re_3 clusters are involved, so the formula is better written Re_3Cl_9.)

In principle, we can use the Born–Haber cycle to predict whether a particular ionic compound should be thermodynamically stable, on the basis of calculated values of U, and so proceed to explain away all of the chemistry of ionic solids. The relevant quantity is actually the *free energy of formation*, ΔG_f°, and this is calculable if an entropy cycle is set up to complement the Born–Haber enthalpy cycle. However, in practice ΔH_f° dominates the energetics of formation of ionic compounds.

The pitfall in making predictions of this kind is that there may be more stable compounds of the elements concerned than the one considered. One can ask, for example, why calcium does not form a stable solid monofluoride, CaF, as well as CaF_2, since the first ionization potential for Ca is only 596 kJ mol^{-1} as against 1748 kJ mol^{-1} for the sum of the first and second *IPs*. To answer this question, we first make an educated guess of the unknown ionic radius of the Ca^+; about 122 pm seems reasonable, by comparison with other monovalent ions, and the radius ratio for CaF is therefore 122/129. Since this is greater than 0.73 (see Table 6.1), the crystal structure should be of the CsCl type. The Madelung constant for the CsCl structure is 1.763, and Eqn. 6.7 gives $U = 865$ kJ mol^{-1}. This last value can be plugged into the Born–Haber cycle to give $\Delta H_f^\circ = -332$ kJ mol^{-1} for CaF(s); it would appear that CaF(s) should be comfortably stable, since ΔH_f° for NaCl is only 79 kJ mol^{-1} more negative than this! The heat of formation of $CaF_2(s)$, however, has been measured to be -1220 kJ mol^{-1}. Thus we have:

$$Ca(s) + F_2(g) \rightarrow CaF_2(s) \qquad \Delta H = -1220 \text{ kJ mol}^{-1} \quad (6.9)$$
$$\underline{2CaF(s) \rightarrow 2Ca(s) + F_2(g) \quad \Delta H = 2(+332) \text{ kJ mol}^{-1} \quad (6.10)}$$
$$\text{Add:} \quad 2CaF(s) \rightarrow Ca(s) + CaF_2(s) \quad \Delta H = -556 \text{ kJ mol}^{-1}. \quad (6.11)$$

The *disproportionation* (self-oxidation/reduction) of CaF(s) is therefore very strongly favored thermodynamically, and so any attempt to make solid CaF from the correct amounts of calcium and fluorine will simply give CaF_2 and unreacted calcium. In this particular example, the conclusion is the same one reached by hand-waving arguments about the special stability of the argon-like, filled-quantum-shell electronic configuration of Ca^{2+}, but we see now that many other factors are involved, and in the transition metals, lanthanides, and actinides the "filled-shell" approach is of little use. The lanthanide elements, for example, are *all* preferentially trivalent in their ionic compounds. This is because the trends in (1st + 2nd + 3rd) IP, ΔH_{sub} and U are influenced in a compensatory manner by the progressive shrinking (*lanthanide contraction*, Section 1.4) of the metal atoms as we go from lanthanum to lutecium. Similarly, the most stable oxidation states of the metals of the first transition period are scandium(III), titanium(IV), vanadium(V), chromium(III), manganese(II), iron(II) or (III), cobalt(II), nickel(II), copper(II), and zinc(II). Only the first three and the last one would be predicted from the full/empty shell argument.

References

1. A. F. Wells, "Structural Inorganic Chemistry," 5th edn., Oxford University Press: London, 1984.

2. D. McKie and C. McKie, "Crystalline Solids," Thomas Nelson and Sons: London, 1974.

3. F. A. Cotton, G. Wilkinson, and P. L. Gaus, "Basic Inorganic Chemistry," 2nd edn., John Wiley & Sons: New York, 1987.

4. "The Mineral Powder Diffraction File," JCPDS, International Center for Diffraction Data.

5. R. C. Weast and M. J. Astle (eds.), "CRC Handbook of Chemistry and Physics," CRC Press: Boca Raton, Florida, 1990 (revised annually).

6. R. D. Shannon, "Revised effective ionic radii and systematic studies of interatomic distances in halides and chalcogenides," *Acta Crystallographica*, 1976, *A32*, 751–767.

7. M. Sayer and K. Sreenivas, "Ceramic thin films: fabrication and applications," *Science*, 1990, *247*, 1056–1060.

8. K. A. Müller and J. G. Bednorz, "The discovery of a class of high-temperature superconductors," *Science*, 1987, *237*, 1133–1139.

9. A. S. Sleight, "Chemistry of high-temperature superconductors," *Science*, 1988, *242*, 1519–1527.

Exercises

6.1 Count the effective numbers of anions and cations contained within the zinc blende, fluorite, and rutile unit cells.

6.2 Predict the crystal structure of magnesium fluoride, given that the appropriate anion and cation radii are 131 and 72 pm, respectively. (The observed structure is of the rutile type.)

6.3 The structure of the sodium selenide crystal may be described as a cubic close-packed array of Se ions, with a sodium ion in every tetrahedral hole.

(a) Sketch the unit cell of this compound.

(b) Name the structure type.

6.4 Estimate the lattice energy of sodium chloride, assuming that the ionic radii are 102 pm (Na^+) and 181 pm (Cl^-) and that the Born exponent is 9. Compare your result with the value given in Eqn. 6.8.

6.5 Set up a Born–Haber cycle for the formation of hematite (α-Fe_2O_3) from its elements.

6.6 Calculate the heat of formation of fluorite, given the following information (units are kJ mol^{-1}):

Electron affinity of fluorine	= −334.4
First ionization potential of calcium	= 596.1
Second ionization potential of calcium	= 1151.7
Lattice energy of fluorite	= 2635.
Heat of sublimation of calcium	= 178.2
Bond energy of fluorine	= 158.0.

[*Answer:* see Section 6.5.]

6.7 Calculate the lattice energy of AlF_3 from the data of Appendix C. The heat of formation of F^-(g) (for example) incorporates $\frac{1}{2}\Delta H_{diss}$ for F_2 and the first electron affinity of F(g), so your task is simplified. [*Answer:* 5922 kJ mol^{-1}.]

Chapter 7

The Defect Solid State

7.1   The Inevitability of Crystal Defects

IN THE PRECEDING chapter, it was tacitly assumed that crystalline solids were perfect, that is, that all of the sites characteristic of a particular structure would be occupied, that the sites that should be vacant in the ideal structure would indeed be unoccupied, and that the atoms or ions making up the lattice were all of the specified kind. In practice, thermodynamics tells us, no crystal can ever be structurally perfect. The equilibrium state of the crystal will be one in which its free energy is minimized, and this would seem to favor a perfect crystal lattice since misplaced or foreign ions will lead to a reduced lattice energy and hence to a less negative heat of formation. Disorder at the atomic level, however, will be reflected in a more positive *entropy* term, and the increase in the product $T\Delta S°$ will tend to compensate for the loss of lattice energy due to disorder, with increasing effectiveness as the temperature rises:

$$\Delta G° = \Delta H° - T\Delta S°. \tag{7.1}$$

Consequently, although hypothetical perfect crystals can be said to exist at the unattainable absolute zero of temperatures, real crystals contain defects that increase in number (thermodynamics) and mobility (kinetics—atoms can move from one site to another on surmounting an Arrhenius-type activation energy barrier) as the temperature rises. Eventually, these defects (e.g., thermal vibrations of atoms around their equilibrium positions, dislocations of atoms or planes of atoms from their ideal sites, creation of vacant sites) become severe enough that long-range atomic ordering breaks down, and the crystal melts. Even in the liquid some transient short-range ordering may persist, especially in ionic melts and in hydrogen-bonded liquids like water. These local ice-like structures that form and decay continually in liquid water have been referred to as *flickering clusters*.

Wherever there is a defect in a crystal lattice, interatomic forces will remain unbalanced and the free energy will be less negative than elsewhere in the crystal, although generally the lattice will deform locally to smooth this out. Nevertheless, defect sites (especially of the extended variety) tend to be more chemically reactive than the bulk crystal and tend to be active sites for crystal growth, dissolution, corrosion, catalytic activity, and so on.

7.2   Main Types of Crystal Defect

Point defects. These are limited to a single point in the lattice, although the lattice will buckle locally so that their influence may spread quite far. A *Frenkel defect* consists of a misplaced interstitial atom and a lattice vacancy (the site it should ideally have occupied). For example, silver bromide, which has the NaCl structure, has substantial numbers of Ag^+ ions in tetrahedral holes in the ccp Br^- array, instead of in the expected octahedral holes. Frenkel defects are especially common in salts containing large, polarizable anions like bromide or iodide.

Defects in which both a cation and sufficient anions to balance its charge (or vice versa) are completely missing from the lattice are called *Schottky defects*. This results in a density that is lower than that calculated on the basis of unit cell dimensions, whereas Frenkel defects do not affect this density. Titanium(II) oxide, for example, also has the NaCl structure, but, even when its composition is $TiO_{1.00}$ (which it rarely is; see Section 7.3), about one-sixth of the Ti^{2+} and O^{2-} sites are vacant.

The existence of Schottky or Frenkel defects, or both, within an ionic solid provides a mechanism for significant electrical conductance through ion migration from site to empty site (leaving, of course, a fresh empty site behind). Solid β-AgI provides a classic example of a non-metallic solid with a substantial electrical conductivity at elevated temperatures. Several such ionically conducting solids have been intensively studied of late as possible solid electrolytes for use in fuel cells or in advanced electrical storage batteries. What is wanted is a solid that allows ions, but not electrons, to

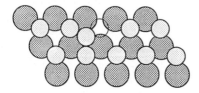 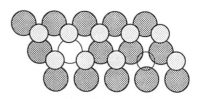

Figure 7.1 Frenkel and Schottky defects. In the Frenkel case (left), a member of the lightly shaded ion sub-lattice is found in the wrong kind of interstice. In the Schottky defect (right), an ion of either kind (here presumed to have equal but opposite charges) is completely missing.

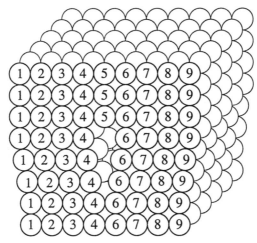

Figure 7.2 Edge dislocation. Layer 5 is incomplete (all atoms are the same).

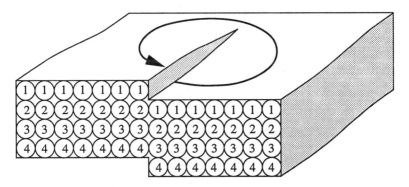

Figure 7.3 Screw dislocation. All the atoms are the same, but layer 1 has become mismatched with layer 2 (etc.). In effect, all the layers become one continuous helicoid surface.

flow through it. The so-called β-*alumina* electrolytes, for example, permit transport of sodium ions in layers of incompletely occupied cation sites separating blocks of aluminum oxide lattice having a spinel-like structure. The term "alumina" (implying Al_2O_3) is a misnomer, since the composition of the β'' variety, for example, would be $NaAl_{11}O_{17}$ if fully stoichiometric. (In reality, it is always non-stoichiometric: $Na_{1+2x}Al_{11}O_{17+x}$; see Section 7.3).

Line defects. These extend in one dimension and may originate in an incomplete layer of atoms (an *edge dislocation*, Fig. 7.2) or from a mismatching of layers (a *screw dislocation*, Fig. 7.3). The latter converts the whole crystal lattice into a single helicoid surface (cf. a multi-storey parkade), but it is a line defect in the sense that the linear axis of the helix defines the

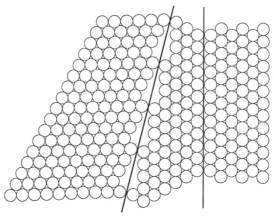

Figure 7.4 Plane defects.

defect. Line defects create lines of weakness in the crystals that may initiate fracture under stress. Where they emerge at the surface of the crystal, they form sites for the initiation of crystal dissolution (revealed as *etch pits* under the microscope) or further crystal growth (e.g., continuing the helicoid structure). These defects, like others, are mobile to some degree, by concerted slippage of layers of atoms.

Plane defects. These are, in effect, grain boundaries. Large crystalline specimens are usually made up of microcrystals, or grains, the lattices of which do not match precisely with those of their neighbors (Fig. 7.4). Impurities tend to be concentrated in these regions of mismatch. Large crystals may also contain sizable inclusions of the solution from which they were grown. Consequently, when precipitating a solid from solution, the chemist or X-ray crystallographer will usually adjust conditions to give small, well-formed crystals. On the other hand, if large crystals tend to form readily from solution, it usually means that the levels of impurities in the system are quite low, so that regular crystal growth is not inhibited. Recent advances in transmission electron microscopy have made it possible to see grain boundaries with individual atoms resolved.[2]

Impurity defects. It is inevitable that some foreign atoms will be present in any macroscopic crystal, usually substituting for the normal lattice atoms (cf. natural chromite, Section 6.4). Few analytical-reagent-grade chemicals are much better than 99.9% pure, which means 0.1% impurities; thus, if the atomic or molecular weights are comparable, there will on average be a foreign atom or molecule in any cubic portion of the lattice of 10 atoms or molecules along the edge. This can have a profound effect upon the bulk properties of the crystal, such as its electrical conductivity, whether this is due to ion migration as in point defects or to electrons hopping from one atom to another.

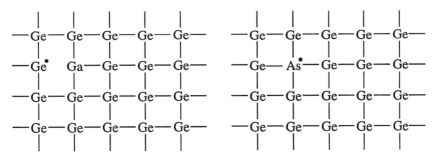

Figure 7.5   Creation of positive (left) and negative defects in a germanium lattice through doping with gallium and arsenic, respectively.

For example, the conductance of elemental germanium, which is tetracovalent and has the diamond structure, is greatly increased by *doping* the crystal with small amounts of its neighbors in the Periodic Table, e.g., with gallium, which has only three valence electrons, or arsenic, which has five. A Ga atom substituting for a Ge therefore leaves the lattice-wide covalent bonding structure one electron short—a so-called *positive hole*, which can move around from one atom (whether Ga or Ge) to another under the influence of an applied electrical field. Gallium-doped Ge is therefore a *p-type semiconductor;* since the holes move more readily at higher temperatures, the conductivity increases markedly with rising temperature (contrast metallic conduction).

Conversely, doping with As places an extra electron in the tetracovalent network, and thus a negative or *n-type semiconductor* is created. If we place *n-* and *p*-type semiconducting crystals in contact (a *p-n* junction), we create a device that conducts electricity preferentially in one direction; this is the basis of action of the semiconductor diodes used in the electronics industry, although specially refined silicon (Section 15.8) is usually employed in place of Ge. Transistors are designed using similar basic principles—typically with *n-p-n* or *p-n-p* junctions.

Crystal surfaces. Crystal surfaces may be looked upon as vast defects inasmuch as the lattice forces are incompletely balanced. The effects of this imbalance are partially offset by distortion of the crystal lattice near the surface, but crystal surfaces still show a strong tendency to adsorb other molecules. Surface areas of powders on the order of 1 to 100 m^2 g^{-1} are customarily measured by the amount of nitrogen gas they adsorb at low temperatures, and release at high temperatures, in forming a multimolecular covering of N_2 (the *Brunauer–Emmett–Teller* or *BET* method). The adsorbed molecules may be activated chemically, as well as held in juxtaposition with other reactants in locally high concentrations, so the surfaces of solids are often active in *heterogeneous catalysis* (Chapter 9).

Stacking faults. If, for example, we have a crystal structure based upon cubic close-packing, the sequence ABCABCABC... of layers may contain occasional errors such as ABCABABC.... These stacking faults are of minor chemical significance.

7.3   Non-Stoichiometry

In introductory chemistry courses, much emphasis is necessarily placed on the concept of stoichiometry, that is, that elements combine in certain definite proportions by weight, proportions that reflect their valences and atomic masses. Throughout much of the chemistry of the main group elements and organic compounds, this concept works extremely well, but in transition metal chemistry in particular it is common for ions of more than one oxidation state to form with comparable ease, and sometimes to occur together in the same ionic solid. The presence of more highly oxidized cations in a metal ion sub-lattice is counterbalanced by vacancies in that sub-lattice. Conversely, the presence of some metal ions of lower oxidation state in the metal ion sub-lattice requires vacant anion sites to balance the charge. In some cases, the charge imbalance is caused by ions of some other element, or, rarely, by multiple valence of the anions. In any event, the empirical formula of a recognizable solid transition-metal compound may be variable over a certain range, with non-integral atomic proportions. Such *non-stoichiometric* compounds may be regarded as providing extreme examples of impurity defects.

For example, iron filings combine with powdered sulfur on heating to form iron(II) (ferrous) sulfide, ostensibly FeS—an easy experiment which has been used on past occasions to introduce students to stoichiometric principles. A worse example would be hard to find among simple reactions between elements. Ferrous sulfide does exist as $Fe^{2+}_{1.000}S^{2-}$; it has the nickel arsenide structure (Table 6.2) and is known as troilite. As usually prepared, however, up to 10% of the Fe ions are missing. Of those that *are* present, two are in the trivalent (ferric) state for every ferrous ion that is missing, so balancing the charges. The S^{2-} array is essentially complete. In some older chemical treatises, it was claimed that reproducible phases such as "Fe_8S_9" existed, but this formula would be better written as $Fe_{1-x}S$ with $x = 0.111$ (no special significance being attached to 0.111), or

$$Fe^{2+}_{0.667}\ Fe^{3+}_{0.222}\ \square_{0.111}\ S$$
$$1-3x \qquad 2x \qquad\quad x$$

where the square symbol represents vacant sites in the iron ion sub-lattice. In general, compositions such as

$$Fe^{2+}_{0.634}\ Fe^{3+}_{0.244}\ \square_{0.122}\ S$$

are just as likely for the product of any given preparation, although the tendency of the vacant sites to become ordered in the lattice can give special stability to certain compositions such as Fe_7S_8. These *iron-deficient* sulfides are known as *pyrrhotite;* an iron-rich material $Fe_{1+x}S$ or *mackinawite* is also known. Finally, the familiar *pyrite* or *"fool's gold"* (FeS_2) is usually stoichiometric, the anion being S_2^{2-} (rather like the peroxide ion, O_2^{2-}). The mechanism of corrosion of steels by aqueous H_2S, e.g., in the *GS (Girdler sulfide) process* for heavy water production, in which the deuterium (^2H) content of water is built up by repeated equilibrations at high and low temperatures with H_2S, can involve several of these often ill-defined sulfides, presenting a complex chemical problem to the corrosion engineer.

In the same way, ferrous oxide (wüstite, Section 6.4) ranges in composition from $Fe_{0.88}O$ to $Fe_{0.95}O$; stoichiometric FeO is not encountered in practice. As an extreme case, δ-TiO, an NaCl-type solid previously mentioned as exhibiting gross Schottky defects even when stoichiometric, ranges widely in composition:

$TiO_{0.69}$ 96% Ti sites, 66% O^{2-} sites occupied

$TiO_{1.00}$ 85% Ti sites, 85% O^{2-} sites occupied

$TiO_{1.33}$ 74% Ti sites, 98% O^{2-} sites occupied.

7.4   Extrinsic (Defect) Semiconductors

Many metal oxides and sulfides exhibit semiconductor properties by virtue of their non-stoichiometry. In Section 7.2, we distinguished positive (*p*-type) and negative (*n*-type) semiconductors.

p-Type oxides gain oxygen on heating in the air, e.g., NiO:

$$NiO + \tfrac{x}{2}O_2 \rightarrow NiO_{1+x} . \qquad (7.2)$$
$$\text{green} \qquad\qquad \text{black}$$

The black phase persists on cooling. It contains $2Ni^{3+}$ in place of $2Ni^{2+}$, plus a vacancy in the Ni^{2+} sub-lattice, for every additional O^{2-}. The Ni^{3+} ions, being smaller and more highly charged than Ni^{2+}, act as charge carriers by skipping through vacant sites under an applied external EMF.

n-Type oxides typically lose O_2 reversibly on heating in the air, e.g., zinc oxide:

$$ZnO(s) \rightleftharpoons ZnO_{1-x}(s) + \tfrac{x}{2}O_2(g). \qquad (7.3)$$
$$\text{white} \qquad\quad \text{yellow}$$

The excess of Zn^{2+} ions in the non-stoichiometric high-temperature phase is counterbalanced by free electrons centered upon the empty O^{2-} sites (*F* or *color* centers; F is for *Farben*, German for color; excitation of these electrons by light causes the yellow color). These act as charge carriers under an applied electrical field.

Similarly, ZnS, on heating to $500\,^{\circ}$C, loses sulfur vapor to give ZnS_{1-x}, which fluoresces strongly in ultraviolet light.

7.5   The Scaling of Metals by Gaseous Oxygen[3-5]

Most metals acquire an oxide film or *scale* on their surfaces on exposure to oxygen or air, especially at elevated temperatures. This film is often protective in the sense of hindering further oxidation, but this is not always the case. Pilling and Bedworth (1923) made an early attempt to rationalize the protective behavior of oxide films on the basis of the volume occupied by the oxide relative to the volume of metal from which it was formed.

If the molar volume V° of oxide per mole of metal is less than the molar volume of the metal, the scale will be under tension as it forms and will tend to crack and so be non-protective. An example would be magnesium:

$$V^{\circ} \text{ of Mg} = \frac{\text{atomic mass}}{\text{density}} = \frac{24.312}{1.74} = 14.0 \text{ cm}^3 \text{ mol}^{-1}$$

$$V^{\circ} \text{ of MgO} = \frac{\text{molar mass}}{\text{density}} = \frac{40.31}{3.58} = 11.3 \text{ cm}^3 \text{ mol}^{-1}.$$

If V° of the oxide per mole of metal content is greater than V° for the metal, the oxide film forms under compression and may be protective, as in the case of nickel:

$$V^{\circ} \text{ of Ni} = 6.6, \quad V^{\circ} \text{ of NiO} = 11.2 \text{ cm}^3 \text{ mol}^{-1}.$$

However, the Pilling–Bedworth approach is of limited applicability, as is shown by the behavior of copper:

$$\frac{V^{\circ} \text{ of CuO}}{V^{\circ} \text{ of Cu}} = \frac{12.5}{7.1}$$

$$\frac{\frac{1}{2}V^{\circ} \text{ of Cu}_2\text{O}}{V^{\circ} \text{ of Cu}} = \frac{11.9}{7.1}$$

which would therefore be expected to form a protective film, be it of copper(I) or copper(II) oxide. (Note, however, that the picture becomes confused where two oxide stoichiometries are involved.) Indeed, oxidation of Cu at 600 to $800\,^{\circ}$C proceeds according to the *parabolic law:*

$$\frac{dy}{dt} = \frac{k}{y} \tag{7.4}$$

or

$$y^2 = 2kt + c \tag{7.5}$$

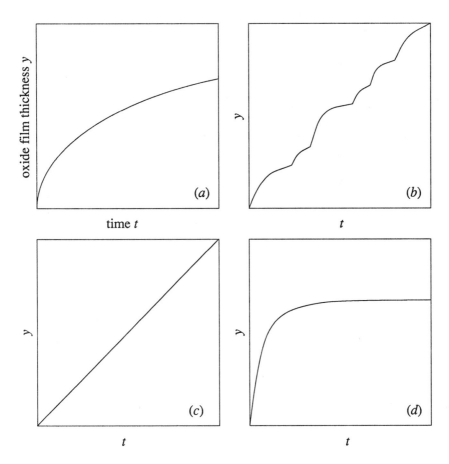

Figure 7.6 Rate laws for the formation of oxide films. (*a*) Parabolic rate law. (*b*) Effect of film cracking (successive parabolic segments). (*c*) Limiting case of (*b*). (*d*) Logarithmic rate law.

where y is the thickness of scale (or weight gain per unit surface area) at time t, and k and c are constants. The parabolic law (Fig. 7.6a) implies protection by the thickening, coherent, oxide film. At $500\,^\circ$C, however, the film is apparently insufficiently plastic to support the Pilling–Bedworth compressional stress, and it cracks intermittently to expose fresh metal (Fig. 7.6b). In the limit of frequent film cracking (as with sodium), Fig. 7.6b reduces to a straight line (*rectilinear rate law*, Fig. 7.6c). The Pilling–Bedworth concept also fails to explain why some metal oxidation processes follow *logarithmic* or *inverse logarithmic rate laws* (Fig. 7.6d, and Eqns. 7.13 through 7.16), which lead to much sharper slowing of oxidation than the intuitively reasonable *parabolic rate law*.

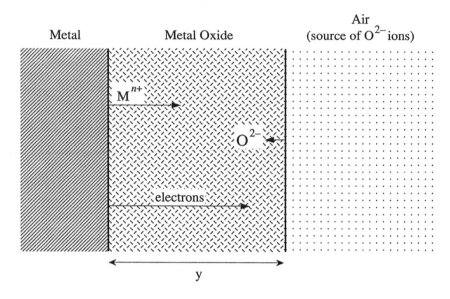

Figure 7.7 Oxidation of a metal by gaseous oxygen, viewed as an electrochemical process.

In 1933, C. Wagner rationalized metal oxidation rates in terms of metal oxide lattice defects. The oxidizing surface was regarded as an electrochemical cell (Fig. 7.7):

$$M \rightarrow M^{n+} + n\,e^- \qquad \text{at anode (metal surface)} \qquad (7.6)$$
$$O_2 + 4e^- \rightarrow 2O^{2-} \qquad \text{at cathode (air-oxide interface)} \qquad (7.7)$$

in which the electrolytic medium is the film (of thickness y) of defective metal oxide. This places "dry" oxidation on a similar basis to "wet" corrosion (Section 14.2). Electron transfer across the film is usually relatively rapid, and, since M^{n+} is usually smaller than O^{2-}, the film grows mainly by diffusion of M^{n+} outwards to meet the O^{2-} that are being added to the oxide sub-lattice at the gas-oxide interface. The M^{n+} ions move through Schottky and/or Frenkel defects under the influence of electric field and concentration gradients (these are equivalent, according to Einstein), which will be inversely proportional to the separation y of the "anode" and "cathode."

The probability that an ion will jump to a neighboring vacancy, in the absence of an external electrostatic field, is $\exp(-E_a/\kappa T)$, where E_a is the activation energy and κ is Boltzmann's constant. In the presence of an electric field gradient ($\propto y^{-1}$), the probability of an up-field jump is $\exp[(-E_a - K/y)/\kappa T]$ and the probability of a down-field jump is $\exp[(-E_a + K/y)/\kappa T]$, where K is a constant. Thus,

$$\frac{dy}{dt} = AK(e^{\Theta} - e^{-\Theta}) = 2AK \sinh \Theta \qquad (7.8)$$

(the "sinh equation"), where $A = \exp(-E_a/\kappa T)$ and $\Theta = K/y\kappa T$. Since

$$e^{\Theta} = 1 + \Theta + \frac{\Theta^2}{2!} + \frac{\Theta^3}{3!} + \cdots \qquad (7.9)$$

we can expand Eqn. 7.8 to:

$$\frac{dy}{dt} = AK\left(2\Theta + \frac{\Theta^3}{3} + \cdots\right). \qquad (7.10)$$

Now, if the temperature T is *high* and the oxide film is *thick* (T and y both being large), Θ can become much less than unity, so

$$\frac{dy}{dt} \approx \frac{(\text{constant})}{y} \qquad (7.11)$$

which is the *parabolic law* (Eqn. 7.4). On the other hand, when T is *low* and the film is *thin*, Θ may be much larger than unity, so that the term in $\exp(-\Theta)$ becomes negligible, whence:

$$\frac{dy}{dt} = AK \exp \Theta \qquad (7.12)$$

which, after further approximations,[3] gives an expression equivalent to the *inverse logarithmic law*:

$$y_0^{-1} - y_t^{-1} = b\ln(k't + 1) \qquad (7.13)$$

or, for a *small* increase of weight w per unit area,

$$w^{-1} = c - k'' \ln t \qquad (7.14)$$

where b, c, k', and k'' are constants. Inverse logarithmic oxidation rates are exhibited by iron at room temperature; the oxide film grows on a freshly exposed metal surface to a thickness of 1.6 nm after one day, and 3.5 nm after one year—a drastic retardation.

If, contrary to our assumption, electron transfer is slower than M^{n+} diffusion and is oxidation rate-controlling, it can be shown that the *direct logarithmic law* applies:

$$y_t - y_0 = a\ln(k^*t + 1) \qquad (7.15)$$

or

$$w = a\ln(k^*t + 1) \qquad (7.16)$$

where a and k^* are constants. Again, this is expected to apply to fairly thin (< 100 nm) films at relatively low temperatures, such as in the oxidation of cobalt at 320 to 520 °C, but in practice it is hard to distinguish direct from inverse logarithmic kinetics.

Finally, there are two special cases in which the *rectilinear law* is observed: when the rate-controlling factor is the rate of supply of O_2, and where the metal oxide is volatile at the temperature of oxidation. The latter case occurs in the high-temperature oxidation of molybdenum, since MoO_3 is quite volatile, and in this case dw/dt is of course negative.

The mechanism of the scaling of iron is so complex as to require special mention. Above 570 °C, wüstite ($Fe_{1-x}O$) is thermodynamically stable and forms the relatively thick basal layer in the oxide film. This is followed by a magnetite (Fe_3O_4) layer which is followed by a final layer of Fe_2O_3. Magnetite itself tends to become non-stoichiometric under oxidizing conditions, with excess Fe^{3+}, so that its composition and color can vary from $Fe_{3.000}O_4$ (black) towards cubic $Fe_{2.667}O_4$ (i.e., γ-Fe_2O_3, chocolate brown). Thus, as outlined in Section 6.4, the oxidation of iron above 570 °C involves mainly the migration of Fe^{2+} and Fe^{3+} outward through a ccp sub-lattice of O^{2-} in a largely non-stoichiometric oxide film. At high temperatures, the final Fe_2O_3 layer must be α-Fe_2O_3, since γ-Fe_2O_3 is unstable with respect to this (and seems to require the elements of water to stabilize it; Section 6.4), but this relatively thin layer has an hcp O^{2-} array and grows by diffusion of O^{2-} inwards. The growth kinetics are further complicated by the tendency for cavities to form within the film. At *low* temperatures, wüstite is unstable with respect to disproportion to Fe and Fe_3O_4, so the oxide film is usually Fe_3O_4 (possibly non-stoichiometric to some degree) and/or γ-Fe_2O_3 (if the stabilizing traces of water are present).

References

1. N. N. Greenwood, "Ionic Crystals, Lattice Defects, and Non-Stoichiometry," Chemical Publishing: New York, 1970.

2. A. L. Robinson, "Spotting the atoms in grain boundaries," *Science*, 1986, *233*, 842-844.

3. U. R. Evans, "The Corrosion and Oxidation of Metals," Edward Arnold: London, 1960, and Supplements 1 (1968) and 2 (1978).

4. U. R. Evans, "An Introduction to Metallic Corrosion," 3rd edn., Edward Arnold: London, 1981, Chapter 1.

5. M. G. Fontana and N. D. Greene, "Corrosion Engineering," 2nd edn., McGraw-Hill: New York, 1978, Chapter 11.

Exercises

7.1 Stoichiometric titanium(II) oxide and magnesium oxide have densities of 4.93 and 3.58 g cm^{-3}, respectively. They both have the NaCl structure, with unit cells of edge 416.2 and 421.2 pm, respectively. Show that only about 84% of the lattice sites in TiO, but essentially all in MgO, are filled.

7.2 On which of the following metals would the oxide film be expected to be protective, according to the Pilling–Bedworth principle?

Metal	Density (g cm^{-3})	Oxide	Density (g cm^{-3})
Be	1.85	BeO	3.01
Ca	1.54	CaO	3.25–3.38
Al	2.70	α-Al$_2$O$_3$	3.97
Ti	4.50	rutile	4.26

[*Answer:* all except calcium.]

7.3 The oxidation of iron by (dry) air at ambient temperature proceeds according to the *inverse logarithmic rate law*, which may be written as:

$$\frac{1}{y} = c - k \ln t$$

where the oxide film thickness y is 1.6 nm after time $t = 1$ day and 3.5 nm after 1 year; c and k are empirical constants. How much longer will a doubling of the oxide film, from 3.5 to 7.0 nm, require?
[*Answer:* 10.5 more years.]

Chapter 8

Silicates and Aluminosilicates

NEXT TO oxygen, silicon and aluminum are by far the most abundant constituents of the crust of the Earth (Table 8.1),[1] where they occur as silica (SiO_2), alumina (Al_2O_3), and a great variety of solid silicates and aluminosilicates. This diversity of silicates originates in the propensity of Si to form strong Si—O—Si links as noted in Section 1.8, leading to very stable chains, rings, sheets, and networks based upon oxygen-sharing (corner-sharing) between $SiO_4{}^{4-}$ tetrahedra. The Al^{3+} ion is capable of substituting for "Si^{4+}" in some of these tetrahedra, giving rise to aluminosilicate frameworks. In all silicates and aluminosilicates, cations of various kinds must be present to counterbalance the negative charge of the anionic units or frameworks. Thus it comes about that four of the five volumes of Deer, Howie, and Zussman's invaluable reference work on mineralogy[2] are taken up with silicates of one kind or another, and these materials constitute increasingly important economic resources.[3]

8.1 Silicate Structures[2-5]

In order to simplify the essential features of these complicated structures, the silicate $SiO_4{}^{4-}$ (or aluminate $AlO_4{}^{5-}$) tetrahedron will be represented by the triangle shown in Fig. 8.1 (left), which is essentially the tetrahedron as viewed down one of the Si—O bonds. Triangle corner-sharing therefore implies O^{2-}-sharing between linked $SiO_4{}^{4-}$ units. Aluminum has the ability (shared by Si, but only at very high pressures; see Section 6.3) to be six- as well as four-coordinate and to link with $SiO_4{}^{4-}$ units by corner-sharing of the $AlO_6{}^{9-}$ octahedron. This octahedron will be represented by the double-triangle symbol in Fig. 8.1, representing an octahedron as viewed

TABLE 8.1
The Commonest Elements of the Earth's Crust

Element	Atom %	Weight %	Element	Atom %	Weight %
O	62.6	46.6	Ti	0.20	0.44
Si	21.2	27.7	F	0.091	0.080
Al	6.47	8.1	P	0.083	0.12
Na	2.64	2.8	C	0.057	0.032
Ca	1.95	3.6	Mn	0.038	0.10
Fe	1.94	5.0	S	0.034	0.052
Mg	1.84	2.1	Cl	0.030	0.048
K	1.42	2.6	Li	0.021	0.007

down a three-fold axis (perpendicular to one of the triangular faces). Some typical silicate frameworks are summarized in Fig. 8.2.

(a) *Discrete silicate ions.* These *orthosilicates* contain isolated SiO_4^{4-} units. Examples include *phenacite* ($Be_2[SiO_4]$) and *willemite* ($Zn_2[SiO_4]$) in which the beryllium and zinc cations are also tetrahedrally surrounded by the oxygens of the silicate anions. A more important case is *olivine* ($(Fe,Mg)_2[SiO_4]$), which can also be regarded as a rumpled hcp array of O^{2-}, with "Si^{4+}" in one-eighth of the T-holes and the divalent cations in one-half of the O-holes. (Compare and contrast the spinel structure [Section 6.4] in which the O^{2-} sublattice is ccp.) The end-member compositions of olivine are known as *forsterite* (Mg_2SiO_4) and *fayalite* (Fe_2SiO_4).

(b) $Si_2O_7^{6-}$. This *pyrosilicate* ion is not very common in solid minerals. *Thortveitite* ($Sc_2[Si_2O_7]$) is a scandium mineral with this anion. Note that the charge on the ion is given by four times the number of silicons, minus twice the number of oxygens.

(c) *Three-silicate rings,* $Si_3O_9^{6-}$. These occur in a few minerals such as *benitoite* ($BaTi[Si_3O_9]$). The ring consists of three silicons linked through three oxygens, so it is actually six-membered and not particularly strained.

(d) *Six-silicate rings,* $Si_6O_{18}^{12-}$. These are known in the precious stone *beryl* ($Be_3Al_2[Si_6O_{18}]$).

(e) *Single-chain silicates,* $(SiO_3)_n^{2n-}$. These are called *pyroxenes.* (Note that n is not intended to mean the chain length, which is "infinite.") Examples are *enstatite* ($Mg[SiO_3]$) and *diopside* ($MgCa[(SiO_3)_2]$).

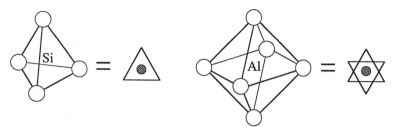

Figure 8.1 Representation of SiO_4^{4-} tetrahedron (left) and AlO_6^{9-} octahedron in terms of the view down three-fold axes.

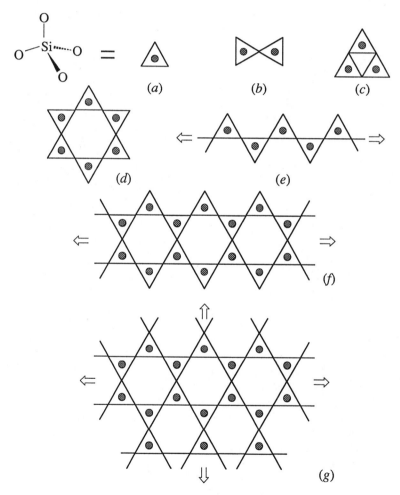

Figure 8.2 Silicate anion structures: (a) orthosilicate, (b) pyrosilicate, (c) three-silicate ring, (d) six-silicate ring, (e) pyroxene, (f) amphibole, and (g) phyllosilicate.

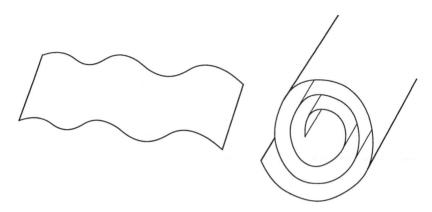

Figure 8.3 Serpentine sheet structures: antigorite (left) and chrysotile.

(f) *Double-chain silicates,* $(Si_4O_{11})_n{}^{6n-}$. These are known as *amphiboles*, such as *tremolite* $(Ca_2Mg_5(OH)_2[(Si_4O_{11})_2])$. These include the true asbestoses, such as *crocidolite* or *blue asbestos*, which owe their fibrous character and heat-insulating properties to the silicate chain structures. Most commercial asbestos, however, is *chrysotile*, which is actually a layer silicate that forms tubular fibers like a rolled carpet [see (g)].

(g) *Sheet silicates,* $(Si_2O_5)_n{}^{2n-}$. These are called *phyllosilicates* (phyllo = leaflike) and are characterized by easy cleavage parallel to the sheets of silicate (where these are flat). This is typified by *talc* $(Mg_3(OH)_2[Si_4O_{10}])$, which has a greasy feel and pearly luster. It is the chief constituent of soapstone (*steatite*), which is used for carvings by Inuit artists. The *serpentines* $(Mg_3(OH)_4[Si_2O_5])$ have sheet silicate structures in which the sheets tend to be curved. *Antigorite* has essentially flat sheets with gentle waves of radius of curvature 7.5 nm, while in chrysotile this curvature is all in one direction, creating tubular fibers of inner and outer diameters about 10 and 20 nm respectively and very great relative length (Fig. 8.3). Chrysotile is a high-melting, non-combustible fibrous solid that was formerly widely used as a fireproof thermal insulator and in construction materials under the name of asbestos. Asbestos fibers have been linked to the high incidence of lung cancer in asbestos workers and are considered to pose a threat to the population at large. Anyone handling commercial asbestos must wear appropriate breathing equipment and clothing, and under no circumstances should asbestos filters be used in preparing foodstuffs such as wines.

8.2   Aluminosilicates

Silica occurs widely in Nature as *quartz*, often in large characteristic crystals but also in the translucent agglomerations of microscopic crystals known as *chalcedony*, which includes cherts and flint. Other natural crystalline varieties of SiO_2 include *tridymite* and *cristobalite* (opal is a semi-precious stone that consists of microcrystalline, hydrous cristobalite). All involve three-dimensional networks of corner-linked SiO_4 tetrahedra.

If we were to replace some of the "Si^{4+}" ions with Al^{3+}, it would still be possible to have a three-dimensional Si—O—Al net, but cations would have to be incorporated in the structure to counterbalance the now anionic framework. Thus, for each substituent Al^{3+}, we must add, say, an Na^+, a K^+, or half a Ca^{2+} ion. The *feldspars*, which, along with quartz and micas (see below), are typical constituents of granites, can be viewed in this way:

$$K[AlSi_3O_8] \text{ orthoclase}$$
$$Na[AlSi_3O_8] \text{ albite}$$
$$Ca[Al_2Si_2O_8] \text{ anorthite.}$$

Plagioclase is a solid solution of albite in anorthite (or vice versa).

The *micas* have layer structures in which silicate sheets are combined with aluminate units; the aluminum ions can be octahedrally as well as tetrahedrally coordinated. For example, the mica *muscovite* contains both octahedral and tetrahedral Al^{3+}:

$$KAl_2(OH)_2[Si_3AlO_{10}].$$
$$\uparrow \qquad\qquad \uparrow$$
$$\text{octahedral} \qquad \text{tetrahedral}$$

The potassium ions are located between the flat aluminosilicate sheets (Fig. 8.4). The crystals cleave very easily parallel to the sheets, and the thin transparent flakes can be used for electrical insulation (e.g., in capacitors) or as furnace windows. *Phlogopite* ($KMg_3(OH)_2[Si_3AlO_{10}]$) has a similar structure but with Mg^{2+} in octahedral environments instead of Al^{3+}.

Many *clay minerals* have aluminosilicate layer structures. For example, in *kaolinite* ($Al_2(OH)_4[Si_2O_5]$; Fig. 8.5), the Al^{3+} are all in octahedral

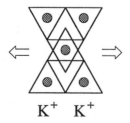

$$K^+ \quad K^+$$

Figure 8.4 Structure of muscovite (sheet viewed edgewise).

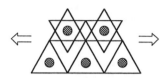

Figure 8.5 Layer structure of kaolinite, $Al_2(OH)_4[Si_2O_5]$ (sheet viewed edge-wise): Al^{3+} are octahedrally coordinated, "Si^{4+}" tetrahedrally.

locations. Clay minerals of the *montmorillonite* (*smectite*) type can absorb large amounts of water between the aluminosilicate layers, with the result that they can swell greatly. In petroleum recovery operations, superheated water is sometimes injected into formations such as the Athabasca tar sands of northern Alberta with the object of reducing the viscosity of the oil content and sweeping it to the surface in a production well. If, however, montmorillonite-type minerals form by hydrothermal reactions within the formation, the swollen products may block the flow path (perhaps to advantage, if regions stripped of oil are thereby sealed off). Typically, kaolinite and dolomite in a tar-sands matrix (which is mainly quartz) will react with high-temperature water to form montmorillonite.

8.3   Zeolites[6]

Zeolites, which merit special consideration in view of their growing importance in chemical engineering, are natural or synthetic aluminosilicates in which the anionic Al—O—Si framework encloses cavities linked by channels (Fig. 8.6). They are represented by:

$$M_{a/n}[(AlO_2)_a(SiO_2)_b] \cdot x H_2O$$

where $1 \leq b/a \leq 5$. Again, when we have n Al^{3+} ions substituting for "Si^{4+}" in what would otherwise be SiO_2, we must have a counterion M^{n+}. Here, however, the M^{n+} occupy the cavity/channel system, along with the x water molecules, and are often easily replaceable with other cations that may diffuse into these *pores* from a solution in which the zeolite may be immersed. Similarly, the water in the pores is driven out on heating, and the name "*zeolite*" comes from the Greek, meaning "*boiling stone.*"

Typical zeolites include the commonly encountered *analcite* (natural, also called *analcime*, $Na[AlSi_2O_6]\cdot H_2O$), *faujasite* (natural, (Na_2,Ca)-$[(AlO_2)_2(SiO_2)_4]\cdot 8H_2O$, the structural analogue of synthetic *zeolites X* and *Y*), and *zeolite A* (synthetic, $M_{12}^I[(AlO_2)_{12}(SiO_2)_{12}]\cdot 27H_2O$). The zeolite-like aluminosilicate framework mineral *sodalite* ($Na_8[(AlO_2)_6(SiO_2)_6]Cl_2$) contains the truncated octahedral structural unit known as the *sodalite cage*, shown in Fig. 8.6a, which is found in several zeolites. The corners of

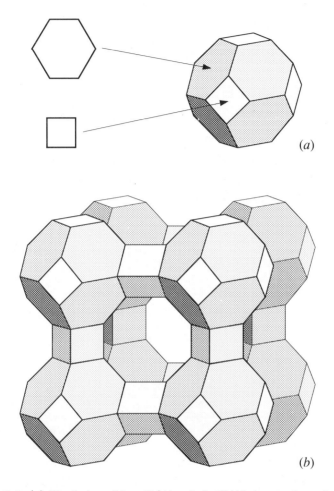

Figure 8.6 (a) The fusion of four-Si/Al and six-Si/Al rings to form the sodalite cage, and (b) the linking of eight sodalite cages to form the cavity-and-channel structure of zeolite-A, $M^I_{12}Al_{12}Si_{12}O_{48}\cdot27H_2O$. Each corner is an Si or Al atom; the edges represent bridging oxygens.

the faces of the cage are defined by either four or six Al/Si atoms, which are of course joined together through oxygen atoms. The zeolite A structure (Fig. 8.6b) is generated by joining sodalite cages through the four-Si/Al rings, as shown, so enclosing a cavity or *supercage* bounded by a cube of eight sodalite cages and readily accessible through the faces of that cube (*channels* or *pores*). The structural frameworks of faujasite and zeolites X and Y are similarly generated by joining sodalite cages together through their six-Si/Al faces. In zeolite A, the effective width of the pores is controlled by the nature of the cation M^+ or M^{2+} (see Table 8.2).

TABLE 8.2
Effective Channel Widths in Zeolite A

Zeolite type	Cation	Channel width(pm)
3A	K^+	300
4A	Na^+ [a]	400
5A	$\frac{1}{2}Ca^{2+}$ [b]	500

[a] smaller than K^+, so pores are wider
[b] size similar to Na^+, but only half as many

Zeolites have many uses, most importantly as *cation exchangers* (e.g., in water softening), as *desiccants* (i.e., drying agents), and as solid *acid catalysts*.

(a) *Zeolites as cation exchangers.* Water that contains significant amounts of Ca^{2+} or Mg^{2+} is said to be "hard." These ions cause soluble soaps (sodium stearate, palmitate, oleate, etc.) to precipitate as insoluble scums, which are an unsightly nuisance as well as a waste of soap:

$$Ca^{2+}(aq) + 2C_{17}H_{35}CO_2^-(aq) \rightarrow (C_{17}H_{35}CO_2)_2Ca(s). \quad (8.1)$$

Synthetic detergents such as alkyl sulfonates may not be precipitated, but Ca^{2+} and Mg^{2+} ions tend to form anion–cation complexes with them in solution and so reduce their effectiveness as detergents. Their detergent power derives from their ability to form *micelles* enveloping the greasy dirt particles with the anionic end of the detergent molecule outwards, so forming a colloidal dispersion in water (Fig. 8.7, cf. Section 12.2). The Ca^{2+} and Mg^{2+} can be replaced by Na^+, which is inoffensive in this context, by treatment with a zeolite in its sodium form, say, Na_2Z:

$$Na_2Z(s) + Ca^{2+}(aq) \rightleftharpoons CaZ(s) + 2Na^+(aq) \quad (8.2a)$$

$$Na_2Z(s) + Mg^{2+}(aq) \rightleftharpoons MgZ(s) + 2Na^+(aq). \quad (8.2b)$$

The equilibria above lie to the right, as the more highly charged cations have greater affinity for the anionic zeolite framework Z, and so a dilute solution of Ca^{2+} or Mg^{2+} (i.e., typical hard water) can be passed through a column of zeolite particles to replace essentially all of the divalent cations in solution with Na^+. All the Na_2Z will eventually be converted to CaZ or MgZ, but, since reaction 8.2 is

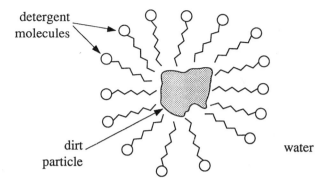

detergent molecules

dirt particle

water

Figure 8.7 Detergent action.

reversible under sufficiently forcing conditions, the Na-form zeolite can be regenerated by back-flushing the column with concentrated brine. The resulting solution of $NaCl/MgCl_2/CaCl_2$ is then rejected.

Alternatively, finely powdered zeolites may be used in the detergent powder itself as a *builder*, since zeolite particles smaller than 10 μm do not stick to clothing. Formerly, sodium polyphosphates were used extensively as detergent builders, that is, as additives to "tie up" the Ca^{2+} and Mg^{2+} in hard water as soluble complexes (or as a precipitate that washes away). However, many communities have found it necessary to ban phosphate detergents because of pollution problems (Section 3.6). This has created a major new market for zeolites in recent years.

(b) *Zeolites as desiccants.* If the water content is driven off (usually by heating to 350 °C in a vacuum), the dehydrated zeolite becomes an avid absorber of *small* molecules—especially water, of course. The size of the molecules that can be absorbed is strictly limited by the pore diameter, which is different for different zeolites (Table 8.2), so that a given zeolite (e.g., zeolite 3A) can be a highly selective absorber of, say, small amounts of water from dimethylsulfoxide (DMSO) solvent. For this reason, dehydrated zeolites are often called *molecular sieves.*

In order to be retained by the charged zeolite framework, a molecule has to be polar, as well as small enough to penetrate it. Thus, the angular (104°) H_2O molecule with its strongly polar bonds has a permanent dipole moment (as well as an ability to form hydrogen bonds to the anionic framework) and is strongly absorbed. However, monatomic helium, tetrahedral methane, and linear CO_2 have no permanent dipole moments and are not held in a typical zeolite, even though they can penetrate it easily. Consequently, zeolites are

used to remove water from the natural gas feedstock used in the cryo-
genic production of helium and in the preparation of liquefied natural
gas (LNG) for shipment. (Helium, from alpha particles produced in
radioactive decay in rocks, tends to accumulate in natural gas.)

(c) *Zeolites as solid acid catalysts.* The effective area of the anionic aluminosilicate framework in the pores of a zeolite is at least 100 times
the external surface area. Consequently zeolites are unusually effective as catalysts for reactions that are favored by aluminosilicate surfaces. Substitution of "Si⁴⁺" by Al³⁺ in a "silica" framework makes
it *acidic* and, potentially, *coordinatively unsaturated.* Suppose, for example, that we heat the NH_4^+ form of a zeolite. Ammonia is driven
off, and one H^+ remains to counterbalance each Al^{3+} that has sub-
stituted for a silicon. The protons, of course, are attached to oxygens
of the aluminosilicate framework:

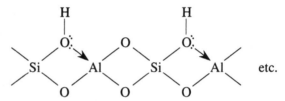

However, further heating drives off the elements of water as water
vapor:

Here, one Al has lost H^+ from the neighboring OH group, and one Si
has lost OH^- and therefore becomes coordinatively unsaturated, as
well as positively charged. This Si can therefore act as catalytic site
for the numerous organic reactions that are *Lewis-acid* (i.e., electron-
pair acceptor) catalyzed. Consider, for example, the dehydration of
alcohols to give olefins:

$$RCH_2CH_2OH + -Si^+ \longrightarrow RCH_2CH_2^+ + -Si-OH \quad (8.3)$$

$$RCH_2CH_2^+ \longrightarrow RCH=CH_2 + H^+ \quad (8.4)$$

$$H^+ + -Si-OH \longrightarrow H_2O + -Si^+ \text{ (regenerated).} \quad (8.5)$$

At higher temperatures, C—H and C—C bonds may be similarly broken. Thus, zeolite catalysts may be used for: (*i*) the alkylation of aromatic hydrocarbons (cf. the Friedel–Crafts reactions with $AlCl_3$ as the Lewis-acid catalyst), (*ii*) the *cracking* (i.e., the loss of H_2) of hydrocarbons, or (*iii*) the isomerization of alkenes, alkanes, and alkyl aromatics.

In (*iii*), zeolites have special merit in their ability to admit straight-chain, but not branched-chain, molecules into their pores. Thus, normal alkanes up to n-$C_{14}H_{30}$ can penetrate the pores of zeolite 5A, where the C—C or C—H bonds may be catalytically broken; the fragments, on re-emerging from the pores, can recombine as isomerized molecules. The reverse process is not possible, since the isomers, having branched chains, cannot enter the pores (Fig. 8.8). This is known as *shape-selective reforming*[7, 8] and is important in the upgrading of the gasoline fraction in petroleum refining. The object is to minimize the amount of n-alkanes (especially n-heptane) in fuel for internal combustion engines, since they promote *engine knock* (violent detonation rather than smooth burning of the air-fuel mixture in the cylinder), and to maximize the branched-alkane content, especially *iso*-octanes. In practice, short-chain alkanes and alkenes are normally used as feedstock for the shape-selective catalytic formation of *iso*-octanes at relatively low temperatures. Until recently, lead alkyls (Section 16.1) were added to most automotive fuels to help suppress engine knock, but they have been largely phased out because of the chronic toxicity of lead and its compounds. The most commonly used non-lead anti-knock additive is now methyl *tert*-butyl ether (MTBE; $CH_3OC(CH_3)_3$), which is made by the reaction of methanol with 2-methylpropene, $(CH_3)_2C{=}CH_2$. The latter is obtained by catalytic cracking of petroleum fractions to give but-1-ene, which is then shape-selectively isomerized on zeolitic catalysts.

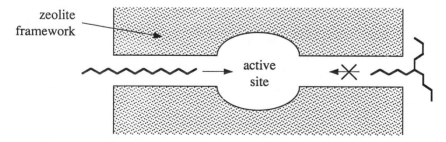

Figure 8.8 Shape-selective reforming. The straight-chain alkane (left) can enter the zeolite pore and penetrate to the catalytic site, whereas the branched-chain isomer cannot.

8.4   Silica and Silicate Glasses[9]

In addition to the crystalline forms of silica (quartz, tridymite, cristobalite, etc.) in which there are regular networks of Si atoms linked through oxygens, there exist amorphous modifications, in which the network is irregular and which are markedly more soluble in water than the crystalline polymorphs (Section 8.5). Furthermore, when quartz is fused at about $1900\,^\circ C$ (preferably under vacuum to remove bubbles from the viscous melt), on cooling it forms a *glass*. Strictly speaking, glasses are not solids but extremely viscous, supercooled liquids. Again, the molecular structure consists of irregular —Si—O—Si— networks. Whereas the quartz structure contains rings consisting of six SiO_4 tetrahedra, linked regularly into other such rings in three dimensions, the rings in silica glass are variable in size and the structure is random.

Fused silica has a very low coefficient of thermal expansion and so (unlike the ordinary soda-lime glass discussed below) does not shatter on shock-heating or -cooling. It can also be very transparent to visible light, if pure silica is made by the reaction of redistilled $SiCl_4$ (a moisture sensitive liquid, bp $57\,^\circ C$; Section 15.8) in the vapor phase with oxygen above $1100\,^\circ C$:

$$SiCl_4(g) + O_2(g) \xrightarrow{1100\,^\circ C} SiO_2(s) + 2Cl_2(g). \qquad (8.6)$$

If care is taken to exclude any hydrogen-containing material (which could produce light-absorbing OH groups in the glass) and traces of transition-metal ions (such as the ubiquitous iron) silica made in this way can be fused and drawn into optical fibers with 95% or better light transmission per kilometer. These are now being used to carry laser light, modulated with digitized telephone signals, over great distances and will undoubtedly revolutionize telecommunications.

Ordinary bottle- or window-glass, however, is a sodium calcium silicate (soda-lime glass) that may contain variable amounts of other minor constituents. A typical composition is 72% SiO_2, 14% Na_2O, 10% CaO, and 3% MgO. The last two constituents are necessary to prevent the glass from being water-soluble; sodium silicates form viscous aqueous solutions called *waterglass*. Even so, ordinary glass is slightly soluble in hot water and is quite quickly attacked by aqueous alkali (e.g., in dishwashers). It is easily made by fusing a mixture of sand (SiO_2), sodium carbonate (soda ash), and lime (CaO—usually containing some MgO if dolomitized limestone was calcined). The most common impurity in glass is iron, which, as Fe^{2+}, imparts a green color.

Soda-lime glass has a large coefficient of thermal expansion and cracks if subjected to sudden changes of temperature. If, instead of lime, the glass is made by fusing SiO_2 (81%) with Na_2CO_3 (equivalent to 4.5% Na_2O), boric oxide (B_2O_3, 12.5%), and Al_2O_3 (2%), a glass is obtained that is higher-melting than soda-lime glass and is very resistant to thermal shock.

This borosilicate glass is popularly known by the Corning Glass Company's tradename, *Pyrex*, and is widely used for kitchenware as well as for laboratory glassware.

Since glasses are supercooled liquids, they will eventually crystallize or *devitrify*, especially if held close to their melting points for extended periods (as novice glassblowers discover to their dismay). Devitrification renders glass brittle, opaque, and useless, but controlled crystallization in the presence of suitable nucleating agents such as TiO_2 can produce extremely tough microcrystalline glass ceramics, which usually include other oxides such as Al_2O_3.

8.5 Soluble Silicates[10, 11]

Silica (and silicates) are soluble in aqueous alkali. The silicate anions produced in solution may be regarded as derived from silicic acid (H_4SiO_4 or $Si(OH)_4$), which has a first pK_a of 9.47 at ionic strength 0.6 mol L^{-1}:

$$H_4SiO_4 \rightleftharpoons H^+ + H_3SiO_4^- \tag{8.7}$$

and a second pK_a of about 12.65:

$$H_3SiO_4^- \rightleftharpoons H^+ + H_2SiO_4^{2-}. \tag{8.8}$$

However, this is an oversimplification, since these monomeric anions are in rapid equilibrium with polymers such as the dimer (cf. Fig. 8.2b), cyclic trimer (8.2c), and a large number of other small silicate polymers, some of which have no counterparts in solid silicate minerals:

$$2(HO)_3SiO^- \rightleftharpoons {}^-O(HO)_2SiOSi(OH)_2O^- + H_2O, \text{etc.} \tag{8.9}$$

Clearly, the higher the concentration of dissolved silica, the more will be present as polymers. High excesses of alkali, however, favor the monomer ($H_3SiO_4^-$ and $H_2SiO_4^{2-}$) and the smaller polymers. On acidification, a gel of silicic acid, which is poorly soluble at ordinary temperatures, is produced. The solubility of silicic acid in water, and hence of solid silica in its various forms, becomes quite important at 200 °C or more (under pressure). This must be considered when, for example, superheated water is injected into oil-bearing formations, since the dissolved silica will be reprecipitated on cooling. This effect is intensified if the water is alkaline, because of the much higher solubility of $H_3SiO_4^-$ (etc.) salts relative to H_4SiO_4, and dissolution followed by reprecipitation of SiO_2 in aqueous alkali along a temperature gradient is used in growing large single crystals of quartz for the electronics industry.

Soluble silicates are usually manufactured by fusing high-purity sand (quartz) with soda ash or caustic soda, with the ratio SiO_2:Na_2O ranging from 0.5 to 3.4, depending upon the end use. Aqueous solutions of the products have a soapy feel, and indeed soluble silicates improve the detergent power of soaps. They have long been incorporated as builders in household wash powders (10 to 15%) and much more so in industrial detergents. Their mode of action is not entirely clear. These silicates are also effective in suppressing the corrosion of iron, probably by precipitating insoluble sodium iron silicates such as *acmite* ($NaFeSi_2O_6$) on the affected surfaces. Sodium silicate solutions are also employed in the hydrothermal production of synthetic zeolites, by reaction with aluminum oxide.

Precipitated silica gels may be dehydrated by heating, and the familiar light, glassy product is widely used as a desiccant and as a carrier for catalysts such as nickel.

References

1. V. M. Goldschmidt, "Geochemistry," Oxford University Press: London, 1954.

2. W. A. Deer, R. A. Howie and J. Zussman, "Rock-Forming Minerals," 5 vols., Longmans: London, 1978. A condensed version of this material is presented in W. A. Deer, R. A. Howie, and J. Zussman, "An Introduction to the Rock-Forming Minerals," Longmans: London, 1966.

3. J. E. Fergusson, "Inorganic Chemistry and the Earth," Pergamon Press: Oxford, 1982, pp. 169–174, and 305 ff.

4. F. A. Cotton and G. Wilkinson, "Advanced Inorganic Chemistry," 5th edn., Wiley-Interscience: New York, 1988.

5. A. F. Wells, "Structural Inorganic Chemistry," 5th edn., Oxford University Press: London, 1984.

6. (*a*) R. P. Townsend (ed.), "Properties and Applications of Zeolites," Special Publication No. 33, The Chemical Society: London, 1980. (*b*) D. W. Breck, "Zeolite Molecular Sieves," Robert E. Krieger Publishing: Malabar, Florida, 1984.

7. S. M. Csicsery, "Shape selective catalysis in zeolites," *Chemistry in Britain*, 1985, *21*, 473–477.

8. J. Haggin, "Shape selectivity key to designed catalysts," *Chemical and Engineering News*, December 13, 1982, 9–15.

9. (*a*) D. Kolb and K. E. Kolb, "The chemistry of glass," *Journal of Chemical Education*, 1979, *56*, 604–608. (*b*) C. M. Melliar-Smith, "Optical fibers and solid state chemistry," *Journal of Chemical Education*, 1980, *57*, 574–579. (*c*) R. L. Tiede, "Glass fibers—are they the solution?" *Journal of Chemical Education*, 1982, *59*, 198–200.

10. R. Thompson (ed.), "The Modern Inorganic Chemicals Industry," Special Publication No. 31, The Chemical Society: London, 1977, pp. 320–352.

11. J. S. Falcone, Jr. (ed.), "Soluble Silicates," A.C.S. Symposium Series No. 194, American Chemical Society: Washington, D.C., 1982.

Exercises

8.1 Show that the empirical formula of the silicate framework of a pyroxene will be $(SiO_3)^{2-}$; of an amphibole, $(Si_4O_{11})^{6-}$; and of a sheet silicate, $(Si_2O_5)^{2-}$.

8.2 How much zeolite A in its Na^+ form would be needed to soften completely one tonne of water, the hardness of which was equivalent to 150.0 ppm (i.e., 150.0 mg/kg) dissolved $CaCO_3$? Assume 100% efficiency of ion exchange. (Your calculated mass will therefore be a minimum, in practice.)

[*Answer:* 547 g.]

Chapter 9

Inorganic Solids as Heterogeneous Catalysts

9.1 Heterogeneous Catalysis

A CATALYST is a substance that increases the rate at which a chemical reaction approaches equilibrium, while not being consumed in the process. Thus, a catalyst affects the *kinetics* of a reaction, through provision of an alternative reaction mechanism of lower activation energy, but cannot influence the *thermodynamic* constraints governing its equilibrium.

There are major advantages to the use of an insoluble, non-volatile solid catalyst for fluid-phase reactions: loss of the catalyst is minimized, it does not significantly contaminate the reaction products, and it stays physically in place in the reaction chamber. Since such heterogeneous catalysts often must operate at quite high temperatures, most are refractory (i.e., high-melting) inorganic materials themselves, or else require a refractory support material such as a metal oxide, often alumina (Al_2O_3). In some special cases, the actual active catalyst is a liquid at reaction temperature and must be used on a refractory support. An important example is the *vanadium pentoxide catalyst* used in the contact process (Section 4.2) for the oxidation of SO_2 to SO_3 at 400 to 600 °C. In practice, the V_2O_5 is used with a K_2SO_4 *promoter* in the pores of an inert support material, and the mixture is molten under the reaction conditions. (V_2O_5 melts at 690 °C, but the fusion point of the mixture is lower.) However, such cases are exceptional.

Although the general definition of a catalyst given above emphasizes the acceleration of the approach to equilibrium, the *selectivity* of a catalyst is often of more importance than its overall catalytic activity (as in the catalytic oxidation of ammonia over platinum; Section 3.4).

As noted in Section 7.2, the activity of solids in heterogeneous catalysis

123

is due to the presence of unbalanced electrostatic attractions or unsatisfied covalence at the surface of a crystalline solid. In recent years, many new and powerful techniques to characterize the surfaces of solids and their adsorbates at the atomic level have become available; these include *low-energy electron diffraction* (LEED), *X-ray photoelectron spectroscopy* (XPS, in which one measures the energies of electrons emitted from surface atoms excited by X-rays), *infrared spectroscopy*, *extended X-ray absorption fine structure analysis* (EXAFS), *thermal desorption spectroscopy* (TDS), and *scanning tunneling microscopy* (STM). The cost of the sophisticated instruments, the complexity of surface phenomena, and the difficulty of relating laboratory observations under clean conditions to what goes on in the "dirty" environment of an industrial catalytic reactor, have hindered progress in placing heterogeneous catalysis on a truly scientific footing, but it is no longer the "black art" of a few years ago.[1] Students will find G. C. Bond's concise monograph[2] a useful introduction to the subject. The Royal Society of Chemistry publishes a continuing series of periodical reports on recent advances in catalysis.[3]

9.2   Physical Adsorption and Chemisorption

The presence of unbalanced attractions at the surface of a solid—let us say a metal such as nickel—means that small molecules will tend to become rather loosely attached to the surface in one or (more likely) several molecular layers with an exothermic adsorption energy ranging to about -20 kJ mol^{-1} for non-polar molecules. (The term *adsorption* is used to denote surface sorption without penetration of the bulk solid, which would be called *absorption*.) No chemical bonds are formed or broken. This state is usually called *physical adsorption* or *physisorption*.

If, however, the adsorbate forms chemical bonds with the surface atoms, the adsorption process is called *chemisorption*. Chemisorption can be quite strongly exothermic (-40 to -800 kJ mol^{-1}) but involves only the first monomolecular layer of adsorbate.

As Fig. 9.1 shows, with reference to the adsorption of hydrogen on nickel, there is no activation energy for physical adsorption, but the transition to the chemisorbed state (in which the H—H bond is broken in favor of two Ni—H) bonds can be made by surmounting an activation energy barrier E_a. This barrier to the breaking of the H—H bond is seen to be small compared with the atomization energy of 435 kJ mol^{-1} for gaseous H_2. This is the key to the activation of H_2 by nickel surfaces—the strong H—H bond is broken for further reactions with only a small investment of energy:

$$2Ni + H_2 \rightleftharpoons 2Ni\cdots H_2 \rightleftharpoons 2NiH. \qquad (9.1)$$

metal	gas	physical adsorption	(surface only) chemisorption

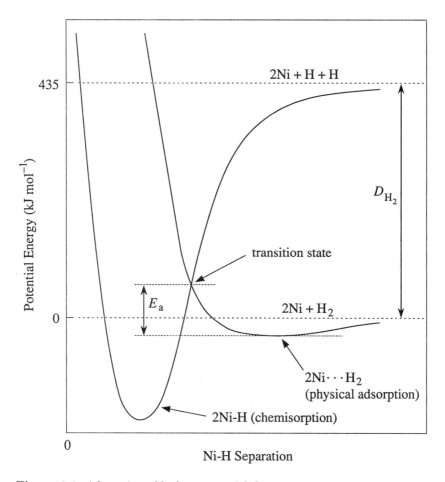

Figure 9.1 Adsorption of hydrogen on nickel.

In the activation of O_2 by a metal M, the double $O\!\!=\!\!O$ bond may be either weakened to a single bond or else broken altogether:

$$
\begin{array}{ccccc}
\underset{\displaystyle\substack{\vdots\\-M-M-\\|\quad\;\;|}}{O\!\!=\!\!O} & \rightleftharpoons & \underset{\displaystyle\substack{|\quad|\\-M-M-\\|\quad\;\;|}}{O\!\!-\!\!O} & \rightleftharpoons & \underset{\displaystyle\substack{\|\quad\;\|\\-M\quad M-\\|\quad\;\;|}}{O\quad O}
\end{array}\,.
\tag{9.2}
$$

If the energy of chemisorption is *too* strongly negative, the adsorbed species will become *un*reactive towards others and will "poison" the catalyst by virtue of using up its reactive sites. The reactant molecules must, however, be sufficiently strongly adsorbed to allow surface reactions to take place at all. Thus, given that chemisorption of the reactants *does* occur, catalytic activity will be *inversely* related to their (negative) energy of adsorption. For a solid to be an effective catalyst for a given reaction, it must interact with the reactants just (but *only* just) firmly enough to cause chemisorption. Since it is unlikely that both reagents in a bimolecular reaction will interact equally strongly with the surface, the critical question is whether the solid can just barely chemisorb the molecule that is the less reactive towards surfaces.[4]

9.3   Transition Metals as Catalysts

Metal catalysts are normally associated with *hydrogenation* (or *dehydrogenation*) reactions such as the Fe-catalyzed Haber process (Section 3.8) or the nickel-catalyzed "hardening" (hydrogenation) of edible vegetable oils to make margarine and similar products. To a lesser extent, they also feature in oxidation–reduction processes such as the oxidation of ammonia to nitric oxide (Section 3.4) and the clean-up of automobile exhausts (Section 2.4), in both of which platinum or its alloys are used. For all metals except gold, the chemisorption bond strength sequence for common gaseous reagents is usually:

$$O_2 > \text{alkynes} > \text{alkenes} > CO > H_2 > CO_2 > N_2. \qquad (9.3)$$

The electronic nature of the interaction between a transition metal and one of these molecules in chemisorption is not easily explained, but it seems that the valence (mainly d) orbitals of the metal should have sufficient electronic vacancies to permit interaction with the electron-rich adsorbate molecules, but not so many as to allow the interaction to become too strong (cf. poisoning). In any event, the strength of the interactions with a given molecule rises as we go from right to left across the transition metals in the Periodic Table. The result is that metals of the Sc, Ti, V, Cr, and Mn groups generally interact so strongly, even with N_2, that they are inevitably "poisoned" by adsorbates. Going further to the right in the Periodic Table, we find that:

(*a*) iron, ruthenium, and osmium *barely* chemisorb N_2 (and, implicitly, the rest of series 9.3 are held more firmly);

(*b*) cobalt and nickel *barely* chemisorb CO_2 and H_2;

(*c*) the Co and Ni analogues rhodium, palladium, iridium, and platinum *barely* chemisorb H_2 but *not* CO_2; and

(*d*) copper, silver and gold *barely* chemisorb CO and ethylene. (Gold, however, does not usually chemisorb oxygen at low temperatures, apparently for kinetic reasons.)

These relative chemisorption strengths enable us to make some simple predictions regarding suitable metal catalysts for specific reactions. For example, a catalyst for the Haber process must chemisorb both N_2 and H_2, but not too strongly. Since N_2 is the less readily bound, we choose Fe, Ru, or Os. The latter two are expensive, so our best choice is iron—usually finely divided, on a suitable refractory support.

Similarly, we can ask what would be the best catalyst for hydrogenating an olefin such as ethylene. Since olefins (alkenes) are more strongly chemisorbed than hydrogen, we choose a metal that just barely chemisorbs H_2—this means Co, Rh, Ir, Ni, Pd, or Pt. In practice, nickel is the least expensive choice. Again, it should be finely divided (maximum surface area) for greatest catalytic efficiency and dispersed on the internal surfaces of a porous support such as alumina with surface area on the order of $200 \text{ m}^2 \text{ g}^{-1}$.

9.4 Defect Oxides and Sulfides in Catalysis

Non-stoichiometric metal oxides are effective catalysts for a variety of oxidation-reduction reactions (as might be expected) since the variable valence of the constituent ions enables the oxide to act as a sort of "electron bank." They resemble metals in that they can also catalyze hydrogenation and alkene isomerization reactions. However, on zinc oxide (for instance) these two processes are independent of each other, whereas hydrogen is necessary for isomerization to occur on metals. The mode of action of semi-conducting metal oxides can be illustrated with reference to their catalysis of the oxidation of carbon monoxide:

$$CO + \tfrac{1}{2}O_2 \rightarrow CO_2. \tag{9.4}$$

p-Type metal oxides, such as $Ni_{1-x}O$, can accommodate excess positive charge in the lattice and can thus be expected to adsorb oxygen to form anions O_2^-, O^-, O^{2-}, etc., on their surfaces. It turns out that $O^-(\text{ads})$ is the most active adsorbed oxygen species in terms of catalytic activity. It is not, of course, a commonly encountered anion under circumstances other than the special environment of a crystal surface. Within the lattice, the electrons are provided by:

$$M^{n+} \rightarrow M^{(n+1)+} + e^-. \tag{9.5}$$

Thus, for NiO, we have:

$$\tfrac{1}{2}O_2(g) + Ni^{2+} \rightarrow O^-(\text{ads}) + Ni^{3+} \tag{9.6}$$
$$O^-(\text{ads}) + CO(\text{ads}) \rightarrow CO_2(g) + e^- \tag{9.7}$$

followed by the regeneration of Ni^{2+}:

$$e^- + Ni^{3+} \rightarrow Ni^{2+}. \tag{9.8}$$

n-Type metal oxides, such as ZnO, tend to give up oxygen and accommodate excess electrons in their lattice defects:

$$CO(g) + 2O^{2-}(lattice) \rightarrow CO_3{}^{2-}(lattice) + 2e^- \tag{9.9}$$

$$\tfrac{1}{2}O_2(g) + 2e^- \rightarrow O^{2-} \tag{9.10}$$

$$CO_3{}^{2-}(lattice) \rightarrow CO_2(g) + O^{2-}(lattice). \tag{9.11}$$

Thus, the two oxides consumed in reaction 9.9 are regenerated in reactions 9.10 and 9.11.

Metal sulfides play an important role in catalyzing a wide variety of hydrogenations (of fats, coal, olefins, etc.) and also desulfurization reactions, the importance of which was stressed in Section 2.5. *Molybdenum disulfide* is important among these defect catalysts and can be made to function as an *n*-type ($Mo_{1+x}S_2$) or *p*-type ($Mo_{1-x}S_2$) semiconductor by exposure to an appropriate mixture of H_2S and hydrogen at temperatures on the order of 600 °C. The equilibrium:

$$H_2S(g) \rightleftharpoons H_2(g) + \tfrac{1}{2}S_2(g) \tag{9.12}$$

then *buffers* the S_2 pressure and hence controls the composition of the solid MoS_2. Alternatively, the concentrations of positive or negative defects can be controlled by *doping* MoS_2 with cobalt(II) or antimony(V), respectively.

Molybdenum disulfide catalysts are usually employed with hydrogen gas, as in the *hydrodesulfurization* of butyl mercaptan, C_4H_9SH. The product is a butane–butene mixture, but Co-doped MoS_2 gives almost exclusively butane. This suggests that butene is the initial product, but is then hydrogenated to butane, since *p*-type MoS_2 phases are excellent catalysts for hydrogenations in general:

$$CH_3CH_2CH_2CH_2SH \xrightarrow{MoS_2} CH_3CH_2CH{=}CH_2 + H_2S \tag{9.13}$$

$$CH_3CH_2CH{=}CH_2 + H_2 \xrightarrow[MoS_2]{p\text{-type}} CH_3CH_2CH_2CH_3. \tag{9.14}$$

A typical commercial hydrodesulfurization catalyst might contain 14% MoO_3 and 3% CoO, on an alumina support. These oxides are converted to Co-doped MoS_2 by exposure to H_2S/H_2 under carefully controlled conditions of temperature and partial pressures.

9.5 Catalysis by Stoichiometric Oxides

Several *insulator* oxides, such as Al_2O_3, MgO, and SiO_2, can hydrate reversibly to give $Al(OH)_3$, AlO(OH), $Mg(OH)_2$, etc. and not surprisingly can catalyze dehydration reactions such as the conversion of alcohols to olefins at elevated temperatures:

$$R\text{—}CH_2OH(g) \xrightarrow[Al_2O_3]{350–400\,°C} R\text{—}CH{=}CH_2(g) + H_2O(g). \qquad (9.15)$$

The Al_2O_3 in reaction 9.15 probably functions as a Lewis acid, since alcohols generally dehydrate via the formation of a carbocation, $R\text{—}CH_2{}^+$. The role of Al_2O_3 as a catalyst for the Claus process (Section 4.1) may be similarly viewed.

Zeolites, which can be regarded as being derived from Al_2O_3 and SiO_2, function catalytically in much the same way (Section 8.3). In addition, they catalyze isomerization, cracking, alkylation, and other organic reactions. Although the true zeolites discussed in Chapter 8 are aluminosilicates, a structurally related class of microporous materials based on aluminum phosphate, $AlPO_4$, has recently been described.[5] These new solids have zeolite-like cavities and channels at the molecular level and can function as shape-selective catalysts.

References

1. (*a*) Thomas H. Maugh, II, "Picture of surfaces begins to emerge," *Science*, 1983, *219*, 944–947. (*b*) Thomas H. Maugh, II, "When is a metal not a metal?" *Science*, 1983, *219*, 1413–1415. (*c*) Thomas H. Maugh, II, "Catalysis; no longer a black art," *Science*, 1983, *219*, 474–477. (*d*) Thomas H. Maugh, II, "Clusters provide unusual chemistry," *Science*, 1983, *220*, 592–595.

2. G. C. Bond, "Heterogeneous Catalysis: Principles and Applications," 2nd edn., Clarendon Press: Oxford, 1987.

3. C. Kemball (ed.), "Catalysis," Specialist Periodical Reports, Vol. 1, Royal Society of Chemistry: London, 1976, and subsequent issues.

4. C. Kemball, "Chemical processes on heterogeneous catalysts," *Chemical Society Reviews*, 1984, *13*, 375–392.

5. J. Haggin, "Aluminophosphates broaden shape selective catalyst types," *Chemical and Engineering News*, June 20, 1983, 36–37.

Exercises

9.1 According to *Chemical and Engineering News* for October 13, 1980,
p. 21, Japanese chemists have succeeded in obtaining good yields
of methane by reaction of H_2 with a mixture of carbon monoxide
and dioxide, at temperatures as low as 270 °C, by use of a special
mixed catalyst containing nickel as the most important metallic constituent. Why is nickel used? In the same vein, why is platinum or
platinum–rhodium alloy (but *not* nickel) used in catalytic converters
for automobile exhausts (see also Section 15.4)?

9.2 Suggest reaction conditions and an appropriate catalyst for the reverse of the water–gas shift reaction (Eqn. 3.7).

9.3 Two important ways in which heterogeneously catalyzed reactions
differ from their homogeneous counterparts are the definition of the
rate constant k' and the form of its dependence upon temperature T.
The heterogeneous rate equation relates the rate of decline of the
concentration (or partial pressure) c of a reactant to the fraction f of
the catalytic surface area that it covers when adsorbed. Thus, for a
first-order reaction,

$$\text{rate} = -\frac{dc}{dt} = k'f$$

if the products are not adsorbed. At low enough temperatures, $f \approx 1$,
and the usual Arrhenius equation gives the true activation energy
for the heterogeneous reaction. Adsorption, however, is invariably
exothermic, so that, as T is increased, f will eventually decrease and,
in the high temperature limit, will approach zero. What will the
apparent activation energy E_{appt} be, in the high temperature limit?
Sketch the dependence of ln(rate) upon $1/T$.

9.4 Show that a substance that is an effective catalyst for a particular
reaction will also catalyze the *reverse* reaction (cf. Section 1.4 and
Exercise 1.2).

9.5 It is possible, by progressively "dealuminating" certain zeolites, to
make solids that approach SiO_2 in composition while still retaining
zeolite architecture. What technological value might such materials
have, and why?
Note: One such substance, the Socony–Mobil catalyst ZSM-5, cat-
alyzes the conversion of methanol to gasoline-type alkanes; the first
step is apparently the formation of dimethyl ether.

Chapter 10

Interstitial Compounds of Metals

10.1   The Nature of Interstitial Compounds

WE HAVE SEEN (Section 1.7) that anions that have large, easily polarized electron clouds will tend not to form stable ionic crystals but, instead, will become involved in covalent bonding. The large, squashy hydride ion H^-, with only two electrons and an ill-defined ionic radius of some 140 to 200 pm, is known in the white solid salts NaH, CaH_2, etc. These salts react rapidly and vigorously with water and other potential sources of H^+ to give H_2 gas; consequently, they are sometimes used as powerful reducing agents. The organic chemists' reductants, *lithium aluminum hydride* and *sodium borohydride*, contain the complex ions (Section 11.2) AlH_4^- and BH_4^- rather than free H^-. In the same way, true salts containing the highly charged monatomic anions boride (B^{3-}), carbide (C^{4-}), and nitride (N^{3-}) are rarely encountered. The familiar *calcium carbide*, once widely used to fuel lamps and still favored for this purpose by cavers, is actually an acetylide, i.e., a salt of the $^-C{\equiv}C^-$ ion, which is made by the reduction of lime with coke in the electric furnace and which is hydrolyzed by water to give acetylene (ethyne) gas, which burns with a luminous flame:

$$CaO + 3C \xrightarrow{2100\,°C} CaC_2 + CO \qquad (10.1)$$

$$CaC_2(s) + 2H_2O(l) \longrightarrow Ca(OH)_2(s) + HC{\equiv}CH(g). \quad (10.2)$$

There are, however, many technologically important compounds of metals called *hydrides*, *carbides*, and *nitrides* in which the non-metal is present more as neutral atoms than as anions and is inserted into the interstices (octahedral and tetrahedral holes) of what may be regarded as the expanded lattice of the elemental metal. Hägg made the empirical observation that this is possible when the *atomic* radius of the non-metal is not greater than 0.59 times the *atomic* radius of the host metal. These interstitial compounds are, therefore, sometimes called *Hägg compounds*.[1, 2] They are, in effect, *interstitial* solid solutions of the non-metal in the metal (as distinct from *substitutional* solid solutions, in which actual lattice atoms are replaced, as in the case of gold–copper and many other alloys). Not surprisingly, they tend to be non-stoichiometric. This picture is obviously naive, in that some degree of bonding based on electron sharing is probably involved, but the nature of this bonding is not well understood.

Atoms of C or N can enter the O-holes of a close-packed metal lattice to give, in the limit, one non-metal for each metal atom (as in the case of TiC and TiN, which have the NaCl structure). Usually, however, there are fewer interstitial than lattice atoms, as with $PdH_{0.6}$, the formation of which is presumably responsible for the ease with which hydrogen gas diffuses through palladium metal. If the non-metal atoms are small enough, as with hydrogen, they may enter the T-holes instead, in which case a limiting composition of two non-metal atoms per lattice atom is theoretically possible, although not usually attained. Titanium and zirconium form hydrides TiH_x and ZrH_x with a distorted fluorite structure (Fig. 6.8), where x ranges up to 1.73 and 1.92, respectively. The tendency of Ti and Zr to hydride, so becoming embrittled, presents a serious technical problem wherever these otherwise tough and highly corrosion-resistant metals are used in contact with hydrogen gas, e.g., in pressurized heavy-water nuclear power reactors in which the fuel elements are usually sheathed in zircaloy (a Zr alloy).

As we go from left to right across the transition metals in the Periodic Table, the metal atoms become smaller, much as in the *lanthanide contraction* (Section 1.5). Furthermore, the atoms of elements of the first transition series are smaller than those of corresponding members of the second and third. Consequently, interstitial carbides are particularly important for metals toward the lower left of the series, as with TiC, ZrC, TaC, and the extremely hard tungsten carbide WC, which is used industrially as an abrasive or cutting material of almost diamond-like hardness. The parallel with trends in chemisorption (Section 9.2) will be apparent.

10.2   Carbon Steels

The carbon–metal atomic radius ratio for the room-temperature form of iron, α-Fe, is 0.60, just in excess of the Hägg limit. Consequently, the ability of α-Fe to accept interstitial C is marginal—only 0.022 weight % or 0.06 atom % C can be accommodated in the random solid solution known as *ferrite*.

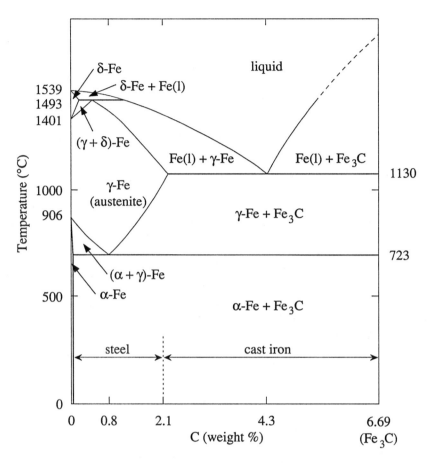

Figure 10.1 The iron–carbon phase diagram.

There is, however, a well-defined solid iron carbide phase known as *cementite*, Fe_3C (6.69 weight % C). Furthermore, as the temperature is increased toward the melting point of 1539 °C, the crystal structure of pure iron changes as follows:

α-Fe (ferromagnetic):   up to   766 °C; bcc
β-Fe (non-magnetic):   766 to   906 °C; bcc
γ-Fe (non-magnetic):   906 to 1401 °C; fcc
δ-Fe (non-magnetic): 1401 to 1539 °C; bcc.

The γ-Fe can accommodate up to 2.11 weight % C in the O-holes of its fcc structure. This solid solution is called *austenite*. The Fe–C phase diagram[3, 4] is shown in outline in Fig. 10.1.

If austenite is cooled *slowly* towards ambient temperature, the dissolved carbon in excess of 0.022 weight % comes out of solid solution as *cementite*, either in continuous layers of Fe_3C (*pearlite*) or as layers of separated Fe_3C grains (*bainite*). In either case, the iron is soft and grainy, as with cast iron. If, on the other hand, the hot austenite is cooled *quickly* (i.e., *quenched*), the γ-Fe structure goes over to the α-Fe form without crystallization of the interstitial carbon as cementite, and we obtain a hard but brittle steel known as *martensite*, in which the C atoms are still randomly distributed through the interstices of a strained α-Fe lattice. Martensite is kinetically stable below 150 °C; above this temperature, crystallization of Fe_3C will occur in time.

If martensite is reheated to between 200 and 300 °C for an appropriate time and is then re-quenched, a *partial* crystallization of Fe_3C will occur, and a tough steel, *sorbite*, with properties intermediate between martensite and pearlite or bainite, will be obtained. This process is known as the *tempering* of steel.[5]

High-carbon austenitic structures can be preserved at ambient temperatures if the iron has been alloyed with sufficient nickel or manganese, since these metals form solid solutions with γ-Fe but not with α-Fe. If over 11% chromium is also present, we have a typical *austenitic stainless steel.* Such steels are corrosion resistant, non-magnetic, and of satisfactory hardness, but, because the α-Fe $\rightleftharpoons$ γ-Fe transition is no longer possible, they cannot be hardened further by heat treatment.

These observations are summarized in Fig. 10.2.

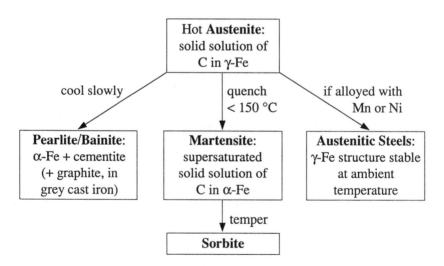

Figure 10.2 Heat treatment of solid iron containing carbon.

10.3 Nitriding

Although nitrogen gas is normally considered to be unreactive, it forms an interstitial nitride with titanium sufficiently easily that the production of titanium metal has to be carried out in an argon atmosphere (the *Kroll process*, Section 15.8). Similarly, oxygen is used in preference to air in modern steelmaking, in part because some nitriding of the steel may otherwise occur, leading to a more brittle product (Section 15.7). Embrittlement of low-alloy steels by nitriding has recently been recognized as a serious problem in ammonia plants after ten years or more of operation.

On the other hand, surface nitriding or *case hardening* of a steel specimen can improve its durability. Both N and C atoms are introduced interstitially to the Fe lattice on the surface of the steel by immersing it in, for example, a solution of sodium cyanide (NaCN) in a molten Na_2CO_3/NaCl mixture at 870 °C.

References

1. F. A. Cotton and G. Wilkinson, "Advanced Inorganic Chemistry," 5th edn., Wiley-Interscience: New York, 1988.

2. A. F. Wells, "Structural Inorganic Chemistry," 5th edn., Oxford University Press: London, 1984.

3. M. G. Fontana and N. D. Greene, "Corrosion Engineering," 2nd edn., McGraw-Hill: New York, 1978, p. 162.

4. R. A. Heidemann, A. A. Jeje and M. F. Mohtadi, "An Introduction to the Properties of Fluids and Solids," University of Calgary Press: Calgary, Alberta, 1984, pp. 106–113.

5. G. J. Long and H. P. Leighly, Jr., "The iron–iron carbide phase diagram," *Journal of Chemical Education*, 1982, *59*, 948–953.

Exercises

10.1 In a much-publicized study in 1989, Pons and Fleischmann claimed to have observed *cold fusion* of deuterium (heavy hydrogen, D) nuclei within palladium electrodes that were being used to electrolyze D_2O. Had this been the case, what other electrode materials might also have shown the same phenomenon?

10.2 Hägg found that metals can accommodate interstitial non-metal atoms of radius up to 59% of that of the metal atoms. Show that, in this limiting case, accommodation of the non-metal atoms in the octahedral holes of a face-centered cubic metal lattice should result in an expansion of the unit cell dimension by 12.4%. (*Hint:* Review the radius ratio rules in Chapter 6.)

Chapter 11

Ions in Solution

11.1   The Born Theory of Solvation[1-4]

WE SAW in Section 6.5 that the dominant factor in the energy balance
governing the existence of ionic solids is the *lattice energy*, U. For ionic
compounds in solution, the corresponding quantity is the *solvation energy*,
which results from the interaction of the ions with the solvent, rather than
with each other. There must, of course, be equal sums of anionic and
cationic charges in the solution (to preserve electrical neutrality), and the
ions will interact with each other significantly except at high dilutions (as
will be considered in Section 11.4). However, it is the *free energy of solvation*
$\Delta G^\circ_{\text{solv}}$ at infinite dilution that replaces U when the Born–Haber cycle is
adapted for solutions.

Suppose we have a spherical conductor of radius R in a vacuum and we
bring up a total charge q from infinite distance, in infinitesimal increments
dq. The work W_0 done in charging the sphere against the charge itself, as
it builds up, will be:

$$W_0 = \int_0^q \frac{q}{4\pi\varepsilon_0 R} dq = \frac{q^2}{2R(4\pi\varepsilon_0)}. \tag{11.1}$$

For the same process in a solvent of *relative permittivity* (*dielectric constant*) D, the corresponding work W_D is given by:

$$W_D = \frac{q^2}{2RD(4\pi\varepsilon_0)}. \tag{11.2}$$

Consequently, if we transfer the charged sphere from a vacuum to the
solvent, its electrostatic self-energy is *lowered* by an amount ΔW:

$$\Delta W = \frac{q^2}{2R(4\pi\varepsilon_0)} \left(1 - \frac{1}{D}\right). \tag{11.3}$$

TABLE 11.1
Relative Permittivities (Dielectric Constants) *D*
and Normal Liquid Ranges
of Some Common Solvents

Solvent	mp (°C)	bp (°C, 1 bar)	D (at 25°C)
Hydrocyanic acid, HCN	−13	25	123.
Sulfuric acid, 100% H_2SO_4	10	290–317	100.0
Water, H_2O	0	100	78.3
Propylene carbonate	−49	242	64.4
DMSO, $(CH_3)_2SO$	18	189	46.6
Acetonitrile, CH_3CN	−45	82	36.2
Methanol, CH_3OH	−98	65	32.6
Acetone, $(CH_3)_2CO$	−94	56	20.7
Sulfur dioxide, $SO_2(l)$	−96	−10	15.4
Ammonia, $NH_3(l)$	−78	−33	16.9
Acetic acid, CH_3COOH	17	118	6.2
n-Hexane, C_6H_{14}	−95	69	1.9

Born pointed out that a mole of gaseous ions of radius r and charge ze would be similarly stabilized on transfer to the solvent, the work difference being $-\Delta G^\circ_{solv}$. Strictly speaking, the foregoing argument is not really applicable to ions, especially if $z = \pm 1$, since one cannot charge up an ion with increments less than $\pm e$, but the result is still valid:

$$- \Delta G^\circ_{solv} = \frac{N(ze)^2(1 - D^{-1})}{8\pi\varepsilon_0 r}. \tag{11.4}$$

For polar solvents like water, DMSO, or 100% sulfuric acid, D^{-1} is quite small compared to unity (Table 11.1) so the electrostatic self-energy of a gaseous ion is almost entirely eliminated on transferring the ion to a polar solvent. For an ionic compound to be freely soluble in a given solvent, the solvation energies of its anions and cations must outweigh the lattice energy sufficiently, otherwise an ionic solid results instead. Ionic solids are therefore not usually very soluble in solvents of low D.

Solubility equilibria can be treated in terms of the free energy of solution (as outlined in Section 1.2), and their temperature dependences can be related to the enthalpy (and entropy) of solution. The Born–Haber cycle for solutions, in terms of enthalpies, can be written as follows:

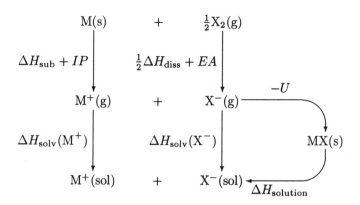

For water, typical values of the heats of hydration (i.e., solvation by water) are:

$$M^+ = -400 \text{ kJ mol}^{-1}$$
$$M^{2+} = -1700$$
$$M^{3+} = -4500$$
$$X^- = -400.$$

The sums of these enthalpies for typical aqueous MX or MX_2 are seen to be comparable with the corresponding $-U$ values given in Chapter 6. The balance, however, is delicate, and solubility equilibria are not accurately predictable from the simple Born model. The data of Table 11.1 may serve as a rough guide to the effectiveness of various solvents for electrolytes: HCN should be excellent, but is too volatile (and, of course, is extremely toxic); H_2SO_4 should be better than water, if its extreme acidity can be tolerated; propylene carbonate is good and offers a very wide liquid range, and so has been extensively investigated as a solvent for electrolytes in advanced fuel cells (Section 13.5); liquid ammonia is generally poor, acetic acid is worse, and hydrocarbons such as hexane are useless.

Table 11.1 does not explain, however, why the ionic solid silver chloride, which is well-known for its poor solubility in water, dissolves readily in liquid ammonia, despite a much less favorable relative permittivity. The reason is that the silver ion interacts strongly with specific ammonia molecules to give a *complex ion*, usually formulated as $Ag(NH_3)_2{}^+$. Indeed, Ag^+ interacts with water in aqueous solution to give a complex believed to be $Ag(OH_2)_4{}^+$, but the energy of the interaction is less than with NH_3. The upshot of this is that the Born theory of solvation fails because it regards the solvent as a continuous dielectric, whereas in fact solute ions (especially metal cations with $z > 1$) often interact in a specific manner with solvent molecules, and in any event the molecular dielectric is obviously very "lumpy" on the scale of the ions themselves.

11.2   Metal Complexes[1, 2, 5]

Cations, particularly those with $z > 1$, are considerably smaller than their parent neutral atoms, and so the electric potential at their surfaces ($\propto z/r$) is very high. Consequently, the electron cloud of a neighboring molecule or anion becomes displaced towards a cation and, if there is an unused pair of electrons (*lone pair*) in the valence shell of the molecule or anion, it may actually become involved in a sort of covalent bond with the cation, using empty valence-shell orbitals on the cation. Such a bond, in which the *donor* molecule (or anion) provides both bonding electrons and the *acceptor* cation provides the empty orbital, is called a *coordinate* or *dative* bond. The resulting aggregation is called a *complex*. Actually, any molecule with an empty orbital in its valence shell, such as the gas boron trifluoride, can in principle act as an electron pair acceptor, and indeed BF_3 reacts with ammonia (which has a lone pair, $:NH_3$) to form a complex $H_3N:\rightarrow BF_3$. Our concern here, however, is with metal cations, and these usually form complexes with from two to twelve donor molecules at once, depending on the sizes and electronic structures of the cation and donor molecules. The bound donor molecules are called *ligands* (from the Latin *ligare*, to bind), and the acceptor and donor species may be regarded as a *Lewis acid* and a *Lewis base*, respectively.

The nature of complex compounds of metals was first made clear by Alfred Werner of the University of Zürich around 1900, largely from his investigations of cobalt(III) complexes, in which ligand substitution is slow—a multitude of mixed-ligand complexes of cobalt(III) can therefore be isolated and characterized. For example, Werner recognized that the various cobalt(III) chloride *ammines* (ammonia complexes—the nomenclature of complexes is explained in Appendix E) that were known all contained a six-coordinate cobalt(III) complex with, say, x chloride ions bound as ligands, $(6 - x)$ ammonia ligands, and $(3 - x)$ chloride ions as such in the crystal lattice. On dissolving one of these ammines in water, the $(3 - x)$ free Cl^- ions could be precipitated at once as solid silver chloride on adding

TABLE 11.2
Cobalt(III) Chloride Ammines

Solid compound	Color	Ionized Cl^-	Formulation as Complex
$CoCl_3 \cdot 6NH_3$	yellow	3	$[Co(NH_3)_6]Cl_3$
$CoCl_3 \cdot 5NH_3$	purple	2	$[Co(NH_3)_5Cl]Cl_2$
$CoCl_3 \cdot 4NH_3$	green	1	*trans*-$[Co(NH_3)_4Cl_2]Cl$
$CoCl_3 \cdot 4NH_3$	violet	1	*cis*-$[Co(NH_3)_4Cl_2]Cl$

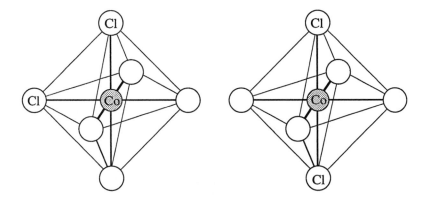

Figure 11.1 *Cis*- (left) and *trans*-dichlorotetraamminecobalt(III) ion isomers. The unmarked spheres represent NH_3 ligands.

Figure 11.2 Square planar *cis*- (left) and *trans*-dichlorodiammine-platinum(II).

excess silver nitrate solution, whereas the remaining x chlorides would be precipitated only on boiling (Table 11.2).

The green and violet tetraammines have the same chemical composition, i.e., they are isomers and are the *only* two isomers with this composition. Werner realized that this was possible only if the six ligands were deployed about the cobalt(III) center in an *octahedral* arrangement (cf. octahedral coordination in solids, Section 6.2); for example, a flat hexagonal complex $Co(NH_3)_4Cl_2{}^+$ would have *three* isomers, like ortho-, meta- and para-di-substituted benzenes. He correctly identified the green compound as the *trans*-isomer (chloro-ligands on opposite sides of the octahedron), and the violet as *cis* (same side), as in Fig. 11.1. In the same way, the existence of two isomers, yellow and orange, of $Pt(NH_3)_2Cl_2{}^0$ showed that these are square planar, rather than tetrahedral, complexes (Fig. 11.2); the yellow one has no electric dipole moment and is therefore the *trans*-isomer, whereas the orange does and is *cis*. A tetrahedral $Pt(NH_3)_2Cl_2{}^0$ would have just one isomer (with a dipole moment). Four-coordinate complexes of plat-inum(II), palladium(II), and gold(III) are virtually always square planar, but tetrahedral complexes such as the purple tetrathiocyanatocobaltate(II) ion in $Hg[Co(NCS)_4]$ are often encountered.

Most commonly, metals in their usual oxidation states such as M^{2+} or M^{3+} (M = a first transition series metal), Li^+, Na^+, Mg^{2+}, Al^{3+}, Ga^{3+}, In^{3+}, Tl^{3+}, and Sn^{2+} form octahedral six-coordinate complexes. Linear two-coordination is associated with monovalent coinage metal (Cu, Ag, Au) ions, as in $Ag(NH_3)_2^+$ or $AuCl_2^-$. Three- and five-coordination are not frequently encountered, since close-packing considerations tell us that tetrahedral or octahedral complex formation will normally be favored over five-coordination, while three-coordination requires an extraordinarily small radius ratio (Chapter 6). Coordination numbers higher than six are encountered amongst the larger transition metal ions (i.e., those at the left of the second and third transition series, as exemplified by TaF_7^{2-} and $Mo(CN)_8^{4-}$) and in the lanthanides and actinides (e.g., $Nd(H_2O)_9^{3+}$, and $UO_2F_5^{3-}$ which contains the linear *uranyl* unit $O{=}U{=}O^{2+}$ and five fluoride ligands coordinated around the uranium(VI) in an "equatorial" plane). For most of the metal complexes dealt with in this book, however, a normal coordination number of six may be assumed.

In the same way, metal ions in polar solvents will form complexes with the solvent molecules. X-ray diffraction, EXAFS, and the visible spectrum show that nickel(II) ion in dilute aqueous solution is present as the green hexaaqua complex, $Ni(H_2O)_6^{2+}$, just as in solids such as $NiSO_4{\cdot}7H_2O$, which is actually $[Ni(H_2O)_6]SO_4{\cdot}H_2O$. In this crystal, the extra water molecule is loosely associated with the sulfate ion independently of the nickel aqua complex; it is sometimes referred to as *lattice* water, as distinct from *complexed* water.

Similarly, the greyish-purple Cr^{3+}(aq) is $Cr(H_2O)_6^{3+}$ just as in solid $Cr(H_2O)_6(NO_3)_3{\cdot}3H_2O$ or $Cr(H_2O)_6Cl_3$ (no lattice water). There is, however, an isomeric solid $CrCl_3{\cdot}6H_2O$ which is green; this is actually *trans*-$[Cr(H_2O)_4Cl_2]Cl{\cdot}2H_2O$, and in solutions of Cr^{3+}(aq) containing fairly high chloride concentrations we find not only $Cr(H_2O)_6^{3+}$ but also dull-green $Cr(H_2O)_5Cl^{2+}$, leaf-green *cis*- and *trans*-$Cr(H_2O)_4Cl^{2+}$, and possibly complexes containing three or more chloro-ligands in place of water. As it happens, ligand substitution at chromium(III) centers is quite slow, occurring on a time-scale of hours at room temperature, just as in the cobalt(III) and platinum(II) cases discussed above, and these various aquachlorochromium(III) species can be separated from each other, e.g., by ion-exchange chromatography (Section 15.4). Not all metal ions undergo ligand substitution quite so slowly, however; in solutions of Ti^{3+}, V^{2+}, V^{3+}, Cr^{2+}, Mn^{2+}, Mn^{3+}, Fe^{2+}, Fe^{3+}, Co^{2+}, Ni^{2+}, Cu^{2+}, Zn^{2+}, Cd^{2+}, Hg^{2+}, and all 1+, 2+, and 3+ ions (including the lanthanides and actinides) outside the transition series, simple ligand substitution normally reaches equilibrium in less than about a second—sometimes in less than a microsecond.

In general, then, metal ions in solution form complexes (frequently six-coordinate) with the solvent molecules, their counter-ions, and other donor molecules that happen to be in the solution. For example, in ammoniacal aqueous solution, Ag^+ forms $Ag(NH_3)_2{}^+$ (as noted above), Cu^{2+} forms a series of aquaammines but most notably the royal blue *trans*-$Cu(NH_3)_4(OH_2)_2{}^{2+}$, and cobalt(II) forms $Co(NH_3)_x(H_2O)_{6-x}{}^{2+}$ which react quite rapidly with the oxygen of the air to give the strawberry-red cobalt(III) complex $Co(NH_3)_5OH_2{}^{3+}$ or (if much Cl^- is present) the $Co(NH_3)_5Cl^{2+}$ ion mentioned above.

11.3 Chelation

Many complexing agents exist that have more than one potential donor atom in the molecule. If the molecular geometry is appropriate, then these can act as ligands that attach themselves to a metal ion through two or more separate points. Such ligands are called *chelating agents* (from the Greek *khele* meaning a claw), and the resulting complexes are known as *metal chelates*. If there are two points of attachment, we speak of *bidentate* chelation; if three, *terdentate;* if four, *quadridentate*, and so on. (Note that the preferred prefixes are Latin, rather than the Greek di-, tri-, tetra-, etc., but both are acceptable).

Suppose, for example, we join two ammonia ligands together with a short hydrocarbon chain; we now have a bidentate chelating agent, for example, H_2N—CH_2—CH_2—NH_2, which is commonly called *ethylenediamine* (abbreviation: en), although it contains no $C{=}C$ double bond. Figure 11.3 shows the structure of $Co(en)_3{}^{3+}$. Note that there are two non-superposable ways in which it can be drawn, these being left-handed and right-handed mirror images. These are known as *optical isomers* of

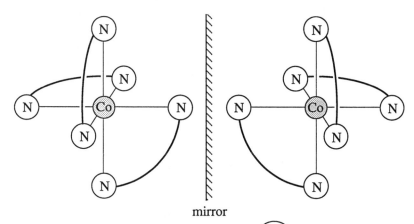

mirror

Figure 11.3 Optical isomers of $Co(en)_3{}^{3+}$. N⌢N represents H_2N—CH_2 —CH_2—NH_2. The two molecules shown are not superposable (cf. left and right hands).

$Co(en)_3{}^{3+}$ because a solution of one will rotate the plane of polarization of light to the left as it passes through, and a solution of the other will rotate it to the right. These isomers will be present in equal amounts as $Co(en)_3{}^{3+}$ salts are usually prepared, but they can be resolved, i.e., separated from each other, by fractional crystallization of a salt of $Co(en)_3{}^{3+}$ from solution with an anion which is itself an optical isomer, for example, the *d*-tartrate anion. The fact that optical isomers of $Co(en)_3{}^{3+}$ and other chelates do exist was used by Werner as incontrovertible proof of his theory of octahedral coordination in cobalt(III) and other complexes.

Bidentate ligands. Besides en, the following bidentate ligands are common:

oxalate (ox)   glycinate   carbonate

2,2′-bipyridine (bpy)   1,10-(or ortho-)phenanthroline (phen)

Note that most of these ligands form a five-membered ring when coordinated to a metal ion. A familiar rule of thumb in organic chemistry is that 6-membered rings are preferred, but this refers to molecules in which the bond angles in the ring are 109.5° (as in the puckered cyclohexane ring) or 120° (as in benzene). Here, we have at least one 90° bond angle (donor atom)—(metal)—(donor atom), if the coordination geometry is octahedral, so five-membered rings are generally more stable. The carbonato-ligand forms a somewhat strained four-membered ring when bidentate and, significantly, is often encountered as a unidentate ligand. Similarly, carboxylates $R—CO_2{}^-$ can be bidentate, with both oxygens coordinated, as in the complex $Th(O_2CCH_3)_6{}^{2-}$, where the coordination number of thorium(IV) is actually 12. In general, though, individual carboxylate groups tend to be unidentate. The glycinate ion, a sort of hybrid of oxalate and ethylenediamine, is the anion of glycine, the simplest member of the α-aminoacids. These compounds are extremely important in that they are the fundamental building-blocks of proteins and other biological molecules. The fact that such compounds are potential chelating agents means that metal ions often interact strongly (and not necessarily beneficially) with biological systems.

Terdentate ligands. By analogy with the discussion of ethylenediamine and glycine above, we can expect *diethylenetriamine* (dien) and *iminodiac-etate* ion (IDA^{2-}) to be effective terdentate chelating agents:

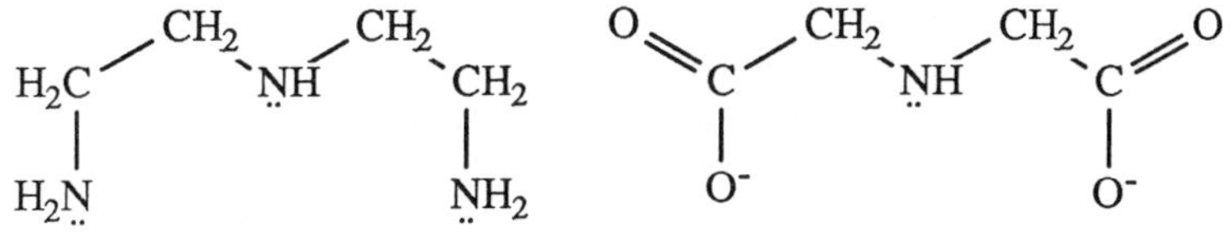

dien IDA^{2-}

Quadridentate ligands. One quadridentate derivative of en and dien is *tren.* It has a carboxylato-analogue in the *nitrilotriacetate* ion (NTA^{3-}):

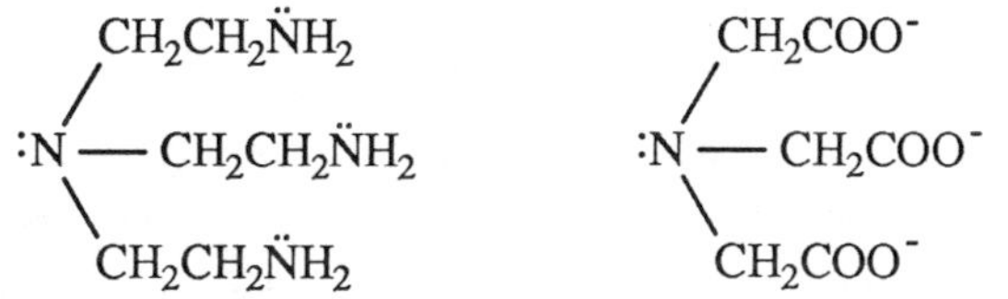

tren NTA^{3-}

These molecules are flexible because rotation about single bonds is free, and so they can wrap themselves around a metal ion to obtain four comfortable donor-atom-to-metal links within five-membered rings. Nitrilotriacetic acid is quite easily synthesized industrially from ammonia, formaldehyde, and hydrocyanic acid (the *Strecker synthesis*) and therefore is potentially a cheap but very effective chelating agent.

There are numerous biologically important quadridentate chelating agents in which four nitrogen donor atoms are locked into a square-planar arrangement in a flat *porphyrin* ring; the basic structure is outlined in Fig. 11.4. These units are found, for example, in the O_2-carrier hemoglobin,

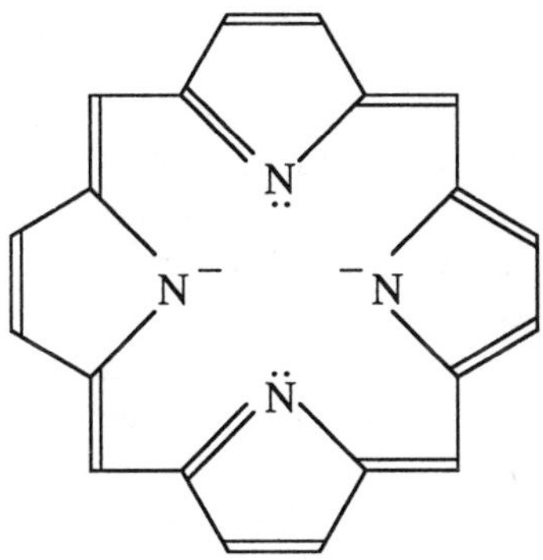

Figure 11.4 The anion of porphine. Porphyrins are porphine rings with peripheral substituents. The porphine ring is flat, and metal ions can be bound within the square planar arrangement of nitrogen atoms.

where the complexed metal ion is iron (see Section 2.2), and in chlorophyll, the green substance that mediates photosynthesis in plants and in which the coordinated metal ion is Mg^{2+}.

Sexadentate chelation. The outstanding example of a powerful and versatile chelating agent is the *ethylenedinitrilotetraacetate* anion, or EDTA^{4-} (sometimes called *ethylenediaminetetraacetate*, or, in medical circles, *versene*):

$$^-O\text{-}CO\text{-}CH_2 \qquad\qquad CH_2\text{-}CO\text{-}O^-$$
$$\diagdown \qquad\qquad \diagup$$
$$: N\text{-}CH_2\text{-}CH_2\text{-}N :$$
$$\diagup \qquad\qquad \diagdown$$
$$^-O\text{-}CO\text{-}CH_2 \qquad\qquad CH_2\text{-}CO\text{-}O^-$$

The successive pK_a values of the parent acid H_4EDTA are 2.0, 2.7, 6.2, and 10.3 at room temperature. The first two of these represent removal of protons from two of the four —COOH groups. It turns out that the other two ionizable protons reside on the nitrogens, so that in Na_2H_2EDTA (which is the form usually supplied by manufacturers) all the carboxyl groups are ionized to —COO$^-$ and the two nitrogens are protonated (a *zwitterion* structure, common among amino acids). When H_2EDTA^{2-} complexes a metal ion, however, all the protons are displaced and the flexible molecule wraps itself around the ion in such a way that, usually, the —COO$^-$ and $\equiv$N: donors occupy six octahedral sites. In a few examples, one carboxyl function is not attached, so that the EDTA ligand is only *quinquedentate*, while in lanthanide complexes, e.g., $La(OH_2)_3EDTA^-$, the EDTA is sexadentate, but the overall coordination number is more than six.

A sexadentate complexing agent which is typical of a rapidly expanding class of new compounds is the flexible *crown ether* 18-crown-6, sketched in Fig. 11.5a. These form strong complexes even with alkali metal ions, which (being large as cations go and of low charge) interact only weakly with most complexing agents. Crown ethers are useful in solubilizing alkali metal salts sufficiently to dissolve in solvents of low relative permittivity. Crown-ether-like molecules consisting of several rings fused to form a cage that can enclose a metal ion are called *cryptands* (Fig. 11.5b), and such complexes are called *cryptates*.

Chelate complexes are much more stable than comparable complexes of unidentate ligands. This *chelate effect* is difficult to express quantitatively because of problems with units (Section 11.4), but we can see qualitatively that, once one donor atom of (say) a sexadentate ligand is attached to a metal ion, the local concentration of five more donors is suddenly very high, and so they readily coordinate with the metal ion. Had the six donors been independently floating around in the solution, no such cooperative action would have been possible. Some authors have pointed out that one

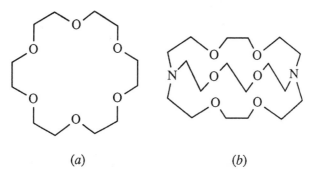

Figure 11.5 (*a*) A crown ether (18-crown-6) and (*b*) a cryptand (2,2,2-crypt).

sexadentate ligand can displace six unidentate ligands, such as aqua or other coordinated solvent molecules, resulting in a positive entropy change (one molecule pinned down for six set loose), and hence a more negative ΔG than for analogous unidentate-unidentate substitutions, but again these arguments turn out to be dependent upon the choice of standard state. The essential points are that multidentate ligands are extraordinarily effective complexing agents, and that the advantage of high *denticity* is especially marked at low concentrations (Exercise 11.5).

11.4 Stability Constants

The equilibrium constant for the *formation* of a metal complex is known as its *stability constant*. (However, some authors present the datum as its reciprocal, the *instability constant* of the complex, by analogy with the dissociation of a weak acid.) There are two kinds of stability constant: *stepwise* $(K_1, K_2, K_3, \ldots K_n)$ and *overall* (β_n). We will assume that there are six aqua ligands to be replaced by some other unidentate ligand X^{x-}, in an aqueous solution of M^{m+}:

$$M(H_2O)_6{}^{m+} + X^{x-} \xrightleftharpoons{K_1} M(H_2O)_5X^{(m-x)+} + H_2O. \qquad (11.5)$$

For clarity, we can ignore the solvent water and charges:

$$\frac{[MX]}{[M][X]} = K_1 \ (\text{L mol}^{-1}) \qquad (11.6)$$

$$MX + X \xrightleftharpoons{K_2} MX_2 \qquad (11.7)$$

$$\frac{[MX_2]}{[MX][X]} = K_2 \ (\text{L mol}^{-1}) \qquad (11.8)$$

$$MX_2 + X \xrightleftharpoons{K_3} MX_3. \qquad (11.9)$$

In principle, this continues until we reach MX_6. Alternatively, we can consider the formation of MX_6 as a single step:

$$M + 6X \xrightleftharpoons{\beta_6} MX_6 \qquad (11.10)$$

$$\frac{[MX_6]}{[M][X]^6} = \beta_6 \ (L^6 \ mol^{-6}) \qquad (11.11)$$

and the connection between the stability constants is therefore

$$\beta_n = K_1 \cdot K_2 \cdot K_3 \ldots K_n. \qquad (11.12)$$

If a sexadentate ligand were to coordinate to M^{m+}, β_n would just be $\beta_1 = K_1$, with units $L \ mol^{-1}$. Consequently, we cannot compare β_1 for the formation of a sexadentate ligand with the overall stability constant β_6 for six comparable unidentate ligands because the units are incompatible.

The stability constants have been defined here in terms of concentrations and hence have dimensions. True thermodynamic stability constants K_n° and β_n° would be expressed in terms of activities (Section 1.1), and these constants can be obtained experimentally by extrapolation of the (real) measurements to (hypothetical) infinite dilution. Such data are of limited value, however, as we cannot restrict our work to extremely dilute solutions. At practical concentrations, the activities and concentrations of ions in solution differ significantly, i.e., their activity coefficients are not close to unity; worse still, there is no thermodynamically rigorous means of separating anion and cation properties for solutions of electrolytes. Thus, single-ion activity coefficients are not experimentally accessible, and hence, strictly speaking, one cannot convert equations such as 11.6 or 11.8 to thermodynamically exact versions.

Modern theories of electrolytes based on the *Debye–Hückel* approach,[3, 4] however, show that ionic activity coefficients are governed mainly by a quantity called the *ionic strength*, I, given by $0.5 \sum c_i z_i^2$ where c_i is the concentration and z_i the charge on ions of the ith kind in the solution. These i kinds include all *ions* in the solution, as well as M^{m+} and X^{x-}. Accordingly, we can vary $[M^{m+}]$ and $[X^{x-}]$ with constant activity coefficients to derive K_n or β_n by maintaining an effectively constant ionic strength I with a swamping concentration of some inert electrolyte. Usually a perchlorate such as $LiClO_4$ or $NaClO_4$ is used because, of all the commoner anions, ClO_4^- has the least tendency to become involved as a ligand itself (Section 5.4).

Consequently, tabulations of stability constants[6, 7] usually list $\log_{10} K$ values for various M^{m+} and X^{x-} (x may be zero, of course) *at specified ionic strengths*, often including extrapolated values for $I = 0$. The temperature is also noted, since ΔH° for complex formation, though rarely large, cannot normally be disregarded.

11.5   Uses of Complexing Agents

Most applications of complexing agents depend upon the fact that they will "tie up" most of the metal ion present in a solution as a complex, leaving only a very low concentration of free metal ion.

Ligands of the NTA and EDTA type (*complexones* or *sequestering agents*, as they are sometimes called) are therefore often able to dissolve deposits of metal oxides, hydroxides, sulfides, carbonates, etc., because they displace solubility equilibria such as reaction 11.13 to the right by reducing the free metal ion concentration to very low levels:

$$M(OH)_m(s) \xrightleftharpoons{K_{sp}} M^{m+}(aq) + mOH^-(aq) \tag{11.13}$$

$$[M^{m+}][OH^-]^m = K_{sp} \tag{11.14}$$

$$M^{m+}(aq) + EDTA^{4-}(aq) \xrightleftharpoons{\beta_1} [M(EDTA)]^{(4-m)-} \tag{11.15}$$

$$\beta_1 = \frac{[M(EDTA)^{(4-m)-}]}{[M^{m+}][EDTA^{4-}]}. \tag{11.16}$$

Thus, if β_1 is large enough, for a given solubility product K_{sp}, and the pH and the free EDTA concentration are appropriate, the solid $M(OH)_m$ of reaction 11.13 will pass entirely into solution (see Exercise 11.6). Equations 11.13 to 11.16 could be applied to the dissolution of brown iron stains or corrosion deposits (nominally $Fe(OH)_3$), and a parallel set of equations could be written for the dissolution of $CaCO_3$ (boiler scale) or of the sulfide tarnish on copper or brass. Many domestic cleaning agents contain chelating agents; bathroom cleaners typically contain 1–5% Na_4EDTA, and Na_3NTA is an excellent detergent *builder* (Section 8.3), acting to keep Ca^{2+} and Mg^{2+} tied up as complexes so they they cannot interfere with the detergent action. In this regard, Na_3NTA offers a biodegradable substitute for $Na_5P_3O_{10}$, a very effective traditional builder but one that presents ecological problems (Section 3.6). Claims (perhaps unjustified) that NTA^{3-} is sufficiently carcinogenic to present a significant public health hazard have led to the present ban on its use as a detergent builder in New York state. Another possible objection to its widespread use is that, by its very nature, it may mobilize normally insoluble toxic heavy metal compounds as NTA complexes in rivers, lakes, etc. Most NTA in waste water would, however, be destroyed by bacterial biodegradation in secondary sewage treatment.

Boiler tubes are often cleaned with EDTA or NTA solutions to remove both $CaCO_3$ scale and corrosion products. In pressurized heavy water nuclear power reactors, radioactive corrosion deposits (in effect, magnetite in which some of the Fe^{2+} has been replaced by radioactive $^{60}Co^{2+}$) can be removed from the coolant water circuits with an aqueous mixture of oxalic and citric acids (both good chelators for Fe^{3+}) and EDTA. In home laundry operations, bloodstains on clothing can be removed by treatment with oxalic acid, which takes up the iron from the hemoglobin (Section 2.2) as $Fe(ox)_3{}^{3-}$. By the same token, oxalates are very toxic when taken internally, as are many other complexing agents. EDTA, for example, is used as a means of removing lead, a cumulative poison, from the body of a person with suspected lead poisoning, but it will strip out many other essential metal ions at the same time; so such "versene" treatments are very risky and should not be undertaken unless the diagnosis is definite. (Unfortunately, the maximum allowable level of lead in blood or urine is only about twice the normal level, so diagnosis requires very careful chemical analysis for Pb.)

On the other hand, the insolubility of the various forms of Fe_2O_3, $Fe(OH)_3$, and $FeO(OH)$ makes it very difficult for plants or animals to get the necessary amount of iron *into* their systems. This is especially true for plants in soils that are alkaline, since OH^- suppresses $Fe(OH)_3$ dissolution (reaction 11.13). Gardeners therefore feed iron to evergreens as soluble chelate (usually EDTA) complexes. Certain bacteria and fungi have evolved to produce iron-chelating substances, in which the donor atoms are usually oxygen, just as in oxalate.[8] In humans, iron reserves are brought into the body by a complicated protein chelating agent (*transferrin*) and are stored in the spleen as $FeO(OH)$ plus Fe phosphates inside a protein envelope (*ferritin*).

Metal ions, even in trace amounts, catalyze many chemical reactions, such as the decomposition of hydrogen peroxide:

$$H_2O_2(aq) \xrightarrow[\text{(etc.)}]{Fe^{3+}} H_2O + \tfrac{1}{2}O_2 \tag{11.17}$$

or the deterioration of foodstuffs, e.g., salad dressing. This catalysis can be suppressed by addition of a small amount of a chelating agent, usually EDTA, to "tie up" any traces of free metal ions.

Although chelating ligands are especially effective, there are many unidentate complexing agents of commercial importance. As noted in Section 11.1, silver ion interacts strongly with ammonia to form $[Ag(NH_3)_2]^+$, so that substantial amounts of AgCl or AgBr, which are very poorly soluble in water, can be dissolved in aqueous ammonia. Thus, at $25\,°C$ and low ionic strength, we have (Exercise 11.4):

$$\frac{[Ag(NH_3)_2{}^+]}{[Ag^+(\text{free})][NH_3(\text{free})]^2} = \beta_2 = 1.07 \times 10^7 \text{ L}^2 \text{ mol}^{-2} \tag{11.18}$$

and the solubility product K_{sp} for AgBr gives us

$$[Ag^+(\text{free})][Br^-] = K_{sp} = 5.3 \times 10^{-13} \text{ mol}^2 \text{ L}^{-2} \tag{11.19}$$

where $[Br^-]$ must equal the total Ag^+ concentration in solution ($=$ $[Ag^+(\text{free})]+[Ag(NH_3)_2{}^+]$) and the common factor $[Ag^+(\text{free})]$ can be eliminated. The tarnish that forms on silverware is mainly Ag_2S, and can similarly be dissolved in aqueous ammonia, which is present in many commercial silver cleaning fluids.

Silver ion also complexes strongly with *thiosulfate* ion (Section 1.8), formerly called *hyposulfate*, and photographers' "hypo" fixing solution makes use of the resulting ability of aqueous $Na_2S_2O_3$ to dissolve solid $AgCl$ or $AgBr$. Briefly, a black-and-white photographic film consists of finely divided silver halide grains in a support material. Grains that are exposed to light (i.e., to the bright parts of the image projected onto the film) decompose very slightly to free silver and halogen, and these minute spots of silver serve as nuclei for reduction of the whole grains to silver metal by subsequent action of a "developer" (typically, hydroquinone with sodium metabisulfite). The unreduced silver halide must then be completely removed by complexation with $Na_2S_2O_3$ solution, otherwise it, too, will ultimately go to silver metal when the film is brought out of the darkroom. The result is a black deposit of Ag wherever light had struck the film inside the camera. The image is therefore a *negative* one, and must be projected onto silver halide coated paper, and the development/fixing process repeated, to secure a permanent positive image. Most color films also use silver halides for capturing the latent image, and interaction of this with various dyes during processing gives a color-positive (for transparencies) or color-negative (for prints) image.[9, 10]

Cyanide ion is a very powerful unidentate complexing agent, especially for transition metal ions such as Fe^{2+} and Fe^{3+}, which form hexacoordinate complexes commonly called *ferrocyanide* ($Fe(CN)_6{}^{4-}$) and *ferricyanide* ($Fe(CN)_6{}^{3-}$), respectively. These complexes (especially ferrocyanide) are of low toxicity despite their cyanide content, because the cyanide is so strongly bound. Indeed, a traditional antidote for swallowed cyanide is to drink freshly mixed aqueous ammonia and iron(II) sulfate, which will tie up the CN^- as $Fe(CN)_6{}^{4-}$. The ammonia serves to increase the pH of the stomach contents, thus minimizing the formation of undissociated HCN which would be absorbed rapidly by the body. The toxicity of cyanide is itself due to its complexing with the iron(III) centers in porphyrin units in cytochrome *c*, which is responsible for the crucial steps in the transfer of electrons to molecular oxygen in the mitochondria (respiratory organelles) of the cells that make up the bodies of the higher animals. Complexation by CN^- takes place rapidly, and completely prevents the reduction of the iron(III) center to iron(II). Since this reduction is essential to the respiratory cycle, death follows very quickly.[11]

Cyanides are widely used in industry, e.g., in gold refining (Section 15.2) and in electroplating (Section 13.6), and must be handled with appropriate care. In particular, contact with acids must be avoided, otherwise

HCN (*prussic acid*, pK_a 9.3), which is highly volatile and can be absorbed through the skin as well as through the lungs, will form. Hydrogen cyanide is also produced when nitrile-based synthetic fibers ("Orlon," "Acrilan," etc.), adhesives, and rubbers burn. Many fire victims who die of "smoke inhalation" are actually killed by HCN. Finally, HCN is found in peach pits and bitter almonds. Their characteristic odor is actually HCN, hence they could be fatally toxic if eaten in large amounts.

11.6   "Hydrolysis" of Aqueous Cations[1, 2, 12]

A water molecule is strongly polarized by coordination to a metal cation, towards which its electrons are attracted:

$$\begin{array}{c} {}^{\delta+}\text{H} \\ \searrow \\ \ddot{\text{O}}{:}\longrightarrow \text{M}^{(z-2\delta)+} \\ \nearrow \\ {}^{\delta+}\text{H} \end{array}$$

Consequently, aqua ligands are quite acidic:

$$M(H_2O)_6{}^{z+}(aq) \overset{K_a}{\rightleftharpoons} M(H_2O)_5OH^{(z-1)+}(aq) + H^+(aq). \qquad (11.20)$$

Thus, a typical aqueous metal ion M^{z+}(aq) can act as a *Brønsted* (i.e., proton-donating) *acid*, of which $MOH^{(z-1)+}$(aq) is the conjugate base. The polarizing power of M^{z+} will be greater, the larger its charge z and the smaller its ionic radius r, so that the greatest acid dissociation or "hydrolysis" *constants* K_a, or smallest pK_a values ($= -\log_{10} K_a$), will be expected for small, highly charged cations:

$$K_a = \frac{[MOH^{(z-1)+}][H^+]}{[M^{z+}]} \qquad (11.21)$$

$$pK_a = pH + \log_{10}\frac{[M^{z+}]}{[MOH^{(z-1)+}]}. \qquad (11.22)$$

From Eqn. 11.22, it is seen that pK_a corresponds to the pH value at which hydrolysis of M^{z+}(aq) is just half complete—assuming no other reactions intervene (but see below). Typical values of pK_a are about 14 for $z = 1$, 9 ± 3 for $z = 2$, and 3 ± 2 for $z = 3$, at least for the lighter elements—pK_a for the lanthanide(III) ions, for example, is on the order of 7 to 9 because of their larger radii:

	Sc^{3+}	Y^{3+}	La^{3+}	
r	81	98	106	pm
pK_a	4.5	7	9	

However, simplistic generalizations based on z and r are only partly success-ful in understanding metal ion hydrolysis. For instance, it is not obvious why K_a for $Fe^{3+}(aq)$ $(r = 64.5 \text{ pm})$ is about 100-fold greater than K_a for $Cr^{3+}(aq)$ $(r = 61.5 \text{ pm})$.

Like stability constants and other thermodynamic properties of metal ions in solution, hydrolysis constants are affected by ionic strength and temperature, and these should be specified when quoting precise pK_a values. For the "ball-park" figures cited here, 25 °C and high dilution are assumed.

Equation 11.21 does not, however, tell the whole story about metal ion hydrolysis. Not only can further proton dissociations occur:

$$\begin{aligned} M^{z+}(aq) &\rightleftharpoons MOH^{(z-1)+}(aq) + H^+ \\ &\rightleftharpoons M(OH)_2{}^{(z-2)+}(aq) + 2H^+ \quad (11.23) \\ &\rightleftharpoons \text{etc.} \end{aligned}$$

until insoluble $M(OH)_z(s)$ comes out of solution, but the conjugate base species $MOH^{(z-1)+}(aq)$, $M(OH)_2{}^{(z-2)+}(aq)$, etc., can polymerize by shar-ing OH^- ligands as bridging groups:

$$2M(H_2O)_{n-1}OH^{(z-1)+} \rightleftharpoons$$

$$(H_2O)_{n-2}M \underset{\underset{H}{O}}{\overset{\overset{H}{O}}{\diagdown\diagup}} M(OH_2)_{n-2}{}^{(2z-2)+} + 2H_2O. \quad (11.24)$$

Condensation (i.e., elimination of bound water or its elements) reactions like reaction 11.24 can build up large, polymeric, highly charged cationic molecules that might be expected to come out of solution as solids contain-ing the available anion. In fact, however, they tend to remain dispersed as *colloids* because mutual electrostatic repulsions (actually, electrical double-layer effects) prevent their *coagulation*. Colloidal solutions or *sols* can be kinetically stable more or less indefinitely and may be mistaken for true solutions, although the colloidal particles may be large enough to scatter light, so that a beam passing through the sol shows up clearly (the *Tyndall effect*). Coagulation can often be induced by adding highly charged ions of the opposite charge—in this case, anions such as sulfate, polyphosphate, silicate polymers, etc.

These phenomena are exemplified in the aqueous chemistry of iron(III). The ion $Fe(H_2O)_6{}^{3+}$ actually has a beautiful pale lilac color, as seen in solid "ferric alum" ($KFe(SO_4)_2 \cdot 12H_2O$) or iron(III) nitrate ($Fe(NO_3)_2 \cdot 9H_2O$). In solution, however, iron(III) is typically yellow (at high acidities) or brown, because of hydrolysis and hydrolytic polymerization. The higher polymers form relatively slowly, so that the properties of the solution change on ageing. At p$H \geq 5$, a dark brown, gelatinous precipitate of $Fe(OH)_3$ forms. This tends to dehydrate to form the yellow-brown solid *goethite* (α-FeO(OH)), and this in turn can be further dehydrated on heating to give red-brown *hematite* (α-Fe$_2$O$_3$). These hydrolytic solids are important in soil science.

The hydrolytic scheme shown in Fig. 11.6 does not take into account complex formation between Fe^{3+} and other solutes, which tends to prevent the iron(III) from precipitating. In natural waters, for example, organic substances known as *humic acids*[13] may be present as a result of the decay of vegetation, especially in streams that drain marshes, muskeg, and peaty moorland. These humic acids have complicated structures but typically have carboxylate and phenolic oxygen atoms (cf. lignin, Section 4.3) and can act as very effective chelating agents for metal ions, particularly iron(III). As a result, surface waters rich in humic acids are usually brown because of the complexed iron(III) they contain.

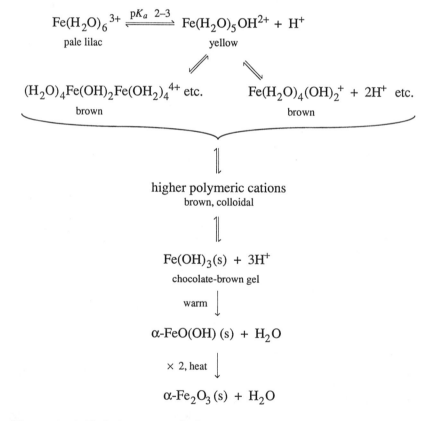

$$Fe(H_2O)_6{}^{3+} \xrightleftharpoons{\ pK_a\ 2\text{--}3\ } Fe(H_2O)_5OH^{2+} + H^+$$

pale lilac   yellow

$(H_2O)_4Fe(OH)_2Fe(OH_2)_4{}^{4+}$ etc. $Fe(H_2O)_4(OH)_2{}^+ + 2H^+$ etc.

brown   brown

higher polymeric cations
brown, colloidal

$Fe(OH)_3(s) + 3H^+$
chocolate-brown gel

warm

α-FeO(OH) (s) + H_2O

$\times$ 2, heat

α-Fe$_2$O$_3$ (s) + H_2O

Figure 11.6 Hydrolysis of iron(III).

Hydrolysis of metal ions of charge 4+ or more can be expected to be extensive, involving the loss of several protons even in very acidic solution so that species such as "$Ti(H_2O)_6{}^{4+}$" are simply not encountered. The hydrolyzed ions, however, tend to be *oxo*- rather than *hydroxo*-species, i.e., $(H_2O)_5 TiO^{2+}$ rather than $(H_2O)_4 Ti(OH)_2{}^{2+}$. Such oxo-cations are usually designated with the *-yl* suffix, as in zirconyl (ZrO^{2+}), vanadyl (VO^{2+}), uranyl ($UO_2{}^{2+}$), and neptunyl ($NpO_2{}^+$) ions. The conception of oxo-complexes as hydrolysis products can be stretched further to include such anionic metal species as chromate ($CrO_4{}^{2-}$), permanganate ($MnO_4{}^-$) or vanadate ($VO_3{}^-$), which may be regarded as fully hydrolyzed "Cr^{6+}(aq)," "Mn^{7+}(aq)," and "V^{5+}(aq)," respectively. Ions such as $HCrO_4{}^-$ can be considered either as incompletely hydrolyzed chromium(VI) ($HOCrO_3{}^-$) or, more traditionally, as manifestations of weakness of the corresponding acid (chromic acid, H_2CrO_4; cf. H_2SO_4).

References

1. J. Burgess, "Metal Ions in Solution," Ellis Horwood: Chichester, 1978.

2. J. P. Hunt, "Metal Ions in Aqueous Solution," W. A. Benjamin: New York, 1963.

3. J. O'M. Bockris and A. K. N. Reddy, "Modern Electrochemistry," Vol. 1, Plenum Press: New York, 1973.

4. R. A. Robinson and R. H. Stokes, "Electrolyte Solutions," 2nd edn. (revised), Butterworths: London, 1965.

5. F. A. Cotton and G. Wilkinson, "Advanced Inorganic Chemistry," 5th edn., Wiley-Interscience: New York, 1988, Chapter 2.

6. L. G. Sillén and A. E. Martell, "Stability Constants," Special Publication No. 17, The Chemical Society: London, 1964, and Supplement (Special Publication No. 25), 1971.

7. A. E. Martell and R. M. Smith, "Critical Stability Constants," 4 vols., 2 suppl., Plenum Press: New York, 1974–1988.

8. T. Emery, "Iron metabolism in humans and plants," *American Scientist*, 1982, *70*, 626–632.

9. G. T. Austin, "Shreve's Chemical Process Industries," 5th edn., McGraw-Hill: New York, 1985, Chapter 23.

10. G. T. Eaton, "Photographic Chemistry," 2nd edn., Morgan and Morgan: Dobbs Ferry, N.Y., 1965 (written for the layman).

11. D. A. Labianca, "On the nature of cyanide poisoning," *Journal of Chemical Education*, 1979, *56*, 788–791.

12. C. F. Baes and R. E. Mesmer, "The Hydrolysis of Cations," Wiley-Interscience: New York, 1976.

13. C. Steelink, "Humates and other natural organic substances in the aquatic environment," *Journal of Chemical Education*, 1977, *54*, 599–603.

Exercises

11.1 Estimate the standard free energy of hydration of one mole of gaseous ionized NaCl. The necessary information is given in Sections 6.5 and 11.1 and in Exercise 6.4. Compare the answer with the corresponding lattice energy (Section 6.5).
[*Answer:* -1051 kJ mol^{-1} .]

11.2 (*a*) Sketch the $Co(en)_3{}^{3+}$ ion (cf. Fig. 11.3) as viewed down a three-fold axis (i.e., looking down onto a triangular face, as in Fig. 8.1) of the octahedron of nitrogens, and show that it can be drawn as a screw-like structure with either a left- or a right-handed "thread."

(*b*) Show that *cis*-$Co(en)_2Cl_2{}^+$, but not its *trans*-isomer, can exhibit optical isomerism.

(*c*) How many isomers of $Cr(H_2O)_3Cl_3{}^0$ are there?

11.3 Silver(I) forms a dithiosulfato complex for which $\beta_2 = 2 \times 10^{13}$ in dilute aqueous solution at 25 °C. Given that the solubility product of AgBr is 5.3×10^{-13} (and assuming that ionic strength effects are negligible), show that the minimal total concentration of $Na_2S_2O_3$ needed to dissolve 1.00 g silver bromide in one liter of water at 25 °C is 0.0123 mol L^{-1}. Note that Na^+ is introduced simply as the counter-ion of $S_2O_3{}^{2-}$ and does not enter into the calculations.

11.4 (*a*) Calculate the minimum total concentration of ammonia that must be present in 1 L of water at 25 °C in order to dissolve 1.000 g silver bromide ($K_{sp} = 5.3 \times 10^{-13}$ mol^2 L^{-2}), assuming that the diammine-silver(I) ion ($\beta_2 = 1.07 \times 10^7$ L mol^{-1}) forms a freely soluble bromide salt.

(*b*) By what factor does the presence of the ammonia increase the solubility of the silver bromide?
[*Answers:* (*a*) 2.247 mol L^{-1}; (*b*) 7316-fold.]

11.5 The sexadentate chelating agent EDTA^{4-} (ethylenedinitrilotetraac-etate, $(^-OCOCH_2)_2NCH_2CH_2N(CH_2COO^-)_2$) can be regarded as being formed by joining together two N-methyliminodiacetate ($MIDA^{2-}$, ($^-OCOCH_2)_2NCH_3$) units through the methyl groups. MIDA itself is an effective terdentate chelating agent.

Suppose we have a solution containing 2 mol L^{-1} $MIDA^{2-}$ and 1 mol L^{-1} $EDTA^{4-}$, i.e., the number of potential ligand atoms (O or N) associated with each of these chelating agents is the same.

(a) How would a small amount of Ni^{II} distribute itself amongst the forms Ni^{2+} (free), $Ni(MIDA)_2{}^{2-}$, and $Ni(EDTA)^{2-}$? (Assume, for the purposes of this exercise, that the MIDA and EDTA ions remain unprotonated.)

(b) How would this distribution change on pouring this solution into 1×10^9 times its volume of pure water? What does this tell you about the effectiveness of chelating agents in very dilute solutions?

Stability constants

	$\log_{10} K_1$	$\log_{10} K_2$
$Ni^{2+} + MIDA^{2-}$	8.73	7.22
$Ni^{2+} + EDTA^{4-}$	18.56	

[*Answer:* (a) $1 : 3.6 \times 10^{16} : 3.6 \times 10^{18}$.]

11.6 Suppose we have two solutions of Ni^{2+} and of Fe^{3+} in water, each 0.001 M in metal ion and 0.1 M in EDTA. If the pH is adjusted to 12, will either or both of the metal ions be precipitated as their hydroxides or stay in solution as the EDTA complexes? (At pH 12, H_4EDTA can be regarded as completely dissociated to $EDTA^{4-}$.) The relevant data are as follows:

For $M^{n+} + EDTA^{4-} = MEDTA^{(4-n)-}$, $\log K = 18.6$ for Ni^{2+}
$= 25.1$ for Fe^{3+}.

The solubility products of $Ni(OH)_2$ and $Fe(OH)_3$ are 6.5×10^{-18} and 1×10^{-36}, respectively.

11.7 The pK_a of the aqueous iron(III) ion in very dilute solution at 25 °C is 2.17. On the basis of this information alone, estimate:

(a) the minimal H^+ concentration required to ensure that 99% of the iron(III) is present as the hexaaqua complex, and

(b) the $[H^+]$ at which you would expect $Fe(H_2O)_6{}^{3+}$ and $Fe(H_2O)_5$-OH^{2+} to be present in equal concentrations.

(c) What factors (polymerization would be one) could interfere with the accuracy of the predictions, and how would you attempt to minimize their effects?

[*Answers:* (a) 0.67 mol L^{-1}; (b) 6.76×10^{-3} mol L^{-1}.]

11.8 Use the thermodynamic data of Appendix C to derive stability constants β_6 for the ferrocyanide and ferricyanide ions at $25\,^\circ$C and infinite dilution.

[*Answers:* 3.5×10^{45}, 2.8×10^{52}.]

Chapter 12

Water Conditioning

12.1 The Importance of Water Treatment[1-5]

WATER IS a vital resource for industry and agriculture, as well as for domestic use. The ever-increasing demands for it are outstripping the supply in many places where abundant water was once taken for granted. Therefore, it is becoming increasingly important to recycle water, whenever possible, to ensure that water returned to rivers or lakes is fit for re-use by someone else (as well as safe for aquatic life) and to learn how to make use of impure but abundant resources such as brackish water or even seawater.

Industry has huge demands for water, but its net consumption is actually relatively small. Some 70% of this water requirement is for cooling. Since it is simply discharged after use, it may cause thermal pollution, but usually it does not cause chemical pollution of natural waterways. Furthermore, much industrial water is now recycled within the plant, so that the gross use-to-intake ratio in North American pulp and paper mills, for example, is now about 5:1 and rising, while the petrochemical industry should reach a ratio of 40:1 by the end of this century. Agricultural irrigation, on the other hand, leads to large water losses through evaporation, and the water that does return to waterways is often badly contaminated with salts leached from the soil, fertilizer residues, and pesticides.

Therefore, objectives of water treatment are: (*a*) to prepare available water for use in boilers, chemical processes, cooling systems, laundry, and for domestic consumption,[1-4] and (*b*) to clean up waste water for discharge into natural waterways.[5, 6] The appropriate treatment of waste water prior to *discharge* is highly dependent on how the water was used. Since water is used in a vast number of ways, we cannot enter into a comprehensive discussion of waste water treatment here. Instead, we shall focus attention upon the less specific question of treating water *intake* for typical contaminants: suspended and colloidal matter, dissolved solids, dissolved gases, bacteria and algae.

159

12.2   Suspended and Colloidal Matter

Suspended matter down to bacterial size can best be removed by subsidence in settling ponds and by skimming off floating material such as grease, which may form a paste that coats heat-exchange surfaces. Filtration would, no doubt, do a better job, but on the scale necessary, it is not always cost-effective.

Colloidal suspensions (typically, of fine particles of clay minerals) or emulsions (i.e., colloidal solutions of liquid in liquid, such as oil in water) cannot be broken by filtration, but, because colloidal particles are electrically charged, they can be brought down from solution by coagulation with a highly charged counterion (Section 11.6). Coagulation alone may not produce a precipitate that will separate by settling; it is usual to employ a *flocculant*—a substance that can form bridges between the coagulated particles and form an open-structured, settlable, hydrolytic precipitate or *floc*. Thus, clay mineral particles generally are negatively charged in water, through dissociation of some alkali-metal cations from the anionic aluminosilicate framework (Section 8.2), and can be coagulated with highly charged cations such as Al^{3+} or Fe^{3+} and precipitated in an $Al(OH)_3$ or $Fe(OH)_3$ floc. Usually, alum ($KAl(SO_4)_2 \cdot 12H_2O$) is added to the water to be treated, whereupon it forms a cationic floc that coagulates the anionic clay colloid. Alternatively, lime and $FeSO_4 \cdot 7H_2O$ (*copperas*) may be added to the water to form a mixed $Fe(OH)_2/Fe(OH)_3$ floc in the presence of oxygen. Sodium aluminate ($Na[Al(OH)_4]$) may be used where an anionic floc is required.

12.3   Origin and Effects of Dissolved Solids

The impact of water hardness due to calcium or magnesium ions on detergents was explained in Section 8.3. The source of most Ca^{2+} and Mg^{2+} in hard water is the dissolution of limestone ($CaCO_3$) or dolomite ($CaMg(CO_3)_2$). Magnesium carbonate is fairly soluble (1.26 mmol L^{-1} at ambient temperature), but $CaCO_3$ is much less so (0.153 mmol L^{-1}). However, if the water contains dissolved CO_2 (as indeed it will if exposed to the air), the relatively freely soluble $Ca(HCO_3)_2$ forms, and the limestone slowly dissolves away:

$$CaCO_3(s) + H_2O(l) + CO_2(aq) \rightleftharpoons Ca(HCO_3)_2(aq). \qquad (12.1)$$

Reaction 12.1 is reversible, however, and, if the solution is boiled, the CO_2 is swept out in the steam, and $CaCO_3$ is reprecipitated until its own solubility is reached. Thus, part of the Ca^{2+}-derived hardness is removed by boiling; this is called *temporary hardness*. The deposit of $CaCO_3$ may be seen as "*fur*" in domestic kettles or as "*scale*" (not to be confused with the oxide *scales* of Section 7.5) or sludge in boilers. It is highly undesirable in that it reduces heat transfer efficiency and may even cause blockages of tubes or valves. Boiler sludge has an unfortunate tendency to form intractable pastes with any oil in the feedwater. Frequent blowdown of boilers, as well as descaling, is, therefore, necessary to minimize sludge accumulation, unless the feedwater is appropriately conditioned.

Boiling does *not* remove magnesium salts, *or* $CaCO_3$ at its own solubility level, *or* "non-carbonate" calcium (i.e., Ca^{2+} that is counter-balanced by Cl^- or other anion of which the Ca^{2+} salt is freely soluble). The hardness that remains after boiling is called *permanent hardness*. Permanent hardness plus temporary hardness is called the *total hardness*. It is incorrect to regard hard water as containing specific salts such as $CaCO_3$, $Ca(HCO_3)_2$, $MgCO_3$, $MgSO_4$, and $CaCl_2$ because these are present in solution as mixtures of ions. The principal ions present in typical surface waters are:

Cations: Na^+, (K^+), Mg^{2+}, Ca^{2+}, Fe^{2+}/Fe^{3+}

Anions: $CO_3^{2-}/HCO_3^-/CO_2$, Cl^-, SO_4^{2-}.

Others, such as Al^{3+} (which hydrolyzes), F^- (CaF_2 is poorly soluble), HPO_4^{2-}, NO_3^-, and NH_4^+ are minor components, usually not exceeding 1 ppm (i.e., 1 mg per kg of water). Aluminum ion is not usually significant in surface waters, except as a consequence of acid rain or the use of alum or sodium aluminate in the treatment of waste water. Iron is fairly soluble as Fe^{2+} salts, which often occur in substantial concentrations in well waters. At pH values typical of surface waters, iron(II) is oxidized by the air to hydrolyzed iron(III) and its polymers. Iron in water imparts a bad taste, causes discoloration of appliances, and leaves iron oxide/hydroxide deposits in pipes or heat exchanges. Iron(III) can be retained in significant concentrations in surface waters through complexation by humic acids (Section 11.6).

Ionic solids apart, the most important dissolved solid in natural waters is silica, SiO_2. Its solubility depends strongly on its solid-state structure: 120 ppm as amorphous SiO_2, but only 10 ppm as quartz, at $20\,^{\circ}C$ and pH 6. The solubility is greater at higher pH (Section 8.5) and also rises markedly with increasing temperature, so that quartz dissolution can cause problems in steam-injection oil recovery (Section 8.2). Silica is also appreciably soluble in high-pressure steam. Consequently, in steam turbines, it can be redeposited on the delicately balanced blades as the steam expands. It is therefore critically important that feedwater for turbine boilers be substantially free of dissolved silica.

12.4   Treatment for Dissolved Solids

The traditional way to free water of dissolved solids is to distil it, either at atmospheric pressure or by multistage flash evaporation at reduced pressure. This removes virtually all solutes but is wasteful of energy unless the low-grade heat can be economically recovered from the condensers. In any case, it makes more sense to remove the relatively small amount of solutes (typically, 0.02%) from the great excess of water, rather than vice versa. (Seawater might be an exception, with about 3.5% dissolved solids.) Water *softening* is concerned primarily with removal of Ca^{2+} and Mg^{2+}, but for some purposes removal of all dissolved solids (*deionization* or *demineralization*) is necessary.

The most widely used techniques for the removal of dissolved inorganic solids are *boiling*, addition of *washing soda, lime–soda softening, complexation, sodium ion exchange, demineralization, reverse osmosis, electrodialysis, adsorption* onto suspended solids, and *aeration*.

(a) *Boiling* will remove temporary hardness, but the resulting precipitation of $CaCO_3$ may be precisely what one is trying to avoid.

(b) *Addition of washing soda* ($Na_2CO_3 \cdot 10H_2O$) precipitates most of the $MgCO_3$ and $CaCO_3$ by the common ion effect, i.e., by forcing up the concentration of CO_3^{2-}(aq):

$$Ca^{2+}(aq) + CO_3^{2-}(aq) \rightleftharpoons CaCO_3(s) \qquad (12.2)$$

$$[Ca^{2+}][CO_3^{2-}] = K_{sp}. \qquad (12.3)$$

This is fairly costly, as substantial excesses of Na_2CO_3 are needed, and furthermore the solution is left quite alkaline (and hence likely to transport silica):

$$CO_3^{2-} + H_2O \rightleftharpoons HCO_3^- + OH^-. \qquad (12.4)$$

(c) *Lime–soda softening* involves removal of the temporary hardness by adding the calculated amount of hydrated lime (Section 5.1):

$$Ca(HCO_3)_2(aq) + Ca(OH)_2(aq) \rightarrow 2CaCO_3(s) + 2H_2O. \qquad (12.5)$$

One can stop at this point (*selective calcium softening*) or add enough further lime to precipitate the Mg^{2+} as $Mg(OH)_2$. In contrast to the calcium analogues, $MgCO_3$ is fairly soluble but $Mg(OH)_2$ is not (see Exercises 12.1 to 12.3):

$$Mg^{2+}(aq) + Ca(OH)_2(aq) \rightarrow Mg(OH)_2(s)_\downarrow + Ca^{2+}(aq). \qquad (12.6)$$

Finally, soda ash or washing soda is added to remove the Ca^{2+} we have put *into* the solution by reaction 12.6, as well as any "non-carbonate" calcium:

$$Ca^{2+}(aq) + Na_2CO_3(aq) \rightarrow CaCO_3(s)\downarrow + 2Na^+(aq). \qquad (12.7)$$

So, we need to add $Ca(OH)_2$ equivalent to the temporary hardness *plus* the magnesium hardness (which is just the total hardness, if "non-carbonate" Ca^{2+} is absent), and Na_2CO_3 equivalent to the permanent (i.e., total minus temporary) hardness. Clearly, if lime–soda softening is to be effective, accurate analyses for Ca^{2+}, Mg^{2+} and temporary hardness are needed, and the lime must be accurately weighed out accordingly.

There are several disadvantages to lime–soda softening. It leaves the water still saturated with 31 ppm $CaCO_3$, unless extra Na_2CO_3 is added (see (*b*), above). Even if no extra soda is added, the solution will be quite alkaline, causing dissolution of silica when hot and also caustic embrittlement of metals in nooks and crannies of pipes or boilers where water so treated may tend to become concentrated by evaporation. Finally, although colloidal $CaCO_3$ and $Mg(OH)_2$ carry opposite charges (negative and positive, respectively) and so tend to coagulate of their own accord, precipitation may need to be helped along by addition of sodium aluminate, powdered magnesium oxide, or recycled sludge, but, even then, filtration may be necessary to remove turbidity.

(*d*) *Complexation* involves the addition of reagents that give *water-soluble complexes* with Ca^{2+} and Mg^{2+}. Sodium triphosphate (e.g., "Calgon"; the complexing anion is $P_3O_{10}^{5-}$), Na_3NTA, or Na_2H_2EDTA may be used (Section 11.3). NTA and the less thermally stable EDTA are used to suppress boiler sludge formation, and so reduce blowdown frequency, but should not be used if fittings of nickel, copper, or their alloys (other than stainless steel) are present, since Ni^{2+} and Cu^{2+} are strongly complexed by NTA or EDTA and the corrosion of these metals will be facilitated (Section 14.5).

(*e*) *Sodium ion exchange* on zeolites (Section 8.3) or on synthetic organic cation exchange resins such as Dowex-50 (a sulfonated polystyrene; Fig. 12.1), in most circumstances, is superior to the above softening methods. The exchange process favors binding of Ca^{2+} or Mg^{2+} over Na^+ in the solid resin phase:

$$RNa_2(s) + M^{2+}(aq) \rightleftharpoons RM(s) + 2Na^+(aq). \qquad (12.8)$$

Thus, if hard water is passed down a column of Na^+-form resin or zeolite, essentially all the Ca^{2+} and Mg^{2+} ions in the water are

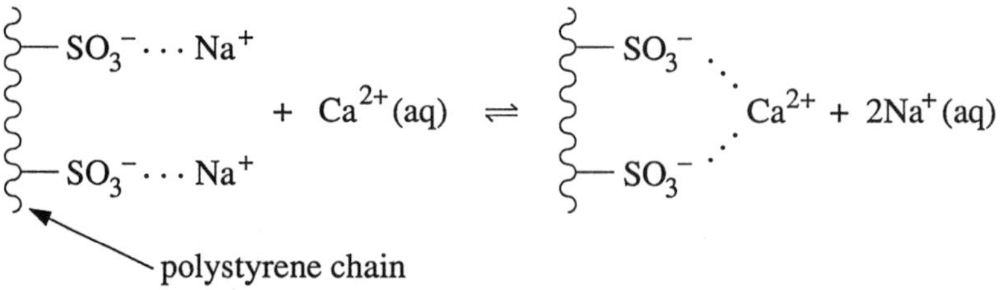

polystyrene chain

Figure 12.1 Exchange of aqueous Ca^{2+} for Na^+ within a bead of a cation-exchange resin in hard water.

replaced by Na^+ until the ion exchange capacity of the resin or zeolite is used up. Furthermore, the resulting softened water is of neutral p*H* (unless it was alkaline to start with). However, reaction 12.8 is reversible, and the ion exchanger can be restored to its Na^+ form by backflushing with sufficiently concentrated brine (NaCl solution). Any iron present in the hard water is likely to be oxidized by the air to Fe^{3+}, which is very strongly absorbed by ion exchange resins or zeolites and cannot easily be removed. Accordingly, unless one can afford to throw away some ion exchanger periodically, any iron should be removed from the water before ion exchange softening (see (*j*), below).

Other problems with ion exchangers include coating of the resin beads or zeolite particles with suspended matter from turbid water (pretreatment with a coagulant may be necessary) or algal growths (chlorination of the water may be required). Zeolites may cause significant silica carry-over and should not be used to treat boiler water for steam turbines. Finally, although Ca^{2+} and Mg^{2+} are objectionable in boiler or laundry operations, they are necessary nutrients in the human diet. However, excessive consumption of Na^+ can contribute to hypertension and other blood circulatory problems. In Canada, for example, the incidence of heart disease and related health problems is lower in areas where the water supply is hard. Accordingly, in homes where a sodium ion-exchanger is installed to provide soft water for washing purposes, a separate supply of hard water should be available for drinking and cooking.

(*f*) *Demineralization* (deionization) is possible if the hard water is first passed down a column of *cation* exchanger in the H^+ form:

$$RH_2(s) + M^{2+}(aq) \rightleftharpoons RM(s) + 2H^+(aq) \qquad (12.9)$$

and then through an *anion* exchanger (usually a hydrocarbon polymer containing quaternary ammonium groups, R_4N^+) in the OH^- form:

$$R_4N^+OH^-(s) + X^-(aq) \rightleftharpoons R_4N^+X^-(s) + OH^-(aq). \qquad (12.10)$$

There are, of course, equal numbers of anionic and cationic charges in the solution, and hence equal numbers of H^+ and OH^- ions are released, forming water:

$$H^+(aq) + OH^-(aq) \rightarrow H_2O(l). \qquad (12.11)$$

Since the predominant anion in hard water is HCO_3^-, anion exchange capacity can be conserved by purging the acidic water coming out of the cation exchanger with air or by vacuum degassing, so sweeping out most of the HCO_3^- as CO_2:

$$H^+(aq) + HCO_3^-(aq) \rightleftharpoons H_2CO_3(aq) \rightarrow CO_2(g) + H_2O(l). \qquad (12.12)$$

If a strongly basic anion exchange resin $R_4N^+OH^-$ is used, silica will also be removed as $(HO)_3SiO^-$ (see Section 8.5), leading to water of a very low total dissolved solids content:

$$R_4N^+OH^-(s) + Si(OH)_4(aq) \rightleftharpoons R_4N^+(HO)_3SiO^- + H_2O. \qquad (12.13)$$

Hydrogen sulfide is similarly removed:

$$R_4N^+OH^-(s) + H_2S(aq) \rightleftharpoons R_4N^+SH^-(s) + H_2O. \qquad (12.14)$$

The chief drawback to demineralization is the relatively high cost of the resins (especially the anion exchanger) and the need to regenerate the cation and anion exchangers with relatively expensive H_2SO_4 and NaOH solutions, respectively.

Dissolved silica can be removed by ion exchange on a strongly basic anion exchange resin. It can also be removed by reverse osmosis, which can reduce colloidal silica levels to 0.1 ppm.

(g) *Reverse osmosis* involves manipulation of the *osmotic pressure* of a solution. If a solution containing n moles of solute particles in a volume V (in m^3) of solution is separated from pure solvent by a semipermeable membrane (i.e., a membrane through which solvent molecules, but not solute particles, can pass), an osmotic pressure Π develops across the membrane, the pure solution being the high pressure side (Fig. 12.2a).

This is because the pure solvent tends to dilute the solution—the solute, however, cannot pass through the membrane to dilute itself. Ideally, the osmotic pressure is given by

$$\Pi = \frac{nRT}{V} = cRT \qquad (12.15)$$

where T is the temperature in kelvins, Π is in pascals, the gas constant $R = 8.3143$ J K^{-1} mol^{-1}, and c is the solute particle concentration

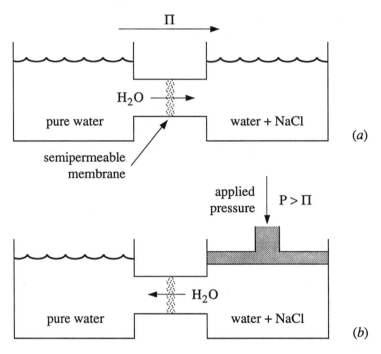

Figure 12.2 (*a*) Osmosis. (*b*) Reverse osmosis.

in mol m^{-3}. Note that, if the solute is an electrolyte $M_m X_x$, there will be $(m + x)$ moles of particles per mole of $M_m X_x$.

The net flow of solvent into the solution will continue unless a pressure equal to Π is applied in the reverse direction—for example, because of the build-up of a hydrostatic head of solution as the dilution process proceeds. Indeed, if a pressure greater than Π is deliberately applied on the solution side, the net flow of solvent will be out of the solution and into the pure solvent (Fig. 12.2b). This is *reverse osmosis*, which obviously can be used to derive pure water from solutions. The semipermeable membrane must be able to withstand a substantial pressure differential, as well as to resist passage of ions and other solutes effectively. Membranes based on cellulose acetate are adequate for use with brackish water, but polyamides are preferred with seawater. Note that the pressurized brine is further concentrated and hence Π is increased by the reverse osmosis process. In practice, then, the plant is typically run at a pressure of about 2Π (6.0 MPa; cf. Exercise 12.4) and the brine is run off after only about 20% of its water content has been extracted.

Reverse osmosis is now extensively used to reduce salt concentrations in brackish waters and to treat industrial waste water, e.g., from pulp mills. Reverse osmosis also holds promise for the economical large-scale desalination of seawater, a proposition of major interest in the Middle East, where almost all potable water is now obtained by various means from seawater or from brackish wells. Thus, at Ras Abu Janjur, Bahrain, a reverse osmosis plant converts brackish feedwater containing 19 000 ppm dissolved solids to potable water with 260 ppm dissolved solids at a rate of over 55 000 m^3 per day, with an electricity consumption of 4.8 kWh per cubic meter of product. On a 1000-fold smaller scale, the resort community on Heron Island, Great Barrier Reef, Australia, obtains most of its fresh water from seawater (36 000 ppm dissolved salts) directly by reverse osmosis, at a cost of about one cent per liter.

(*h*) *Electrodialysis*[3, 6] utilizes the fact that cations, but not anions, can readily pass through a cation exchange membrane, while the reverse is true of anion exchange membranes. If, for example, salt water is made to flow through a cell that is divided longitudinally into sections by alternating cation and anion exchange membranes and is subjected to a transverse electric potential gradient, the Na^+ and Cl^- will become concentrated in alternate compartments and depleted in the others (Fig. 12.3).

Note that, for n pairs of anion and cation membranes, the passage of one faraday causes transfer of n moles of NaCl, but only one mole NaCl is discharged at the electrodes. Ordinary polystyrene ion exchange resins (as under (*e*) and (*f*) above) will perform satisfactorily in electrodialysis, although fluorocarbon membrane materials such as "Nafion" (Section 5.3) are preferred if the budget permits. For best effect, the process is normally repeated in several stages.[5] Electrodialysis can help recover process chemicals from dilute wastes, while at the same time purifying the excess water for discharge; for example, NaOH and H_2SO_4 are regenerated in the cathode and anode compartments when waste Na_2SO_4 solution is electrodialyzed. (This is called *salt splitting.*)

(*i*) *Adsorption* onto suspended solids affords an inexpensive way of removing dissolved or colloidal silica from boiler feedwater. Silica may be adsorbed on solid magnesium oxide or hydroxide, or crushed dolomite. This is *not* a stoichiometric reaction involving the whole solid phase, but occurs to a variable extent on its surface only. The extent of adsorption is expressed by the *Freundlich adsorption isotherm:*

$$\frac{\text{mass SiO}_2 \text{ adsorbed}}{\text{mass MgO (etc.)}} = k[\text{SiO}_2(\text{aq})]^{1/n} \qquad (12.16)$$

where k and n are empirical constants at a given temperature. This

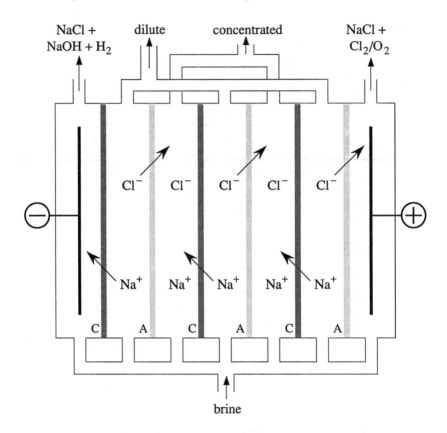

Figure 12.3 Schematic drawing of an electrodialysis cell. C: cation exchange membranes. A: anion exchange membranes.

procedure can be used in conjunction with lime–soda softening (see (c) above), since this results in a precipitate of $Mg(OH)_2$. It may, however, be necessary to add $MgSO_4$ to get enough $Mg(OH)_2$ for the silica adsorption step.

(j) *Aeration* will remove iron(II) from most natural waters. It sweeps excess CO_2 out (so increasing the p*H*), and the Fe^{2+} then reacts relatively rapidly with excess oxygen to precipitate as $Fe(OH)_3$:

$$Fe^{2+}(aq) \xrightarrow[\text{p}H > 6]{O_2} Fe(OH)_3{\downarrow}. \qquad (12.17)$$

Iron(II) reacts only slowly with oxygen in more acidic conditions, and in any event the product iron(III) may then remain in solution. To ensure a high p*H*, lime may be added:

$$Fe^{2+}(aq) + 2OH^-(aq) \rightleftharpoons Fe(OH)_2 \xrightarrow[\text{rapid}]{O_2} Fe(OH)_3. \qquad (12.18)$$

If the iron is present as humic acid complexes,[7] these can be co-agulated with alum (Section 12.2). Instead of trying to precipitate the iron, it may be better to keep it in solution, in which case it can be complexed with a chelating agent such as NTA or EDTA. As a last resort, Fe^{2+} or Fe^{3+} can be removed by cation exchange, but the absorption on the zeolite or resin is usually irreversible.

Organic matter in natural waters is usually in the form of humic acids (Section 11.6), which may discolor drinking water, foul ion exchange resins, transport toxic metal ions, or generate carcinogenic chlorocarbons if the water is chlorinated (see below). Humic acids may be removed by coagulation. Low levels of organic solutes can be removed by sorption on activated charcoal.

12.5 Sewage Treatment

The processing of sewage is mainly microbiological, and the details lie outside the scope of this book. In outline, there are three levels to which sewage may be treated.

(*a*) *Primary treatment* amounts to the mechanical removal of suspended solids and the removal of colloidal matter by coagulation.

(*b*) *Secondary treatment* involves one of two biological processes that decompose (biodegrade) organic matter.[3, 4, 7] In the trickling filter method, the sewage percolates through a pebble bed on which the microorganisms that feed on its organic content accumulate. In the activated sludge process, which is preferred nowadays, the sewage is mixed with some sludge from a previous batch (to provide a source of aerobic bacteria) and is oxygenated with air or pure oxygen to the required degree determined by the *biochemical oxygen demand* (BOD). The solids are then separated, dewatered, and incinerated or otherwise disposed of. In either case, the issuing water is often chlorinated (Section 12.7) to kill escaping microorganisms before it is released into rivers or lakes. In principle, the dewatered sludge could be a valuable fertilizer, but it usually contains too many heavy metal ions to be safe for use in growing foodstuffs. Methane is a by-product that can be used to drive generators for the sewage plant's electrical requirements. The use of oxygen, rather than air, in the activated sludge process keeps the working volume down, and the treatment can then be carried out odor-free in closed tanks.

(*c*) *Tertiary treatment* refers to the removal of specific pollutants from the discharge of secondary treatment plants. Most commonly, this means removal of phosphates, which can be done by adding lime to precipitate hydroxyapatite (cf. Sections 3.6 and 5.5):

$$5Ca(OH)_2 + 3HPO_4{}^{2-} \rightarrow Ca_5(PO_4)_3OH_\downarrow + 6OH^- + 3H_2O. \quad (12.19)$$

Alternatively, addition of alum precipitates $AlPO_4$, while $FeCl_3$ brings down $FePO_4$.

12.6   Dissolved Gases

Carbon dioxide is quite soluble in cold water, largely because it hydrates to form carbonic acid:

$$CO_2(aq) + H_2O(l) \rightleftharpoons H_2CO_3(aq) \quad (12.20)$$

and mineral springs often contain much more dissolved CO_2 than would be present if the water were in equilibrium with air (0.5 ppm). Carbon dioxide concentrations in excess of 10 ppm can cause greatly accelerated corrosion of metals (Chapter 14), but thorough aeration will obviously reduce the CO_2 content to acceptable levels. Carbon dioxide in steam can be neutralized with organic amines.

Oxygen is soluble in water to the extent of 9.4 ppm from air (1 bar) at $20\,^\circ C$ and is the oxidant responsible for most metallic corrosion. Consequently, deaeration of water by purging with nitrogen or vacuum degassing may be desirable in some circumstances; this should not be undertaken without circumspection, since deoxygenation may cause activation of otherwise passive metals or cause cathodic areas to become anodic (Chapter 14). At high temperatures, aqueous oxygen is consumed quite rapidly by hydrazine or sodium sulfite (Section 14.7).

Hydrogen sulfide enters natural waters from the decay of organic matter (e.g., in swamps), the bacterial reduction of sulfate ion, or underground sour natural gas deposits. It can be removed by aeration, anion exchange (Eqn. 12.14), or oxidation by chlorine to elemental sulfur:

$$Cl_2(g) + H_2S(aq) \rightarrow S(s) + 2HCl(aq). \quad (12.21)$$

12.7   Bacteria and Algae

Bacteria in water are usually thought of in terms of human disease. Indeed, until quite late in the 19th century, disastrous outbreaks of water-borne diseases such as cholera, dysentery, and typhoid fever were common in the major cities of the world and sometimes re-emerge even today when water treatment is disrupted. Bacteria are also responsible, however, for the destruction of wood, for example in cooling towers,[8] by breaking down the cellulose fibers. Certain bacteria derive their metabolic energy from the iron(II)–iron(III) redox cycle. These "iron bacteria" can proliferate to the extent that they block pipes. In any case, they will discolor water. In addition, objectionable growths of algae can occur in water tanks or circuits, given even minimal supplies of nutrients. Consequently, biocidal agents are widely used in the treatment of industrial, as well as municipal, water supplies.

Chlorination is the longest-established and cheapest means of disinfecting water on a large scale. Chlorine is available at low cost as a by-product of caustic soda production (Section 5.3) and reacts quickly with water to give hypochlorous acid, HOCl (Section 5.4), which is the actual bactericidal agent. Alkaline solutions of sodium hypochlorite are less effective, inasmuch as HOCl penetrates bacterial cell walls much more readily than OCl$^-$ ion. Generally, less than 10 ppm Cl$_2$ will suffice to disinfect water. Aqueous chlorine, however, reacts with other possible solutes such as H$_2$S (Eqn. 12.21), NH$_3$ (giving chloramine, NH$_2$Cl), and organic matter, so the chlorination plant operator arranges for 0.2 to 0.5 ppm chlorine-equivalent to remain in the water five minutes after treatment, as this is enough for continued bactericidal action en route to the user.

Chlorination can result in unacceptable taste intensification, where potable water is concerned. This often originates in the chlorination of phenols present in trace amounts from industrial pollution. If economics permit, the use of chlorine dioxide (Section 5.4) or ozone (Section 2.3) in place of chlorine will minimize taste intensification and will also prevent the formation of chlorocarbon carcinogens, notably chloroform. These carcinogens may form from chlorination of contaminants such as acetone, a commonly used solvent that finds its way into water supplies:

$$(CH_3)_2CO + 3HOCl \rightarrow CH_3COOH + CHCl_3 + 2H_2O. \qquad (12.22)$$

However, the chief source of chloroform is probably chlorination of naturally formed humic acids, especially in the tropics and subtropics. The World Health Organization has set a limit of 30 μg L^{-1} as the acceptable chloroform concentration in drinking water. Overzealous use of chlorine to sterilize sewage-plant effluent has also led to major fish kills in rivers. The use of ozone for water sterilization is therefore gaining much favor, but ozone does not persist long enough to give continued protection of the water once it leaves the treatment plant. A compromise solution is to use ozone as the primary disinfectant and to add some chlorine for more lasting sterilization.

Other biocidal agents that can be used in closed industrial water systems include copper(II) salts (which, however, can cause corrosion of metals), chlorinated phenols such as sodium pentachlorophenate (NaOC$_6$Cl$_5$—this is toxic), and quaternary ammonium salts (R$_4$N$^+$X$^-$).

References

1. W. H. Betz and L.D. Betz, "Handbook of Industrial Water Conditioning," 8th edn., Betz Laboratories, 1980.

2. F. N. Kemmer (ed.), "The Nalco Water Handbook," McGraw-Hill: New York, 1988.

3. G. T. Austin, "Shreve's Chemical Process Industries," 5th edn., McGraw-Hill: New York, 1985, Chapter 3.

4. J. A. Kent (ed.), "Riegel's Handbook of Industrial Chemistry," 8th edn., Van Nostrand-Reinhold: New York, 1983, Chapters 2 and 26.

5. R. M. Harrison (ed.), "Pollution: Causes, Effects and Control," Special Publication No. 44, The Royal Society of Chemistry: London, 1983, Chapters 2–7.

6. D. Pletcher, "Industrial Electrochemistry," Chapman and Hall: London, 1984, Chapter 11.

7. J. E. Fergusson, "Inorganic Chemistry and the Earth," Pergamon Press: Oxford, 1982.

8. W. L. Marshall, "Cooling water treatment in power plants," *Industrial Water Engineering,* February–March, 1972, 38–42.

9. S. D. Faust and O. M. Aly, "The Chemistry of Water Treatment," Butterworths: London, 1983.

Exercises

12.1 (*a*) Suppose that the temporary and total hardnesses of municipal tap water were equivalent to 72.0 and 150.0 mg $CaCO_3$ per kg water (i.e., 72 and 150 ppm). Calculate the weights of lime (as CaO) and soda ash (as Na_2CO_3) needed to soften one cubic meter of water by the lime–soda process, assuming that the permanent hardness is due to magnesium ion alone.

(*b*) How would the calculations be changed if 20 mg kg^{-1} $CaCO_3$-equivalent of the permanent hardness were actually due to non-carbonate calcium hardness?

[*Answers:* (*a*) 84.1 and 82.7 g, respectively.]

12.2 In water at ambient temperature, the solubilities of $MgCO_3$, $CaCO_3$, $Mg(OH)_2$, and $Ca(OH)_2$ are about 106, 15.3, 9.0, and 1600 mg kg^{-1}, respectively. Show that the stoichiometric application of the lime–soda process should leave a residual hardness of some 31 ppm $CaCO_3$-equivalent.

12.3 Consider again the tap water described in Exercise 12.1a. Suppose, instead of lime–soda treatment, we add $EDTA^{4-}$ exactly equivalent to the total concentrations of Ca^{2+} and Mg^{2+}.

(a) What will be the residual $CaCO_3$-equivalent hardness, given that the stability constants β_1 of $CaEDTA^{2-}$ and $MgEDTA^{2-}$ are 5.0×10^{10} and 4.9×10^8 L mol^{-1}, respectively? (Ignore pH and ionic strength effects.)

(b) What would the residual hardness be, if all of the original 150 ppm $CaCO_3$-equivalent hardness were of the temporary type?

(c) Compare your answers with those for the lime–soda method.

[Answers: (a) 0.13 ppm; (b) 0.017 ppm $CaCO_3$-equivalent.]

12.4 Seawater typically contains 35 g kg^{-1} dissolved salts. Assuming this to be effectively all NaCl, and ignoring the interionic interactions (Debye–Hückel, ion pairing, etc.), arrive at an order-of-magnitude estimate of the back-pressure necessary to obtain pure water from seawater by reverse osmosis.

[Answer: 3 MPa or more, i.e., 30 bars.]

12.5 Compare the advantages and disadvantages of zeolites with those of synthetic organic ion exchange resins in the context of water treatment.

Chapter 13

Oxidation and Reduction in Solution

13.1 Galvanic Cells[1-3]

CONSIDER FIRST the dissolution of zinc metal in a strong aqueous acid to give $Zn^{2+}(aq)$ ions and hydrogen gas:

$$Zn(s) + 2H^+(aq) \rightarrow Zn^{2+}(aq) + H_2(g). \tag{13.1}$$

The zinc is oxidized, i.e., loses electrons to the hydrogen ions, which are reduced. If zinc metal is simply placed in direct contact with aqueous acid, the transfer of electrons takes place within the solution. If, however, zinc oxidation and hydrogen reduction are made to take place in different compartments (half-cells), the electrons can be transferred via an external circuit and become available for doing external electrical work. There are two other requisites to make this *galvanic cell* (Fig. 13.1) work: a *salt bridge* or other means of allowing inert ions to flow between the half-cells so as to balance the flow of electrical charge through the external circuit, and an inert conducting surface (electrode) such as platinum metal to supply the electrons from the external circuit to the hydrogen ion reduction reaction. Thus, the complete galvanic cell:

$$Zn(s) \mid Zn^{2+}(aq) \parallel 2H^+(aq) \mid H_2(g),Pt(s)$$

$\qquad$ solution-gas interface on Pt metal

$\qquad$ salt bridge or porous partition

$\qquad$ metal-solution interface

$$\tag{13.2}$$

is a combination of two half-cells:

$$Zn(s) \mid Zn^{2+}(aq) + 2e^- \tag{13.3}$$

$$2e^- + 2H^+(aq) \mid H_2(g),Pt(s) \tag{13.4}$$

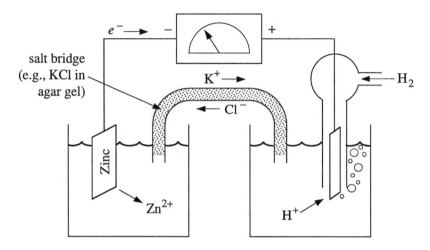

Figure 13.1   A galvanic cell, composed of $Zn/Zn^{2+}(aq)$ and $H_2/H^+(aq)$ electrodes.

in which occur the two half-reactions:

$$Zn(s) \rightarrow Zn^{2+}(aq) + 2e^- \qquad (13.5)$$

$$2e^- + 2H^+(aq) \rightarrow H_2(g). \qquad (13.6)$$

The electrode in the half-cell in which oxidation is occurring is said to be the anode (here, the zinc metal), whereas the other is the cathode (here, the platinum). In principle, we could connect any pair of feasible half-cells to form a galvanic cell; the electromotive force (EMF, in volts) and polarity (which electrode becomes the cathode and which the anode?) of such a cell will depend upon the identity of the half-cells, the temperature and pressure, the activities of the reacting species, and the current drawn.

If we choose a set of standard conditions (cf. Section 1.2) and one convenient half-cell to serve as a reference for all others, then a set of standard half-cell EMFs or *standard electrode potentials* E° (Appendix D)[1-4] can be measured while drawing a negligible electrical current, i.e., with the cell working *reversibly* so that the equations of reversible thermodynamics (Chapter 1) apply. The standard reference half-cell is reaction 13.6, the *standard hydrogen electrode* (SHE), and the standard conditions are those listed in Section 1.2, although for our purposes the molar concentration scale (mol L^{-1}) can generally be used without significant loss of precision. We will simplify matters further, for illustrative purposes, by *equating activities with molar concentrations;* our numerical results will therefore be only approximate, except where these concentrations are very low.

By international agreement, the algebraic sign of $E°$ for a half-cell is chosen to be the same as its electrical sign relative to the SHE. This means, in effect, that we must write the half-reactions with the electrons on the left-hand side, i.e., $E°$ values are taken to be *reduction potentials*.* Consequently, a reagent such as chlorine that is more oxidizing than aqueous H^+ $(\rightarrow \frac{1}{2}H_2)$ under standard conditions will have a positive $E°$:

$$\tfrac{1}{2}Cl_2(g) + e^- \rightarrow Cl^-(aq) \quad E° = +1.36\,V \tag{13.7}$$

and a reagent such as chromium(II) $(\rightarrow Cr^{3+})$ that is more reducing than H_2 $(\rightarrow H^+)$ will be associated with a negative $E°$:

$$Cr^{3+}(aq) + e^- \rightarrow Cr^{2+}(aq) \quad E° = -0.42\,V. \tag{13.8}$$

Note that *reducing reagents* (i.e., suppliers of electrons) appear on the right in a half-reaction.

Consideration of electrode potentials in the context of Born-Haber-type cycles (cf. Sections 6.5 and 11.1) leads to the following generalizations:[†]

(*a*) Strongly *reducing* reagents (very negative $E°$ values) result from low ionization potentials (e.g., alkali metals), low heats of sublimation (e.g., alkali metals again), and high heats of hydration (e.g., small ionic radii: Li^+ more than Na^+);

(*b*) Strongly *oxidizing* reagents (very positive $E°$ values) are associated with negative electron affinities (e.g., halogens), low heats of dissociation (e.g., halogens again), high heats of hydration (e.g., small ionic radii: F^- rather than Cl^-).

As noted in Section 1.2, $E°$ is a measure of the available free energy *per mole of electrons* to be transferred, under standard conditions:

$$\Delta G° = -nFE° \tag{13.9}$$

where n is the number of (moles of) electrons transferred and F is the Faraday (i.e., $96\,487$ C mol^{-1}). The more positive $E°$ is, the greater is $-\Delta G°$ and hence the greater the tendency for that particular half-cell to go in the direction indicated, with respect to oxidizing H_2 to $H^+(aq)$. More generally, a half-reaction should proceed spontaneously in the sense written by *reversing* another half-reaction of less positive $E°$. Iron(III) ions, for

---

*In some of the older American literature, such as the venerable and still useful books by Wendell M. Latimer,[5] the reverse convention (*oxidation potentials*) was used; the electrons appear on the right, and the sign of $E°$ is reversed.

†As an exercise, check these conclusions and compare them with actual tabulated $E°$ values (Appendix D).

example, should (and do) oxidize aqueous iodide to iodine:

$$Fe^{3+}(aq) + e^- \rightarrow Fe^{2+} \qquad\qquad E° = +0.771\,V \quad (13.10)$$

$$\tfrac{1}{2}I_2(s) + e^- \rightarrow I^- \qquad\qquad E° = +0.536\,V \quad (13.11)$$

$$\text{Subtract:} \quad Fe^{3+} + I^- \rightarrow Fe^{2+} + \tfrac{1}{2}I_2 \qquad \Delta E° = +0.235\,V. \quad (13.12)$$

We can extract the equilibrium constant K for reaction 13.12 from this information (cf. Section 1.2).

$$nF(\Delta E°) = -\Delta G° = RT \ln K \qquad\qquad (13.13)$$

$$
\begin{aligned}
K &= \exp \frac{nF(\Delta E°)}{RT} \qquad\qquad\qquad (13.14)\\[2mm]
&= \exp \frac{1 \times 96\,487 \times 0.235}{8.3143 \times 298.15}\\[2mm]
&= 9.4 \times 10^3\\[2mm]
&= \frac{[Fe^{2+}][I_2]^{1/2}}{[Fe^{3+}][I^-]}.
\end{aligned}
$$

For non-standard conditions, the *Nernst equation* is necessary:

$$
\begin{aligned}
E &= E° - \frac{RT}{nF} \ln Q \qquad\qquad\qquad (13.15)\\[2mm]
&= E° - \frac{0.05916}{n} \log_{10} Q
\end{aligned}
$$

for a half-reaction at $25\,°C$, or

$$\Delta E = \Delta E° - \frac{RT}{nF} \ln Q \qquad\qquad (13.16)$$

for a balanced reaction, where

$$Q = \frac{[\text{products}]^p}{[\text{reactants}]^r} \qquad\qquad (13.17)$$

which, in the case of reaction 13.12, means:

$$Q = \frac{[Fe^{2+}][I_2]^{1/2}}{[Fe^{3+}][I^-]}. \qquad\qquad (13.18)$$

When reaction 13.12 is allowed to run down to equilibrium, ΔE (the EMF of a galvanic cell based on reaction 13.12) runs down to zero, all concentrations assume their equilibrium values, and so Q becomes K. To put it another way, if all the reagents are present at standard-state concentrations (unity, by definition), Q becomes unity and E is just $E°$.

There are some complications. As we know from Section 11.6, iron(III) ions tend to hydrolyze in aqueous solution unless the pH is very low. Accordingly, it is understood that, unless otherwise stated, $E°$ values refer to measurements in *acidic* solution, even if the hydrogen ions do not explicitly appear in the balanced redox equation, e.g., reaction 13.12. Secondly, iodide ion actually reacts with iodine in water to give brown $I_3{}^-$:

$$I_2(s) + I^-(aq) \rightarrow I_3{}^-(aq). \tag{13.19}$$

Since the equilibrium constant of reaction 13.19 is 0.93, purple solid iodine is significantly more soluble in aqueous iodide solutions than in pure water. For dilute solutions, however, this can be ignored, as can a similar reaction involving chlorine and chloride in connection with reaction 13.7. Finally, we should remind ourselves that we have assumed that all the activity coefficients are unity (although here again the discrepancy so introduced can be ignored for simplicity).

13.2 Manipulation and Use of Electrode Potentials

As an illustration of the use of electrode potentials, consider the classical method of analysis of copper in brass which involves dissolving the weighed sample in nitric acid to obtain $Cu^{2+}(aq)$, adjusting the pH to a weakly acidic level, allowing the Cu^{2+} to react completely with excess potassium iodide to form iodine and the poorly soluble CuI, and then titrating the iodine with sodium thiosulfate solution that has been standardized against pure copper by the same procedure:

$$Cu^{2+} + I^- + e^- \rightarrow CuI(s) \qquad E° = +0.86\,V \quad (13.20)$$
$$S_4O_6{}^{2-} + 2e^- \rightarrow 2S_2O_3{}^{2-} \qquad E° = +0.08\,V \quad (13.21)$$
$$NO_3{}^- + 2H^+ + 2e^- \rightarrow NO_2{}^- + H_2O \qquad E° = +0.94\,V. \quad (13.22)$$

It follows from reactions 13.11 and 13.20 that copper(II) ion will oxidize iodide to iodine, and from reactions 13.11 and 13.21 that thiosulfate will reduce I_2 back to I^- while forming the tetrathionate ion (Section 1.8). Because the relevant $\Delta E°$ values are several hundred mV, the equilibrium constants are large, and these reactions go to essential completion.

We must, however, ensure that there is no contaminant such as free Fe^{3+} in the solution before the potassium iodide is added; otherwise, more I^- will be oxidized than there is Cu^{2+} in the sample (reaction 13.12). The effective concentration of Fe^{3+} can be reduced to negligible levels by adding sodium fluoride to complex it. (The divalent copper ion is little affected.)

From the $E°$ of half-reaction 13.22, it would seem that the nitrate ion present from the dissolution of the brass should also oxidize iodide ion. This $E°$ value, however, refers to *standard* conditions, which implies 1 mol L^{-1} H$^+$, whereas we have adjusted the pH to near-neutrality. Suppose the pH is adjusted to 7, i.e., $[H^+] = 1 \times 10^{-7}$ mol L^{-1}, while $[NO_2{}^-]$ and $[NO_3{}^-]$ retain their standard-state values of unity; the corresponding EMF for the half-reaction 13.22 is then:

$$E_{\text{pH } 7} = E° - \frac{0.05916}{2} \log_{10} \frac{[NO_2{}^-]}{[NO_3{}^-][H^+]^2} \qquad (13.23)$$

$$= +0.94 - \frac{0.05916}{2}(2 \times 7)$$

$$= +0.53 \text{ V}$$

so that the oxidation of I$^-$ by NO$_3{}^-$ is no longer favored. Even in slightly acidic solutions, the oxidation of I$^-$ by NO$_3{}^-$, though marginally favored thermodynamically, is so *slow* that it does not occur, in practice. This illustrates an important limitation to the use of $E°$ data to predict the outcome of an electrochemical reaction; we can say definitely what its equilibrium position is, but *not* whether it will proceed to that equilibrium at a detectable rate.

The oxidation states of manganese in 1.0 mol L^{-1} aqueous acid afford a colorful illustration of how $E°$ data can be manipulated:

$$
\begin{array}{lll}
\text{MnO}_4{}^-\text{(aq)} & \text{permanganate ion (intensely purple)} & \text{Mn}^{\text{VII}} \\
\quad\downarrow +e^- \;\; +0.56\text{ V} & & \\
\text{MnO}_4{}^{2-}\text{(aq)} & \text{manganate ion \quad (dark bottle-green)} & \text{Mn}^{\text{VI}} \\
\quad\downarrow \;{}^{+2e^-}_{+4H^+} \;\; +2.26\text{ V} & & \\
\text{MnO}_2\text{(s)} & \text{pyrolusite \qquad (brown-black solid)} & \text{Mn}^{\text{IV}} \\
\quad\downarrow \;{}^{+e^-}_{+4H^+} \;\; +0.95\text{ V} & & \\
\text{Mn}^{3+}\text{(aq)} & \text{"manganic" ion \quad (red-violet)} & \text{Mn}^{\text{III}} \\
\quad\downarrow +e^- \;\; +1.51\text{ V} & & \\
\text{Mn}^{2+}\text{(aq)} & \text{"manganous" ion \quad (pale pink)} & \text{Mn}^{\text{II}} \\
\quad\downarrow +2e^- \;\; -1.18\text{ V} & & \\
\text{Mn(s)} & \text{manganese metal} & \text{Mn} \\
\end{array}
$$
$$(13.24)$$

(a) *Calculation of $E°$ for multi-electron half-reactions.* Equation 13.9 makes it clear that $E°$ is proportional to the (negative) free energy change *per mole of electrons transferred.* Consequently, we can obtain EMFs of cells by simply adding the $E°$ values of the two constituent half-cells, and we can predict if a given half-reaction will go by reversing another by direct comparison of the $E°$ data (Appendix D), since each of these is on a "per electron" basis. We *cannot*, however, obtain $E°$ for the direct reduction of $MnO_4^-(aq)$ to MnO_2 simply by adding $E°$ for the MnO_4^-/MnO_4^{2-} couple to that for the MnO_4^{2-}/MnO_2, since different numbers of electrons are transferred. It is the *free energies,* and not the $E°$ values, that are additive. So, to evaluate $E°$ for

$$MnO_4^- + 4H^+ + 3e^- \rightarrow MnO_2(s) + 2H_2O \qquad (13.25)$$

we can convert $E°$ to the appropriate free energies, add these, and then convert back to $E°$. To take a short cut, the procedure reduces to multiplying $E°$ for each step by the number of electrons in that step, adding all these products algebraically, and dividing this sum by the total number of electrons.

For the specific half-reaction 13.25:

$$E° = \frac{(1 \times 0.56) + (2 \times 2.26)}{3} = +1.69V. \qquad (13.26)$$

(*b*) *Identification of species likely to disproportionate.* According to thermodynamics, the half-reaction

$$Mn^{3+} + e^- \rightarrow Mn^{2+} \qquad E° = +1.51\,V \qquad (13.27)$$

can proceed spontaneously by reversing some other reaction with $E° < +1.51\,V$. One such reaction is

$$MnO_2(s) + 4H^+ + e^- \rightarrow Mn^{3+} + 2H_2O \qquad E° = +0.95\,V \qquad (13.28)$$

which means that Mn^{3+} is capable of oxidizing (or reducing) *itself,* to Mn^{2+} and MnO_2 (disproportionation):

$$2Mn^{3+} + 2H_2O \rightarrow Mn^{2+} + MnO_2 + 4H^+ \qquad \Delta E° = +0.56\,V. \qquad (13.29)$$

Indeed, manganese(III) solutions do decompose slowly, putting down brown MnO_2 on the vessel walls, although reaction 13.29 tells us that this disproportionation reaction is less thermodynamically favored at high $[H^+]$, and the *kinetics* of decomposition are also slower in highly acidic media, so that $Mn^{3+}(aq)$ chemistry *can* be studied at very low pH, so long as the time scale is kept short. In a reduction-sequence display such as sequence 13.24, any species that has a more positive $E°$ following than preceding it will be thermodynamically unstable with respect to disproportionation.[‡]

[‡]As an exercise, you may check this and note that manganate ion is a case in point.

(c) *pH effects.* Whenever hydrogen ions appear in a half-reaction, $E°$ must be pH-dependent. There is no *direct* pH effect on the MnO_4^-/MnO_4^{2-} couple, since reaction 13.30 is balanced as it is:

$$MnO_4^- + e^- \rightarrow MnO_4^{2-}. \qquad (13.30)$$

However, the MnO_4^{2-}/MnO_2 couple is much less oxidizing in 1.0 mol L^{-1} aqueous alkali ($[H^+] = 1 \times 10^{-14}$ mol L^{-1}) than in aqueous acid:

$$MnO_4^{2-} + 4H^+ + 2e^- \rightarrow MnO_2(s) + 2H_2O \quad E° = +2.26\,V \quad (13.31)$$

$$E_{pH\ 14} = E° - \frac{0.05916}{2} \log_{10} \frac{[MnO_2]}{[MnO_4^{2-}][H^+]^4}. \qquad (13.32)$$

which works out to $2.26 - [(0.5916/2) \times 4 \times 14] = +0.60\,V$, since MnO_2 is a solid phase of unit activity and $[MnO_4^{2-}]$ in the standard state is set to unity. It is therefore possible to minimize the tendency of manganate ion to disproportionate by working in strongly alkaline solution.

The Mn^{3+}/Mn^{2+} couple is ostensibly pH-independent, but we must bear in mind that manganese(III), in particular, will hydrolyze unless the acidity is very high, and both $Mn(OH)_3$ and $Mn(OH)_2$ will come out of solution in alkaline media. Small degrees of hydrolysis in solution have very little impact on $E°$, but precipitation of hydroxides (and all metal hydroxides except those of the alkali metals and Ca, Sr, Ba, and Ra are poorly soluble in water) affects $E°$ profoundly. Thus, in alkaline media, the electrode potentials (often called E_h by geologists, wherever $[H^+]$ is not the standard value) of Mn^{2+} and Mn^{3+} are controlled by the solubility products (K_{sp}) of $Mn(OH)_2$ and $Mn(OH)_3$, respectively. In practice, $Mn(OH)_3$ tends to dehydrate to $MnO(OH)$, so we shall consider the $Mn^{2+}(aq)/Mn(s)$ couple:

$$Mn^{2+}(aq) + 2e^- \rightarrow Mn(s) \quad E°_{pH\ 0} = -1.180\,V. \qquad (13.33)$$

In 1 mol L^{-1} NaOH solution (pH 14),

$$Mn(OH)_2 \rightleftharpoons Mn^{2+}(aq) + 2OH^-(aq) \quad K_{sp} = 3 \times 10^{-13} \quad (13.34)$$

so

$$[Mn^{2+}]_{pH\ 14} = \frac{3 \times 10^{-13}}{[OH^-]^2} = 3 \times 10^{-13} \qquad (13.35)$$

$$E_{h(pH\ 14)} = -1.180 - \frac{0.05916}{2} \log_{10} \frac{1}{[Mn^{2+}]} = -1.55\,V. \qquad (13.36)$$

(*d*) *Effects of complexation.* Just as K_{sp} of an insoluble phase can control the concentration of free metal ion and hence E_h, so control by complexing the metal ion can alter $E°$. For example, $E°$ for the half-reaction

$$Mn^{III}EDTA^- + e^- \rightarrow Mn^{II}EDTA^{2-} \qquad (13.37)$$

can be calculated from

$$Mn^{3+} + e^- \rightleftharpoons Mn^{2+} \qquad E° = +1.51\,V \qquad (13.38)$$

given that the stability constants β^{III} and β^{II} for $Mn^{III}EDTA^-$ and $Mn^{II}EDTA^{2-}$ are $10^{25.3}$ and $10^{13.8}$, respectively. Ignoring pH effects, we can write

$$[Mn^{3+}] = \frac{[Mn^{III}EDTA^-]}{\beta^{III}[EDTA^{4-}]} \qquad (13.39)$$

and

$$[Mn^{2+}] = \frac{[Mn^{II}EDTA^-]}{\beta^{II}[EDTA^{4-}]} \qquad (13.40)$$

from which $[EDTA^{4-}]$ can be eliminated and, for the standard-state conditions relevant to reaction 13.37, the concentrations of the EDTA complexes can be set to unity (the degree of their dissociation is very small):

$$\frac{[Mn^{2+}]}{[Mn^{3+}]} = \frac{[Mn^{II}EDTA^{2-}]\beta^{III}}{[Mn^{III}EDTA^-]\beta^{II}} \qquad (13.41)$$

$$= \frac{\beta^{III}}{\beta^{II}}.$$

The standard electrode potential $E°_{EDTA}$ for half-reaction 13.37 is then derived from the Nernst equation for 13.38:

$$E°_{EDTA} = +1.51 - \frac{0.05916}{1}\log_{10}\frac{[Mn^{2+}]}{[Mn^{3+}]}\,V \qquad (13.42)$$

$$= +1.51 - 0.05916 \times (25.3 - 13.8)\,V$$

$$= +0.83\,V.$$

Notice that since complexation stabilizes Mn^{III} much more than Mn^{II}, $E°$ is reduced to a more moderate value. For the couple

$$Mn(CN)_6{}^{3-} + e^- \rightarrow Mn(CN)_6{}^{4-} \qquad E° = -0.24\,V \qquad (13.43)$$

Mn^{III} is obviously very strongly stabilized by complexing with cyanide.

(e) *Redox of solvent water.* Any half-reaction with a negative electrode potential can, in principle, be reversed by

$$2H^+(1 \text{ mol } L^{-1}) + 2e^- \rightarrow H_2(g) \quad E^\circ = 0.000 \text{ V} \quad (13.44)$$

and, if E° is more negative than -0.414 V, by neutral water (pH 7):

$$2H^+(1 \times 10^{-7} \text{ mol } L^{-1}) + 2e^- \rightarrow H_2(g) \quad E_h = -0.414 \text{ V} \quad (13.45)$$

or, if E° is more negative than -0.828 V, by 1.0 mol L^{-1} aqueous alkali:

$$2H^+(1 \times 10^{-14} \text{ mol } L^{-1}) + 2e^- \rightarrow H_2(g) \quad E_h = -0.828 \text{ V} \quad (13.46)$$

Metallic manganese certainly dissolves readily in 1 mol L^{-1} aqueous acid, as expected, but in aqueous base a coating of insoluble $Mn(OH)_2$ would form immediately, stopping the predicted reaction. In cases such as aluminum (E° for $Al^{3+}/Al(s) = -1.66$ V), the product Al^{III} is soluble in alkali as the aluminate ion $Al(OH)_4^-$, but not in neutral water, in which the aluminum oxide/hydroxide layer is protective. A further factor which may prevent expected H_2 evolution is *overpotential*. This will be discussed in detail in Section 13.4, but in essence it is the result of the kinetic slowness of one or more steps in a thermodynamically feasible electrochemical reaction.

Similarly, any half-reaction with E° greater than $+1.229$ V should proceed by oxidizing water to O_2, under standard conditions (1.0 mol L^{-1} H^+):

$$O_2(g) + 4H^+(aq) + 4e^- \rightarrow 2H_2O(l) \quad E^\circ = +1.229 \text{ V}. \quad (13.47)$$

Since we have shown (Eqn. 13.26) that E° is $+1.69$ V for the $MnO_4^-/MnO_2(s)$ couple, we can expect aqueous acidic $KMnO_4$ solution to evolve oxygen and deposit MnO_2. Indeed, it does, though slowly, and solutions of $KMnO_4$ used for volumetric analysis must be restandardized if kept for more than a few hours.

According to reaction 13.47, for which the Nernst equation gives $E_h = +0.401$ V at pH 14, the oxidation of water is clearly favored by alkaline conditions (see Exercises). At the same time, however, many oxidation half-reactions also have lower E_h values in basic media. For example, E_h for the manganate/manganese dioxide couple

$$MnO_4^{2-} + 2H_2O + 2e^- \rightarrow MnO_2(s) + 4OH^- \quad E^\circ = +0.61 \text{ V} \quad (13.48)$$

is only 0.21 V more positive than that required for water oxidation in 1.0 mol L^{-1} aqueous alkali, as against 1.03 V more in aqueous acid (compare reactions 13.24 and 13.47), because *two* H^+ ions are consumed (two OH^- liberated) per electron transferred in reaction 13.48 as compared with *one* in reaction 13.47. So, quite apart from the problem of disproportionation discussed above, aqueous MnO_4^{2-} is expected to be tractable only in highly alkaline solutions. Once again, any such predictions may be subject to modification by overpotential effects.

13.3   Pourbaix (*E_h*-p*H*) Diagrams

The ranges of E_h and p*H* over which a given aqueous species is thermodynamically stable can be displayed graphically as *stability fields* in a *Pourbaix diagram*.[6-9] These are constructed with the aid of the Nernst equation, together with the solubility products of any solid phases involved, for certain specified activities of the reactants. For example, the stability field of liquid water under standard conditions (partial pressures of H_2 and O_2 of 1 bar, at 25 °C) is delineated in Fig. 13.2 by:

$$E_h = 1.229 - 0.05916\,\text{p}H\ \text{V (for } O_2 \text{ evolution)} \tag{13.49}$$

and

$$E_h = 0.000 - 0.05916\,\text{p}H\ \text{V (for } H_2 \text{ evolution)}. \tag{13.50}$$

For non-standard partial pressures of the gases, these boundaries will be slightly displaced, but their slopes will remain the same. In Fig. 13.2, the stability field of water is only slightly narrowed by considering gas pressures a million-fold lower.

The E_h-p*H* relations for the important iron–water system at 25 °C are summarized in Fig. 13.3 with some simplifications. First, it is assumed that no elements other than Fe, O, and H are involved; in a natural water system, the presence of CO_2 would oblige us to include $FeCO_3$ (siderite), and any sulfur compounds could lead to precipitation of iron sulfides in certain E_h-p*H* regimes, etc. As it is, the only solids considered are Fe metal, $Fe(OH)_2$, and $Fe(OH)_3$, whereas in practice magnetite (Fe_3O_4), hematite (α-Fe_2O_3), goethite (α-$FeO(OH)$), and other Fe–O–H phases could be present. Indeed, our choice of solubility products for $Fe(OH)_2$ and $Fe(OH)_3$ is somewhat arbitrary, since these materials become less soluble as they "age," i.e., become better crystallized as time passes; here, mid-range values are chosen. Furthermore, the hydrolytic sequence given in Section 11.6 for iron(III) has been ignored. A small modification[8] is needed to take this into account, but for simplicity we will represent the hydrolysis of Fe^{3+}(aq) as leading directly to $Fe(OH)_3$(s). Finally, certain of the boundary lines in Fig. 13.3 refer to a specific activity of aqueous iron ions. This is arbitrarily set at 10^{-3} mol L^{-1} (and, for convenience, activity

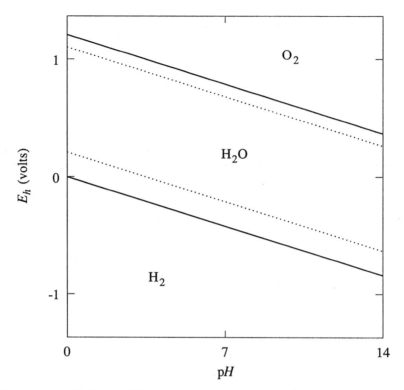

Figure 13.2 Redox stability field of liquid water at 25 °C. Solid lines refer to gas partial pressures of 1 bar (standard state), broken lines to 10^{-6} bar.

is equated with concentration), but a comparison with Fergusson's version[9] for an iron ion activity of 10^{-5} shows that the qualitative features of the diagram are not very sensitive to this variable.

Figure 13.3 is constructed as follows. The boundary between the stability fields of solid iron and 1×10^{-3} mol L^{-1} $Fe^{2+}(aq)$ is given by the Nernst equation (13.52) for:

$$Fe^{2+}(aq) + 2e^- \rightarrow Fe(s) \quad E° = -0.44\,V \quad (13.51)$$

$$E_h = -0.44 - \frac{0.05916}{2} \log_{10}[Fe^{2+}]^{-1} = -0.53\,V. \quad (13.52)$$

Since this is independent of pH, it is simply a horizontal line at the lower left of Fig. 13.3. The value of $E°$ used in Eqn. 13.52 is somewhat more negative than that derived from U.S. National Bureau of Standards thermodynamic data ($-0.409\,V$; Appendix C) and widely cited. In fact, since various values between -0.409 and $-0.475\,V$ have been reported by very experienced experimentalists using several different methods, the value

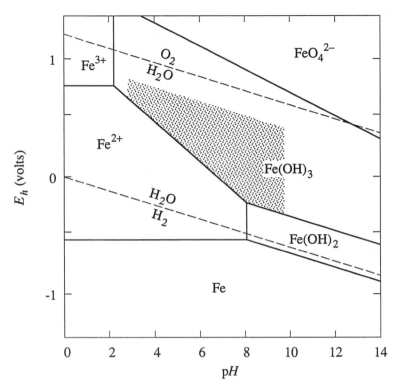

Figure 13.3 E_h-pH (Pourbaix) diagram for the iron–water system at 25 °C for a maximum dissolved-iron concentration of 0.001 mol L^{-1}. The shaded area shows the E_h-pH range of natural waters.

selected here is intended to be a conservative compromise.[4] The moral is that one should not place blind faith in tabulated data, even from the most authoritative sources.

The boundary between the Fe^{2+}(aq) and Fe^{3+}(aq) stability fields will be simply the pH-independent $E°$ value for reaction 13.53, if hydrolysis is ignored, since (by definition) the concentrations of the two ions will be equal at the boundary:

$$Fe^{3+}(aq) + e^- \rightarrow Fe^{2+}(aq) \quad E° = +0.77\,V \quad (13.53)$$

$$E_h = E° - 0.05916 \log_{10} \frac{[Fe^{2+}]}{[Fe^{3+}]} = +0.77\,V. \quad (13.54)$$

The vertical Fe^{3+}(aq)/$Fe(OH)_3$(s) boundary at top left involves pH through Eqns. 13.55 and 13.56, but not E_h, since no electrons are transferred in reaction 13.55.

$$Fe(OH)_3(s) \rightleftharpoons Fe^{3+}(aq) + 3OH^-(aq) \quad K_{sp} = 10^{-38.1} \quad (13.55)$$

$$[H^+][OH^-] = K_w = 1 \times 10^{-14} \qquad (13.56a)$$

$$\log_{10} K_{sp} = \log_{10}[Fe^{3+}] - 3(14 - pH) \qquad (13.56b)$$

$$pH = 14 + \frac{-38.1 + 3.0}{3} = 2.3. \qquad (13.56c)$$

The $Fe^{2+}(aq)/Fe(OH)_2(s)$ dividing line is similarly calculated to stand at pH 7.95 for 10^{-3} mol L^{-1} Fe^{2+}, given $\log_{10} K_{sp} = -15.1$ for Fe(OH)$_2$:

$$Fe(OH)_2(s) \rightleftharpoons Fe^{2+}(aq) + 2OH^-(aq) \quad K_{sp} = 10^{-15.1}. \qquad (13.57)$$

The boundary for the Fe(s) and Fe(OH)$_2$(s) stability fields, however, is given by a combination of reactions 13.51 and 13.57:

$$Fe(OH)_2(s) + 2e^- \rightarrow Fe(s) + 2OH^-(aq). \qquad (13.58)$$

That is, reaction 13.57 controls [Fe^{2+}] in the half-reaction 13.51:

$$
\begin{aligned}
E_h &= -0.44 - \frac{0.05916}{2} \log_{10}\left(\frac{K_{sp}}{[OH^-]^2}\right)^{-1} \qquad (13.59) \\
&= -0.44 - 0.02958[15.1 - 2(14 - pH)] \\
&= -0.06 - 0.059\,pH.
\end{aligned}
$$

This boundary line therefore depends on both E_h and pH and has a slope of $-(\ln 10)(RT/F)$ (bottom right of Fig. 13.3). Parallel to this and above it is the Fe(OH)$_2$/Fe(OH)$_3$ boundary, which is defined by a combination of reactions 13.53, 13.55 and 13.57; the concentrations of Fe^{2+}(aq) and Fe^{3+}(aq) in reaction 13.53 are controlled by the pH-dependent solubilities of Fe(OH)$_2$ and Fe(OH)$_3$, so that what appears to be a solid-solid reaction:

$$Fe(OH)_3(s) + e^- \rightarrow Fe(OH)_2(s) + OH^-(aq) \qquad (13.60)$$

is actually mediated by a solution-phase redox process in an aqueous environment. Development as above leads to:

$$E_h = +0.24 - 0.05916\,pH. \qquad (13.61)$$

The $Fe^{2+}(aq)/Fe(OH)_3(s)$ boundary, which cuts through the center of Fig. 13.3, represents the situation in which the Fe^{3+}(aq) concentration in half-reaction 13.53 is controlled by the solubility of Fe(OH)$_3$, given that [Fe^{2+}] is 1×10^{-3} mol L^{-1}. Proceeding as before, we obtain:

$$
\begin{aligned}
E_h &= 0.77 - 0.05916 \log_{10} \frac{[Fe^{2+}][OH^-]^3}{K_{sp}} \qquad (13.62) \\
&= 1.00 - 0.177\,pH - 0.05916 \log_{10}[Fe^{2+}] \\
&= 1.18 - 0.177\,pH.
\end{aligned}
$$

As a final embellishment, the oxidation of iron(III) to the aqueous ferrate(VI) ion, FeO_4^{2-}, is included:

$$FeO_4^{2-}(aq) + 8H^+(aq) + 3e^- \rightarrow Fe^{3+}(aq) + 4H_2O \quad E^\circ = +1.9\,V. \quad (13.63)$$

As the finished diagram 13.3 will make obvious, we need only consider the ferrate(VI) ion in alkaline environments, in which the iron(III) concentration is controlled by the poor solubility of $Fe(OH)_3$, as in reaction 13.55:

$$
\begin{aligned}
E_h &= 1.9 - \frac{0.0592}{3} \log_{10} \frac{K_{sp}}{[OH^-]^3[FeO_4^{2-}][H^+]^8} \quad (13.64) \\
&= 1.7_9 + 0.0197 \log_{10}[FeO_4^{2-}] - 0.0986\,pH \\
&= 1.7_3 - 0.0986\,pH.
\end{aligned}
$$

where again the total dissolved iron concentration has been set at 1×10^{-3} mol L^{-1}.

In addition to the above information, Fig. 13.3 includes the standard-state stability field of liquid water itself (broken lines) from Fig. 13.2, and the E_h-pH ranges found in natural waters (shaded area). The usefulness of the Pourbaix diagram now becomes apparent:

(a) We see that solid $Fe(OH)_2$ will be oxidized by O_2 in aqueous media, no matter what the pH, and, of course, bare iron metal will always be susceptible to corrosion by aqueous oxygen but may become coated with potentially protective oxide/hydroxide films at pH greater than about 9.

(b) Aqueous Fe^{2+} (1×10^{-3} mol L^{-1}) will be oxidized by O_2 to aqueous Fe^{3+} below, and to solid $Fe(OH)_3$ above, pH 2.3. Mechanistically, this process is complex, but since it is strongly retarded by hydrogen ions, it is possible to handle aqueous iron(II) salt solutions in the open air for limited periods without significant oxidation, if they kept acidic. Thus, kinetic factors may limit the utility of Pourbaix diagrams, like any other thermodynamic tool.

(c) Iron metal can, in principle, reduce water to hydrogen at any pH, although the margin in E_h is very narrow and so the rate of H_2 evolution is likely to be negligible, except in acidic media.

(d) Oxidation of iron(III) to iron(VI) is unlikely to be significant except in strongly alkaline media, since the solvent water will tend to be oxidized instead (to oxygen). Overpotential effects (Section 13.4), however, may suppress O_2 evolution kinetically. It is possible to produce purple solutions of FeO_4^{2-} by anodic oxidation of iron in concentrated aqueous alkali.

(e) The $Fe^{2+}(aq)/Fe(OH)_3(s)$ couple provides an E_h-pH buffer in natural
waters. Except for some extreme examples of *acid rain* (Section 2.5),
the pH of natural waters ranges from an upper limit of 9 to 10, set
by $CO_3{}^{2-}/HCO_3{}^-$ and $H_4SiO_4/H_3SiO_4{}^-$ buffers, down to around 3.
The stability field of water is a somewhat narrower band than is shown
in Fig. 13.2, since the relevant partial pressures of O_2 and H_2 may be
quite low and other factors may intervene,[8] but, even so, the effect of
the ubiquitous and abundant iron in the Earth's crust (cf. Table 8.1)
is clear. The upper limit of dissolved iron in natural waters is around
1×10^{-3} mol L^{-1}, the value chosen in constructing Fig. 13.3. The
lower limit of pH will be buffered by Fe^{III} hydrolysis,[8] but the details
of this have been omitted for simplicity in Fig. 13.3.

One final comment on the Fe/H_2O system: had Fig. 13.3 been extended beyond pH 14, we would have had to include the stability fields of
$Fe(OH)_4{}^-(aq)$ and $Fe(OH)_3{}^-(aq)$.[8] In other words, iron, like aluminum,
chromium, zinc, and many other metals, exhibits *amphoteric behavior* (i.e.,
has both acidic and base-like properties), but only if a sufficiently wide
range of pH is considered. Amphoterism, like many other chemical properties, is not so much something that a given element does or does not have,
but rather a trait that different elements display to different extents.

13.4   Kinetic Aspects of Electrochemistry: Overpotential

The foregoing considerations are based upon the concepts of *reversible
thermodynamics*—the electrochemical cells are considered to be operating
reversibly, which means in effect that no net current is drawn. Real cell
EMFs, however, can differ substantially from the predictions of the Nernst
equation because of electrochemical *kinetic* factors that emerge when a
non-negligible current is drawn. An electrical current represents electrons
transferred per unit of time, that is, it is proportional to the extent of electrochemical reaction per unit of time, or *reaction rate*. The major factors
that can influence the cell EMF through the current drawn are:

(a) the ohmic voltage drop across the cell itself, since cells necessarily have
some internal electrical resistance—this loss in EMF will be proportional to the current drawn, if the internal resistance remains constant
(Ohm's law);

(b) depletion of reactants, or build-up of products, near the electrode surfaces (the so-called *concentration polarization*)—diffusion of molecules
or ions to or from these surfaces through the solutions is not instanta-
neous and may slow down the electrochemical reaction significantly;

(c) slowness of one or more steps in the electrochemical reactions themselves (*activation polarization*), which is a matter of chemical reaction
mechanism and manifests itself as an *overpotential* η, which reduces
the EMF predicted by the Nernst equation.[1, 10, 11]

Factor (a) can be minimized by making the internal resistance of the
cell as small as possible, e.g., by having a high concentration of an inert
electrolyte in the cell. Factor (b) can be reduced by stirring the cell contents
vigorously. Factor (c), since it originates in chemical reaction kinetics, can
often be modified by catalyzing one or more of the electrochemical reactions
or half-reactions that make up the cell (see below).

Contributions to overpotential for the $H_2(g)/H^+(aq)$ electrode include
the following:

(a) A layer of water molecules is inevitably adsorbed on the electrified
metal surface (cf. Chapter 9), so creating an electrical double layer,
since the water molecules are dipolar. Furthermore, the $H^+(aq)$ ion itself is solvated by several water molecules in dilute solution (cf. Chapter 11) and can be regarded as having two tightly coordinated water
molecules to which it forms a linear hydrogen bond

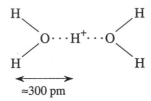

≈ 300 pm

plus several further water molecules hydrogen-bonded to the four
outer protons in turn. Consequently, even if, in the approach of
$H^+(aq)$ to the metal, substantial desolvation of both the metal surface and the proton were to occur (and this itself would cost energy
and so contribute to η), the electron would still have to tunnel out
some 300 pm through a potential energy barrier in order to reduce
$H^+(aq)$ to the hydrated hydrogen atom, $\dot{H}(aq)$.

(b) This hydrogen atom is higher in energy than the parent $H^+(aq)$ and
the electron residing on the metal. In quantum mechanical parlance,
the empty $1s$ orbital on the $H^+(aq)$ ion is high in energy relative to the
electronic conduction bands on the metal, so the electrical potential of
the electrons on the metal must be increased to match the $1s$ energy.

(c) Even if electron transfer to $H^+(aq)$ does occur, the newly created
$\dot{H}(aq)$ species must find another $\dot{H}(aq)$, and desolvate, to form H_2
gas. If it does not, then either the electron returns to the electrode,
or the $\dot{H}$ atom diffuses into the metal causing *hydriding* (and usually
embrittlement) of the metal, as explained in Section 10.1.

These considerations refer to the formation of $H_2(g)$ from $H^+(aq)$, but any forward pathway for a reaction is necessarily a reverse pathway too (cf. the principle of microscopic reversibility; Section 1.4), so the same factors create an overpotential for the oxidation of $H_2(g)$ to $H^+(aq)$. In particular, other reactions involving diatomic gases will be subject to constraints similar to the above. Indeed, high overpotentials are associated with most electrodes involving these gases.

The quantitative treatment of overpotential and related phenomena goes back to 1905, when Tafel showed that, for an electrochemical half-cell from which a net electrical current I is being drawn, an excess potential ΔE away from the equilibrium potential will inevitably exist, and that ΔE will be a linear function of the logarithm of the *current density* i ($i = I/$area of interface):

$$\Delta E = a + b \log_{10} i. \qquad (13.65)$$

Figure 13.4 illustrates the Tafel plots for the hydrogen electrode, operating in both the conventional forward direction (current density i_f, b negative):

$$2H^+(aq) + 2e^- \rightarrow H_2(g) \qquad (13.66)$$

and the reverse direction (reversed current density i_r):

$$H_2(g) \rightarrow 2H^+(aq) + 2e^- \qquad (13.67)$$

$$\Delta E = a' + b' \log_{10} i_r \qquad (13.68)$$

where b' is opposite in sign but often very similar in magnitude to b. This is illustrated in Fig. 13.4.

For an electrode process involving only a single electrochemical step, the Tafel slope b for the forward reaction is given by $-2.303RT/\beta nF$, where β is a symmetry factor of value between 0 and 1 but usually close to 0.5, and the other symbols have the same meaning as in the Nernst equation (Eqn. 13.15). One can regard β as being the ratio of the distance of the top of the potential energy barrier for electron transfer from the metal surface to the total distance between the metal surface and the electron acceptor in the solution ($H^+(aq)$, in the present discussion). For electron transfer in the *reverse* direction, the corresponding factor is therefore $(1 - \beta)$, so that b' is given by $2.303\ RT/(1 - \beta)nF$. If, however, *several* electrochemical steps are involved (and we have seen that this is the case for the $H_2(g)/H^+(aq)$ electrode reaction), we must write $b = -2.303RT/\alpha_f nF$ and $b = -2.303RT/\alpha_r nF$ where α_f and α_r are the *transfer coefficients* for the forward and reverse reactions, respectively.[§]

[§] Bockris and Reddy[1] (p. 1007) explain the relationship between transfer coefficients and reaction mechanism; for our purposes, it is enough to note that this theoretical foundation exists and to take experimental values of b and b'.

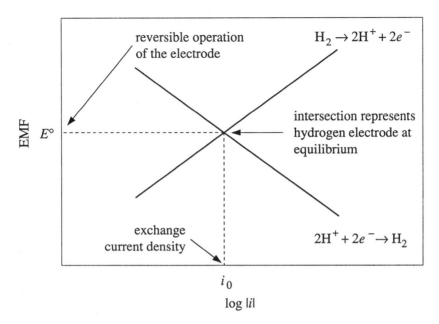

Figure 13.4 Tafel plot for the two half-reactions of the reversible hydrogen electrode.

In any operating condition of the hydrogen electrode, there will in principle be some reduction of H^+ going on together with some oxidation of H_2. The predominance of the first would mean that the electrode is working as a cathode, while the predominance of the second would correspond to operation as an anode. If, however, the electrode is at equilibrium (the situation for which $E°$ applies), the forward and reverse reaction rates are exactly equal, and the corresponding values of i_f and i_r are therefore also equal but opposite in direction, so that no net current flows. The value of i_f or i_r that corresponds to this is called the *exchange current density*, i_0, and the corresponding ΔE value relative to SHE is just $E°$, the standard electrode potential, if standard conditions apply. (If not, this voltage is given by the Nernst equation.) If the constants in the Tafel Eqns. 13.65 or 13.68 are known, $\log_{10} i_0$ is given by a/b or a'/b'.

Figure 13.5 shows the net current density i as a curve defined by the sum of i_f and i_r, which goes to zero when $i_f = i_r = i_0$. This curve merges asymptotically with the two Tafel lines when substantial currents are drawn in either direction, so that the intersection point of these lines, which defines $E°$ and i_0, is obtainable experimentally by extrapolation of the linear (Tafel) portions of EMF versus $\log i$ plots. We can now define the overpotential η quantitatively. It is the excess in EMF of the electrode relative

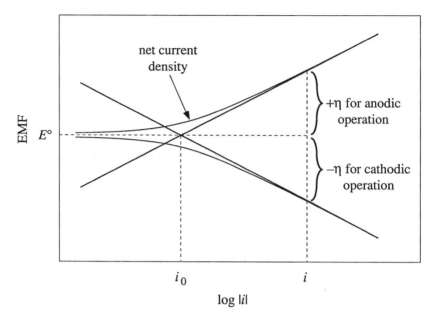

Figure 13.5 Overpotential η associated with a particular current density i. The net current density curve is given by the Butler-Volmer equation.

to the reversible value E°, for a particular value of the current density:

$$\eta = b\log_{10}\frac{i}{i_0}. \qquad (13.69)$$

For the forward reaction, the sign of b is negative, so η reduces the EMF. Equations 13.65, 13.68, and 13.69 can be combined and rearranged to give the *Butler-Volmer equation* (Eqn. 13.70) for the net current density, i, of an electrode process involving a single electrochemical step:

$$i = i_0\left(\exp\frac{(1-\beta)\eta nF}{RT} - \exp\frac{-\beta\eta nF}{RT}\right). \qquad (13.70)$$

If i is not too large, the overpotential η can be made negligible by making i_0 as large as possible. This is achieved by catalyzing the electrode reactions. Catalysis affects only the reaction rate—it does not affect the position of the equilibrium (cf. Section 9.1). In other words, catalysis changes i_f and i_r without changing the EMF of the intersection point, so that i_0 increases while E° remains constant. This serves to emphasize a general chemical principle—catalysis affects *both* the forward *and* the reverse rates of a chemical reaction to the same extent. For example, the metallic contact surface chosen for the hydrogen electrode in cells such as that shown

TABLE 13.1
Exchange Current Densities for the
Half Reaction $2H^+ + 2e^- \rightleftharpoons H_2$

Electrode	i_0 $(A\ cm^{-2})$
Pb	2×10^{-13}
Zn	5×10^{-11}
Cu	2×10^{-7}
Fe	1×10^{-6}
Pt	1×10^{-3}
platinized Pt	1×10^{-2}

in Fig. 13.1 is usually platinized platinum, that is, platinum foil on which finely divided platinum has been deposited to obtain the maximum effective surface area for catalysis of the H_2/H^+ reaction. In chloralkali cells (Section 5.3), it is now usual to use ruthenium-dioxide-coated titanium for the anodes, as this material has a lower overpotential for chlorine evolution than does the graphite formerly used.

Since different surfaces will catalyze a given reaction to different degrees, it follows that i_0 values are specific for a particular electrode surface. The exchange current densities for the evolution of hydrogen from 1 mol L^{-1} HCl, for example, range over some 11 powers of 10 (Table 13.1).

To illustrate the significance of this, let us consider the dissolution of zinc and, separately, iron in 1 mol L^{-1} acid. We might naively expect the zinc to dissolve (corrode) more rapidly than the iron, since the corresponding $E°$ values are -0.76 and -0.44 V, against 0.00 V for hydrogen evolution. In fact, the iron dissolves faster than the zinc. Figure 13.6 shows that $i(Zn)$ for the dissolution of zinc:

$$Zn \rightarrow Zn^{2+} + 2e^- \tag{13.71}$$

will match $i(H_2)$ for the evolution of hydrogen

$$2H^+ + 2e^- \rightarrow H_2 \tag{13.72}$$

at the point $i(H_2) = i(Zn) = I_{corr}$ (the corrosion current density) and $E = E_{corr}$ (the corrosion potential). This point corresponds to the dissolution reaction:

$$Zn + 2H^+ \rightarrow Zn^{2+} + H_2 \tag{13.73}$$

if the anodic and cathodic areas are the same (*general corrosion*, Section 14.5), since the rate of delivery of electrons (the corrosion current) by half-reaction 13.71 must equal the rate of their consumption by half-reaction 13.72. The corrosion current is a direct measure of the rate of

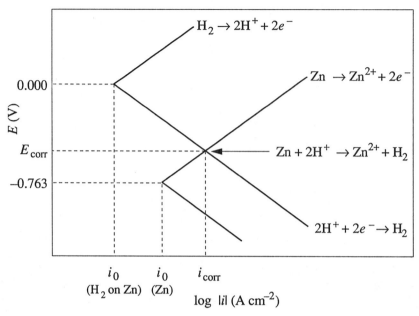

Figure 13.6 Mixed-potential representation of the dissolution of zinc metal in 1.0 mol L^{-1} aqueous acid.

reaction 13.71. Figure 13.7 shows the same kind of *mixed-potential* plot for the dissolution of iron in 1 mol L^{-1} acid, superimposed on that for zinc. It will be seen that the corrosion current density, and hence the dissolution rate, of the iron is somewhat higher than for the zinc, and that this is a consequence of a much higher exchange current density for H_2 evolution on iron relative to zinc.

13.5   Fuel Cells[1, 10]

A conventional fossil-fuel-fired power plant converts the chemical energy of combustion of the fuel first to heat, which is used to raise steam, which in turn is used to drive the turbines that turn the electrical generators. Quite apart from the mechanical and thermal energy losses in this sequence, the *maximum* thermodynamic efficiency ε for any heat engine is limited by the relative temperatures of the heat source (T_{hot}) and heat sink (T_{cold}):

$$\varepsilon = \frac{T_{\text{hot}} - T_{\text{cold}}}{T_{\text{hot}}} \times 100\%. \qquad (13.74)$$

Even with an optimistic heat sink temperature of 300 K, an almost uncontainable heat source temperature of 3000 K would be needed to have

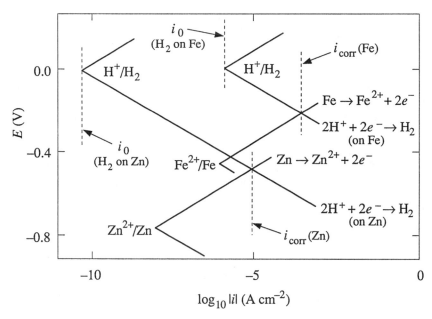

Figure 13.7 Influence of exchange current density for different surfaces on the rate of metal dissolution by hydrogen evolution.

a maximum theoretical efficiency of 90%. This refers to the conversion of the *enthalpy* of combustion, ΔH.

If, on the other hand, we could convert the *free energy change* of combustion ΔG directly to electricity in a suitable galvanic cell, the maximum extractable energy would be:

$$- \Delta G^\circ = nF \Delta E^\circ \tag{13.75}$$

and the efficiency, in terms of what one could extract by simply burning the fuel (ΔH), would be:

$$\varepsilon = \frac{\Delta G^\circ}{\Delta H^\circ} \times 100\% \tag{13.76}$$

which, because ΔH° is usually the major component of ΔG°, can be close to 100%. Indeed, consideration of Eqn. 1.8 shows that there will be some reactions for which ε (defined as above) could exceed 100%!

Such cells are called *fuel cells* and were discovered by Sir William Grove as long ago as 1839, but they have attracted major industrial interest only recently. For reasons that will emerge later, the reaction usually chosen is:

$$H_2 + \tfrac{1}{2}O_2 \rightarrow H_2O. \tag{13.77}$$

In the future, hydrogen gas may be distributed widely for domestic and industrial use from a central electrolysis facility (using the off-peak surplus generating capacity of a nuclear, hydroelectric, or geothermal power plant) or possibly one in which water is photolyzed. The gas would be piped to consumers much as methane is today. At present, however, hydrogen for fuel cells is usually generated on site by reforming propane or natural gas. This is an excellent way of providing electricity in remote communities where bottled propane is available. The cost of long-distance transmission of electricity from a conventional power plant large enough to be economical is eliminated, and the silent, non-polluting fuel cell has a better response to load fluctuations than the noisy, smelly diesel or gasoline generators presently used. The advantage in efficiency ($> 40\%$) over conventional electrical generating systems is most evident for fuel cell installations in the 10 kW to 10 MW range. The electricity delivered is low voltage direct current, but higher voltages can be generated using batteries of fuel cells in series, and an inverter can be installed to give alternating current.

The efficiency ε of a fuel cell operating at standard conditions, in which the hydrogen is oxidized to liquid water,

$$H_2 + \tfrac{1}{2}O_2 \rightarrow H_2O(l) \quad \Delta H = -285.83 \text{ kJ mol}^{-1} \tag{13.78}$$
$$\Delta E^\circ = +1.229 \text{ V}$$

is given by

$$\varepsilon = \frac{\Delta G^\circ}{\Delta H^\circ} = \frac{-2 \times 96\,487 \times 1.229}{-285\,830} \tag{13.79}$$
$$= 0.830 = 83.0\%.$$

If the product is water *vapor*, ε becomes 98.1%. Such theoretical efficiencies would require intractably high temperatures in a heat engine. In practice, of course, there are several factors that reduce ε, but even in the worst case ε is higher than the 25 to 30% efficiency realized by conventional or nuclear power stations. These factors include the following:

(a) Parasitic reactions (e.g., of hydrocarbons in the fuel supply) may reduce the energy output of the cell.

(b) As with all galvanic cells, the internal electrical resistance will reduce the EMF, and so ε.

(c) Concentration polarization (Section 13.4) may reduce the EMF. Porous electrodes are used to improve gas-electrolyte-electrode contacts, and some of the energy output of the cell can be used to stir the electrolyte.

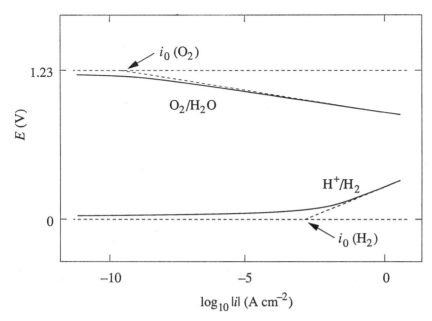

Figure 13.8 The oxygen reduction and hydrogen oxidation electrode reactions (standard conditions, on shiny platinum).

(d) Activation polarization (overpotential) is usually significant, especially at the cathode, since the exchange current density for O_2 reduction is even lower than i_0 for H_2 oxidation and must be catalyzed (Fig. 13.8, and cf. Section 13.4).

For this reason, cathodes are usually made of finely divided platinum on a porous support, for aqueous electrolytes. The catalytic surfaces of the anodes are particularly susceptible to poisoning by CO, olefins, sulfur compounds, etc., in the fuel. These lie above H_2 in the chemisorption series (Section 9.3). For the same reason, the *direct* use of hydrocarbon gases as fuel is impractical.

Figure 13.9 shows how the standard potentials of the anode and cathode in a practical aqueous-medium H_2/O_2 fuel cell change with the current drawn. The horizontal scale is linear in current, not logarithmic in current density as in Figs. 13.4 to 13.8, so even the overpotential-limited Tafel region is nonlinear. Note that most of the voltage loss occurs at the oxygen electrode. At higher currents, the internal resistance of the cell is the chief limiting factor, and ultimately mass transport (i.e., the rate of supply of reactants and ions to the electrodes) becomes limiting. Figure 13.10 gives the same information in terms of the net EMF of the cell (anode versus cathode) and the efficiency ε.

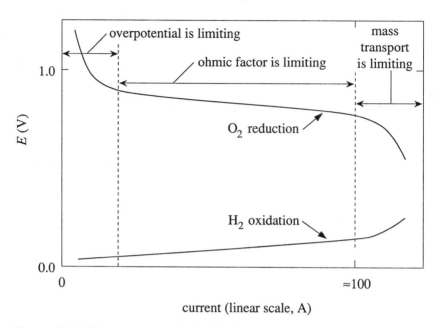

Figure 13.9 Voltage-current relationship for a typical oxygen-hydrogen fuel cell. The vertical separation between the anode and cathode curves represents the cell output voltage.

It is usual to operate an aqueous-medium fuel cell under pressure at temperatures well in excess of the normal boiling point, as this gives higher reactant activities and lower kinetic barriers (overpotential, etc.). The United Technologies PC-11 fuel cell used by Hydro-Québec derives hydrogen by reforming propane and oxidizes it in an aqueous acidic medium at 150 °C. Other designs have used concentrated phosphoric acid as the electrolyte on a solid support at 200 °C.

An alternative to reliance on catalytic reduction of overpotential is the use of much higher temperatures than can be reached with aqueous cells. Molten carbonates are the favored electrolytes in this case, as in a Ferranti-Packard design in which heat from the propane reformer is used to melt a mixture of carbonates:

$$C_3H_8 \xrightarrow{\text{limited air}} 3CO + 4H_2. \tag{13.80}$$

The anode reaction is then:

$$H_2 + CO_3{}^{2-}\,(\text{solvent}) \xrightarrow{700\,°C} H_2O + CO_2 + 2e^-. \tag{13.81}$$

The issuing gases contain some carbon monoxide, which is burned to CO_2 before they are fed to the cathode compartment:

$$CO_2 + \tfrac{1}{2}O_2 + 2e^- \rightarrow CO_3{}^{2-}\,(\text{solvent}). \tag{13.82}$$

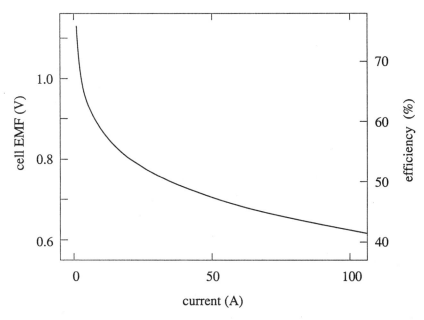

Figure 13.10 Hydrogen/oxygen fuel cell voltage and efficiency as a function of the current drawn.

Thus, the net reaction is just Eqn. 13.77.

The ultimate limitations of any fuel cell are the thermal and electrochemical stabilities of the electrode materials. Metals tend to dissolve in the electrolyte or to form electrically insulating oxide layers on the anode. Platinum is a good choice for aqueous acidic media, but is expensive and is subject to poisoning. Nickel or silver electrodes work well in the aqueous alkaline media (e.g., 6–12 mol L^{-1} KOH at temperatures up to 80 °C) used in cells fuelled with alcohols or hydrazine. However, hydrogen from a hydrocarbon reformer cannot be used as fuel in such cells because it contains CO_2, which will react with the alkali to form a carbonate.

13.6 Electrochemical Energy Storage Cells

The term "fuel cell" is usually reserved for electrochemical cells to which the reagents are fed from an external source. Storage cells (often called "batteries," although this term really refers to a collection of interconnected cells) already contain all the necessary reagents, and may be classed into two groups, *rechargeable* and *disposable* cells. In this section, we will survey briefly the chemical characteristics of some common storage cells; Pletcher[10] gives a good concise account of their technical aspects. As with fuel cells, the maximum theoretical (*reversible* or *no-load*) EMF of a storage cell is reduced in practice by the voltage drop across the internal resistance of the cell and by the overpotentials developed by the anode and cathode reactions, and the cell design therefore aims largely to minimize these effects.

The Leclanché cell. This inexpensive disposable flashlight-type cell has been on the market for over 100 years; yet, its chemistry is still not completely understood. The cell consists of an outer zinc shell that acts as the anode (seen by the external circuit as the source of electrons and hence the *negative* terminal) and oxidizes away during operation of the cell, a carbon rod or disk that serves as the cathodic current collector (*positive* terminal), and a moist paste of manganese dioxide, ammonium chloride, and zinc chloride that fills the cell and acts as both the electrolyte and the source of the cathodic reaction (reduction of Mn^{IV}). Usually, graphite in the form of carbon black is added to the paste to increase its electrical conductivity. The basic reactions are:

Cathode:

$$2MnO_2(s) + 2NH_4^+(aq) + 2e^- \rightarrow 2MnO(OH)(s) + 2NH_3(aq) \quad (13.83)$$

Anode:

$$Zn(s) \rightarrow Zn^{2+}(aq) + 2e^- \quad (13.84)$$
$$Zn^{2+}(aq) + 2NH_3(aq) + 2Cl^-(aq) \rightarrow Zn(NH_3)_2Cl_2(aq)$$

Ammonium chloride plays a key role in the formation of a soluble complex of zinc(II), which would otherwise precipitate as $Zn(OH)_2$ on the anode. The cell EMF, which ideally is 1.55 V, may fall by several tenths of a volt because of concentration polarization if large currents are drawn continuously, but it tends to recover (though slowly and incompletely) on breaking the circuit, as reaction products diffuse into the bulk paste. Leclanché cells cannot be recharged. The small 9 V batteries used in transistor radios, etc., typically consist of six shallow Leclanché cells stacked and connected in series.

An alternative to the Leclanché cell is the *alkaline manganese cell,* in which the electrolyte is a strongly alkaline paste and the anode and cathode reactions in their respective compartments are:

$$Zn(s) + 4OH^-(aq) \rightarrow Zn(OH)_4^{2-}(aq) + 2e^- \quad (13.85)$$

$$2MnO_2(s) + 2H_2O + 2e^- \rightarrow 2MnO(OH)(s) + 2OH^-. \quad (13.86)$$

The alkaline cell gives superior performance, but, because only high-grade electrolytically manufactured MnO_2 can be used, its cost is much higher than that of the Leclanché cell.

Nickel-cadmium cells. The design of the rechargeable alkaline Ni-Cd cell varies widely according to its intended use, which ranges from large storage batteries for power system backup to button cells for miniature electronic devices. However, the essential features are a cadmium anode (negative terminal in the external circuit) and a nickel current collector, which are in contact with the respective metal hydroxides and the electrolyte, aqueous KOH paste. A separator (usually of porous polyethylene) keeps the anode and cathode regions apart. The chemistry is complicated, especially at the anode (where various oxides/hydroxides of nickel(II), -(III) and -(IV) are involved), but can be simplified for illustrative purposes as follows:

Cathode:

$$2NiO(OH)(s) + 2H_2O + 2e^- \rightarrow 2Ni(OH)_2(s) + 2OH^-(aq) \quad (13.87)$$

Anode:

$$Cd(s) + 2OH^-(aq) \rightarrow Cd(OH)_2(s) + 2e^-. \quad (13.88)$$

The reaction products remain in place near the current collectors, and the cell is rechargeable, if suitably designed. The maximum EMF is 1.48 V. Although it is more expensive, the cell is superior to the Leclanché cell in almost all respects.

The *silver oxide* and *mercuric oxide button cells* used in cameras and other devices requiring a miniature source of EMF consist of a zinc disk, which serves as the anode, and, on the other side of a porous separator, a paste of Ag_2O or HgO. The reaction products are zinc hydroxide and metallic silver or mercury. Inert metal caps serve as the current collectors.

The *lead/acid battery* used in automobiles consists of six 2.05 V cells connected in series (for a 12 V electrical system) in which the current collectors are lead grids filled, when in the charged condition, with powdered lead (anode) and a lead/lead(IV) oxide mixture (cathode), and the electrolyte is aqueous sulfuric acid. During discharge, the following reactions occur:

Cathode:

$$PbO_2(s) + 2H^+(aq) + H_2SO_4(aq) + 2e^- \rightarrow PbSO_4(s) + 2H_2O \quad (13.89)$$

Anode:

$$Pb(s) + H_2SO_4(aq) \rightarrow PbSO_4(s) + 2H^+(aq) + 2e^-. \quad (13.90)$$

The solubility of lead(II) sulfate in water is low, and, as a result of the common ion effect, virtually nil in aqueous sulfuric acid, so the newly formed $PbSO_4$ remains inside the lead grids and the electrolyte becomes more dilute as the sulfuric acid is consumed. Thus, the state of discharge of the battery may be monitored through the falling density (*specific gravity*) of the liquid phase. Conversely, since all the electroactive materials are solids and remain localized within the grids, the battery is readily rechargeable; there is always some powdered lead remaining in the electrodes, so that their conductivity remains high. Care should be taken not to over-charge lead/acid batteries, especially those of the sealed "maintenance-free" type, since hydrogen and oxygen will then be liberated at the electrodes, with an attendant explosion hazard. Lead/acid batteries are well suited to deliver currents on the order of 100 A over several seconds, such as are needed to start an automobile engine.

13.7   Electrolysis, Electroplating, and Electroforming

Crude hydrogen for industry is most economically made today by the cracking of hydrocarbons (Section 8.3) or by the water–gas reaction (Section 3.3), but if very pure hydrogen and/or oxygen are required, or if electricity is very cheap, water may be electrolyzed instead. As noted in Section 13.5, electrolytic hydrogen generation offers a means of converting surplus electrical generating capacity to an alternative, readily stored energy source during periods of low demand.

A water electrolysis cell is like a hydrogen/oxygen fuel cell run in reverse, and factors, such as overpotential, which limit fuel-cell efficiency are also important in electrolysis. The electrolyte, which must be unchanged by electrolysis, is usually aqueous NaOH or KOH, as these are not aggressively corrosive toward metals such as the nickel-on-steel commonly used for the electrodes. Platinum would be the best electrode material, in terms of both corrosion resistance and lowered overpotentials for gas evolution, but it is very expensive, and overpotentials on nickel are satisfactory if it is deposited so as to present a high surface area. The operating temperature is typically 80 °C (again to minimize overpotential effects), but temperatures up to 150 °C can be used if the cell is pressurized—this has the additional advantage that the gas bubbles that tend to coat the electrodes and increase their effective electrical resistance are kept small. The anode and cathode compartments need to be kept separated by a diaphragm of some sort (cf. chloralkali diaphragm cells; Section 5.3), although in H_2/O_2 production the purpose is just to keep the gaseous products separated, as the electrolyte undergoes no net change.

Other commercially important inorganic chemicals that can be made electrolytically include caustic soda and chlorine, chlorate and perchlorate salts (Section 5.4), potassium dichromate ($K_2Cr_2O_7$), manganese dioxide, and potassium permanganate.[10]

The electrolytic purification of metals will be considered at length in Chapter 15. In essence, metals can be deposited in high purity from solution on a cathodic surface, by careful control of the voltage and other parameters. The anode can be a billet of the impure metal, and the impurities will either stay in solution or form an insoluble "anode slime"; here, both dissolution and reprecipitation of the desired metal are accomplished in a single electrolytic step. Alternatively, a crude solution of the metal ion might be prepared by some other means, and the pure metal deposited on a cathode with an anode of some inert material; the product of electrolysis at the anode will normally be oxygen gas.

Electroplating of one metal onto another is widely used for protection against corrosion and wear or for cosmetic purposes.[10] Again, the source of metal for deposition could be anodic dissolution or a prepared solution with an inert anode. In contrast to electrolytic refining, only a very thin layer (typically on the order of 1 to 10 μm) of the plating metal is wanted, but usually this layer must be uniform, cohesive, non-porous, and often shiny in appearance. To understand the roles of some of the variables in electroplating, it is useful to consider the electrodeposition process as having at least three steps:

(*a*) diffusion of a metal ion to the cathodic surface;

(*b*) reduction of the metal ion to form an *adatom;* and

(*c*) migration of the adatom to its final site in the metal lattice.

After the initial deposition of a few atomic layers of the plating metal on the substrate, the quality of the thickening film will depend largely on whether process (*c*) is completed before more adatoms are created. For example, if the adatoms have time to migrate to join the step of a screw dislocation (Fig. 7.3), the step will grow laterally and rotate helically about the screw axis to create a continuous layer of new metal atoms. This kind of situation requires that the reduction step (*b*) be slow compared to migration (*c*)—that is, the current density must be *low*. High current densities may lead to the formation of local irregularities, culminating in the extreme case with the growth of whiskers and feathery outgrowths called *dendrites*.

High concentrations of electrolyte are needed to ensure high conductivity of the solution and hence a more uniform deposition potential over the surface of the specimen. At the same time, however, adatom formation (*b*) may be too rapid unless the *free* metal ion concentration is kept low by complexing it with appropriate ligands, such as silver ion with cyanide:

$$Ag^+ + 2CN^- \rightleftharpoons Ag(CN)_2^- \qquad \beta_2 \approx 1 \times 10^{20}. \qquad (13.91)$$

Another factor that may prevent the occurrence of excessive local potentials (and hence excessive deposition rates) on the metal surface is the incursion of a different electrochemical reaction, such as hydrogen evolution, at such places as an alternative to the metal deposition that predominates at lower potentials.

Finally, metal objects can actually sometimes be fabricated in their entirety by electrodeposition (*electroforming*), with much the same considerations as electroplating. Conversely, portions of a metal specimen can be selectively electrolyzed away (*electrochemical machining*). This technique is especially useful where the metal to be shaped is too hard, or the shape to be cut too difficult, for conventional machining. The sample is made the anode, a specially shaped tool the cathode, and electrolyte solution (e.g., aqueous NaCl) is fed rapidly but uniformly over the surface to be machined. Current densities may reach several hundred amps per square centimeter across the electrolyte gap of a millimeter or so. Excellent tolerances can be achieved in favorable circumstances.[10]

References

1. J. O'M. Bockris and A.K.N. Reddy, "Modern Electrochemistry," 2 vols., Plenum Press: New York, 1970.

2. R. A. Robinson and R. H. Stokes, "Electrolyte Solutions," 2nd edn. (revised), Butterworths: London, 1965.

3. P. H. Rieger, "Electrochemistry," Prentice-Hall: Englewood Cliffs, N.J., 1987.

4. A. J. Bard, R. Parsons, and J. Jordan (eds.), "Standard Potentials in Aqueous Solution," Marcel Dekker: New York, 1985.

5. (*a*) W. H. Latimer, "Oxidation Potentials," 2nd edn., Prentice-Hall: Englewood Cliffs, N.J. 1952. (*b*) W. H. Latimer and J. H. Hildebrand, "Reference Book of Inorganic Chemistry," 3rd edn., Macmillan: New York, 1951.

6. M. Pourbaix, "Atlas of Electrochemical Equilibria in Aqueous Solution," Pergamon Press: Oxford, 1966.

7. R. M. Garrels and C. L. Christ, "Solutions, Minerals and Equilibria," Harper and Row: New York, 1965, Chapter 7.

8. D. W. Barnum, "Potential-p*H* diagrams," *Journal of Chemical Education,* 1982, *59,* 809–812.

9. J. E. Fergusson, "Inorganic Chemistry and the Earth," Pergamon Press: Oxford, 1982, pp. 53–61.

10. D. Pletcher, "Industrial Electrochemistry," Chapman and Hall: London, 1982.

11. M. G. Fontana and N. D. Greene, "Corrosion Engineering," 2nd edn., McGraw-Hill: New York, 1978, Chapter 9.

Exercises

13.1 The standard pressure for thermodynamic calculations has recently been changed from one "atmosphere" (101 325 Pa) to one bar (100 kPa). What correction should be applied to old $E°$ values for half-reactions *not* involving gases, relative to the H^+/H_2 standard electrode, to bring them to the new scale?
[*Answer:* Subtract 0.169 mV. Note that this is less than the typical uncertainty of ± 1 mV in $E°$.]

13.2 Calculate the stability constant β_2 of $Au(CN)_2^-$ at $25°C$ from the following standard electrode potentials:

$$Au^+ + e^- \rightleftharpoons Au \qquad\qquad E° = +1.68\,V$$
$$Au(CN)_2^- + e^- \rightleftharpoons Au + 2CN^- \qquad E° = -0.6\,V$$

[*Answer:* 3.3×10^{38}.]

13.3 Given that $E° = +1.229\,V$ for the half-reaction

$$O_2 + 4H^+ + 4e^- \rightleftharpoons 2H_2O$$

in acid solution (1 mol L^{-1}), calculate the EMF of the same half-cell in neutral water (pH 7) and in aqueous NaOH (1 mol L^{-1}), under standard conditions.
[*Answers:* +0.815, +0.401 volt.]

13.4 From the table of standard electrode potentials given in Appendix D, select reagents that should be thermodynamically capable (in 1 mol L^{-1} acid, standard conditions) of: (*a*) oxidizing HCl to chlorine gas; (*b*) reducing aqueous chromium(III) to chromium(II); (*c*) reducing solid silver chloride to metallic silver; (*d*) oxidizing water; and (*e*) reducing water.

13.5 Consider these electrode potentials (in volts, for 1 mol L^{-1} H^+, at $25°C$):

$$UO_2^{2+} \xrightarrow{+0.05} UO_2^+ \xrightarrow{+0.62} U^{4+} \xrightarrow{-0.61} U^{3+} \xrightarrow{-1.85} U$$
$$PuO_2^{2+} \xrightarrow{+0.91} PuO_2^+ \xrightarrow{+1.17} Pu^{4+} \xrightarrow{+0.98} Pu^{3+} \xrightarrow{-2.03} Pu$$

(*a*) Identify (with rationale) any of these species that would be unstable with respect to disproportionation under these conditions. (*b*) Would a change in pH affect your answer to (*a*)? (*c*) Calculate $E°$ for the half-reactions:

$$U^{VI} + 3e^- \rightarrow U^{III}$$
$$U^{VI} + 2e^- \rightarrow U^{IV}$$
$$U^{VI} + 6e^- \rightarrow U^0$$
$$Pu^{VI} + 3e^- \rightarrow Pu^{III}$$

(*d*) With the aid of the standard electrode potentials listed in Appendix D, select a reagent that should reduce plutonium(VI) to plutonium(III) in 1 mol L^{-1} acidic solution, but leave uranium(VI) untouched, assuming standard state conditions.

[*Answers:* (*c*) +0.02, +0.335, −0.915, +1.01 V; (*d*) Any reagent on the right-hand side of a reaction with $E°$ lying between 0.335 and 0.91 volts—iron(II) sulfate solution, perhaps. Considerations such as these can be developed as a basis for separating uranium and plutonium in the reprocessing of spent nuclear fuel elements.]

13.6 Which of the stability field boundaries in Fig. 13.3 are dependent upon the arbitrary choice of $[Fe^{n+}(aq)]$ (here, 1×10^{-3} mol L^{-1})?

13.7 The operation of the Ni-Cd alkaline flashlight cell involves the higher oxidation states of nickel. Assuming (for simplicity) that the only such material present is the insoluble solid NiO_2, construct a Pourbaix (E_h-pH) diagram for Ni in aqueous systems, showing the Ni(s), Ni^{2+}(aq), $Ni(OH)_2$(s), NiO_2(s), and H_2O (i.e., $O_2/H_2O/H_2$) stability fields for $[Ni^{2+}] = 1 \times 10^{-3}$ mol L^{-1} (where this is not limited by solubility, etc.). Relevant data:

$$Ni^{2+} + 2e^- \rightleftharpoons Ni(s) \qquad\qquad E° = -0.257\,V$$
$$Ni(OH)_2(s) \rightleftharpoons Ni^{2+} + 2OH^- \qquad K_{sp} = 1.6 \times 10^{-16}$$
$$NiO_2(s) + 4H^+ + 2e^- \rightleftharpoons Ni^{2+} + 2H_2O \qquad E° = +1.593\,V$$
$$O_2(g) + 4H^+ + 4e^- \rightleftharpoons 2H_2O \qquad E° = +1.229\,V$$
$$2H^+ + 2e^- \rightleftharpoons H_2(g) \qquad\qquad E° = 0.000\,V.$$

Under what circumstances is breakdown of the water theoretically possible? What could prevent this from occurring?

13.8 C. G. Vayenas and R. D. Farr (*Science,* **208**, 593 (1980)) describe a solid electrolyte (Section 7.2) fuel cell in which ammonia is the fuel and is catalytically converted at 1000 K with oxygen (or air) to nitric oxide. The idea is that the energy released in this step in industrial nitric acid production could be recovered directly as electricity.

Given that $\Delta G°$ for the oxidation of one mole of NH_3 exclusively to NO is −269.9 kJ mol^{-1} at 1000 K and 1 bar, calculate: (*a*) the maximum no-load EMF that the cell could (in theory) deliver, and (*b*) the maximum theoretical efficiency of the cell, given the following data (25 °C, 1 bar) and assuming $\Delta C_p°$ to be independent of temperature:

	$\Delta H_f°\,(kJ\ mol^{-1})$	$\Delta C_p°\,(J\ K^{-1}\ mol^{-1})$
O_2	0	29.35
NO	90.25	29.84
NH_3	−41.66	35.06
$H_2O(g)$	−241.82	33.58

[*Answers:* (*a*) 0.56 V; (*b*) 120%.]

13.9 A hydrogen/oxygen fuel cell, operating at near-standard conditions to produce *liquid* water as the reaction product, delivers 50 A at 0.70 V. What is the efficiency of the cell?
[*Answer:* 47%.]

13.10 Consideration of the relevant $E°$ data might suggest that electrolysis of aqueous NaCl should give O_2 rather than Cl_2 at the anode. In fact, O_2 is produced only in minor quantities. Suggest possible reasons for this.

13.11 Under what circumstances could the theoretical efficiency of a fuel cell (as defined in this Chapter) exceed 100%? Suggest an appropriate combination of fuel and oxidant to achieve this.

13.12 From the data in Appendix C, calculate the theoretical maximum EMF of a methane/oxygen fuel cell with an acidic electrolyte under standard conditions. Assume the products to be liquid water and aqueous CO_2. (*Hint:* You need to know the number of electrons transferred per mole CH_4 consumed. Write a balanced equation for the net reaction, and obtain the number of electrons from Eqn. 13.47).
[*Answer:* 1.05 V.]

Chapter 14

Corrosion of Metals
in Aqueous Systems

WE HAVE seen (Section 7.5) that the "dry" oxidation of metals by oxygen
or air can be viewed as an electrochemical process in which the *electrolyte*
of the cell is the developing solid oxide layer itself. If liquid water is present,
the diffusion of the ions and molecules involved in the electrochemical corrosion process is greatly facilitated, and consequently aqueous corrosion of
metals is much more important than "dry" oxidation at near-ambient temperatures. Although most of the corrosion problems encountered in practice
involve only a single metal, aqueous electrochemical corrosion can be especially severe, and its principles most clearly illustrated, in cases where two
different metals are in electrical contact with each other.

14.1   Bimetallic Corrosion

Consider a zinc strip immersed in water. At equilibrium, a small number
of Zn^{2+} ions will pass into solution per unit time, leaving twice as many
electrons behind, while an equal number of Zn^{2+} ions already in the water will be redeposited as elemental zinc (reaction 14.1). The rate of this
process, in terms of the electrons transferred per unit surface area of the
metal, is the exchange current density i_0 for equilibrium 14.1, as explained
in Section 13.4:

$$Zn^{2+}(aq) + 2e^- \rightleftharpoons Zn(s). \tag{14.1}$$

For a strip of copper immersed in water, a similar equilibrium 14.2 will
be set up:

$$Cu^{2+}(aq) + 2e^- \rightleftharpoons Cu(s). \tag{14.2}$$

The driving force for copper deposition (reaction 14.2) is much greater
than for zinc precipitation (reaction 14.1), as the corresponding standard

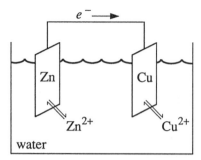

Figure 14.1 The basis of the galvanic (bimetallic) corrosion cell. Zinc, which is more readily oxidized than copper, becomes the anode.

electrode potentials of $+0.340$ and $-0.763\,\mathrm{V}$ indicate. Consequently, if the copper and zinc strips are placed in the same aqueous medium and are in electrical contact with each other as in Fig. 14.1, reaction 14.2 will proceed in the direction left to right by driving reaction 14.1 in the sense right to left. The zinc will therefore dissolve while the copper ions in the aqueous phase are redeposited.

This situation cannot persist for long—the concentration of copper(II) ions in water will initially be extremely small, unless some other source is also involved, and will quickly be depleted. The important point is that, as soon as electrical contact is made, the zinc becomes an anodic electrode, and the copper a cathode. If another cathodic reaction besides reaction 14.2 is possible, however, then dissolution (i.e., corrosion) of the zinc will continue, while the copper will serve merely as an electrically conducting surface to deliver electrons for the alternative cathodic reaction. In pure water, the obvious alternative reaction is hydrogen evolution (reaction 14.3) for which E_h is $-0.414\,\mathrm{V}$ at pH 7:

$$2\mathrm{H}^+(\mathrm{aq}) + 2e^- \rightarrow \mathrm{H}_2(\mathrm{g}). \tag{14.3}$$

In practice, the overpotential for hydrogen evolution on copper is moderately high (i.e., i_0 is fairly low; Table 13.1), though not as high as on zinc itself. Furthermore, the zinc will tend to become coated with zinc hydroxide as the pH begins to rise as a result of the progress of reaction 14.3. The zinc-copper couple will, however, produce hydrogen from hot water, although zinc alone will not, because of the greater overpotential for hydrogen evolution on zinc. Clearly, if the medium is acidic, H_2 will be evolved much more readily—from the copper surface, and not from the zinc, even though it is the zinc that is corroded away. In general, the same considerations apply to other metals of moderately negative $E°$ values: corrosion with hydrogen evolution as the cathodic process is important only at low pH.

If, however, the water contains dissolved oxygen, as it usually does if it is or has been in contact with air, the oxygen absorption reaction 14.4 becomes a likely cathode reaction:

$$O_2(g) + 4H^+(aq) + 4e^- \rightarrow 2H_2O(l) \quad E° = +1.229\,V. \tag{14.4}$$

In neutral water or even 1 mol L^{-1} alkali, E_h for reaction 14.4, better written for basic media as reaction 14.5, is still quite positive (+0.815 and +0.401 V for 1 bar O_2 at pH 7 and 14, respectively, or about 0.01 V less for air at 1 bar), so that oxygen absorption remains an important cause of aqueous metallic corrosion, even in alkaline media, although it is clear from reactions 14.4 and 14.5 that it is favored by low pH:

$$O_2(g) + 2H_2O(l) + 4e^- \rightarrow 4OH^-(aq). \tag{14.5}$$

Oxygen absorption, then, is the usual complementary cathodic reaction for the anodic dissolution of metals other than the most reactive ones (Li, Na, K, Ca, etc.), in near-neutral aerated water. Regardless of whether the cathodic reaction is 14.3 or 14.5, however, the following observations regarding bimetallic corrosion apply:

(a) The cathodic metal is *protected* from corrosion—it has a *surfeit* of electrons (for delivery to reaction 14.3 or 14.5) and hence is not going to be oxidized so long as the metal with which it is in electrical contact is corroding and so supplying the electrons.

(b) The electrical current flowing between the two metals (the corrosion current) is a direct measure of the corrosion rate, since each mole of zinc that dissolves away releases two moles ($2F$) of electrons for consumption at a relatively remote site, i.e., the cathode.

(c) Either cathodic reaction involves a diatomic gas, and hence high over-potentials (Section 13.4); therefore, the cathodic reaction is usually the major factor in controlling the corrosion *rate*. This rate is determined by the corrosion current, that is, by the corrosion current *density*, i_{corr} of Fig. 13.6, multiplied by the cathodic surface area. Thus, the larger the cathodic surface area, the faster the corrosion occurs in terms of moles of metal lost per unit time. Furthermore, this same rate of metal loss will be borne by the anode regardless of its size. A small anode will soon be completely dissolved away, with possible catastrophic consequences in engineering applications. One must therefore never allow small metallic structural components to remain in contact with a less anodic metal; otherwise, they will corrode away very rapidly. Iron nails or rivets must not be used with copper or stainless steel sheets, for example.

(d) Lowered oxygen concentrations (*deaeration*) or increased pH (alkaline conditions) reduce the EMF of the cathodic half-reactions 14.3 and 14.4 (or 14.5) and so will ordinarily help reduce the corrosion rate.

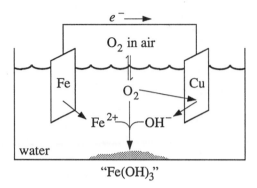

Figure 14.2 Bimetallic corrosion of iron. The immediate anode product is $Fe^{2+}(aq)$, which reacts with dissolved oxygen and OH^- (from the cathode) in bulk solution to form iron(III) oxide/hydroxide which precipitates away from the anodic surface.

In the corrosion of iron by neutral aerated water, in a bimetallic couple with (e.g.) copper or platinum as the cathode (as in Fig. 14.2), the initial product at the anode is $Fe^{2+}(aq)$:

$$Fe^{2+}(aq) + 2e^- \rightarrow Fe(s) \quad E° = -0.44\,V. \tag{14.6}$$

Iron(II) is only slowly oxidized further to iron(III), partly because E_h for oxygen reduction at pH 7 ($+0.815\,V$) is only marginally more positive than that for the iron(II)/iron(III) couple:

$$Fe^{3+}(aq) + e^- \rightarrow Fe^{2+}(aq) \quad E° = +0.771\,V \tag{14.7}$$

but mainly because the rate equation

$$-\frac{d[Fe^{2+}]}{dt} = k[Fe^{2+}][O_2][OH^-]^2 \tag{14.8}$$

(where $k = 5 \times 10^{13}$ L^3 mol^{-3} s^{-1} at 25 °C) shows that the half-period for the oxidation of $Fe^{2+}(aq)$ to Fe^{III} would be over an hour at pH 7 and ≈ 9 ppm dissolved O_2. Consequently, the Fe^{2+} ion has ample time to diffuse away from the anode before it is oxidized further.

The example just given assumed a constant pH of 7. Simultaneously with $Fe^{2+}(aq)$ release at the anode, however, hydroxide ion is formed by reduction of oxygen at the cathode (reaction 14.5), increasing the local pH, and diffuses out to mix with Fe^{2+} in the bulk aqueous phase. The solubility product of $Fe(OH)_2$ is about 8×10^{-16}, which means that concentrations of several μmol Fe^{II} per liter could build up before $Fe(OH)_2$ is precipitated. At the same time, increased pH will accelerate the oxidation by O_2 of Fe^{2+}

to Fe^{3+}, which is therefore precipitated as its very insoluble hydroxide—nominally $Fe(OH)_3$, although in practice the brown deposit familiar to us as rust is the partially dehydrated form, γ-$FeO(OH)$ (lepidocrocite). This substance will be deposited from bulk solution rather than on the anodic or cathodic surfaces, and consequently is not protective, in marked contrast to the protective film of iron oxides that forms in the "dry" oxidation of iron (Section 7.5). The E_h-pH relationships that govern this process are summarized in the Pourbaix diagram for the iron-water system (Fig. 13.3).

The products of the corrosion of iron by well-oxygenated hot water are usually goethite (α-$FeO(OH)$) and ultimately hematite (α-Fe_2O_3). If the oxygen concentration is low, however, the product is usually magnetite (Fe_3O_4) or some partially oxidized non-stoichiometric oxide Fe_xO_4 ($2.667 \leq x \leq 3.000$; cf. Section 7.5).

14.2   Single-Metal Corrosion

In bimetallic corrosion, the anodic and cathodic surfaces are well defined, being different metals, and are established instantly on placing these metals in electrical contact. The corrosion that results can be very vigorous, but one can usually arrange either to avoid using dissimilar metals together or else to place an electrically insulating gasket between them.

Even single metals, however, are subject to aqueous corrosion by essentially the same electrochemical process as for bimetallic corrosion. The metal surface is virtually never completely uniform; even if there is no pre-existing oxide film, there will be lattice defects (Chapter 7), local concentrations of impurities, and, often, stress-induced imperfections or cracks, any of which could create a local region of abnormally high (or low) free energy which could serve as an anodic (or cathodic) spot. This electrochemical differentiation of the surface means that local galvanic corrosion cells will develop when the metal is immersed in water, especially aerated water.

A more effective source of anodic and cathodic regions on a single metal is variation in the thickness, or even local absence, of a protective oxide film. The oxide film on an iron specimen, for example, is usually quite thin everywhere (Section 7.5), but there will usually be some spots where it is particularly thin or has been abraded away mechanically. These spots become anodes when the specimen is placed in neutral aerated water, and the areas with a more protective oxide film become cathodic. Iron(II) release (reverse of reaction 14.6) then starts up at the anodic sites, and the rate-controlling oxygen reduction reaction 14.5 will commence at the relatively large cathodic areas. Iron metal at the cathodic areas remains negatively charged and thus is protected from corrosion, while metal loss at the anodic sites causes them to deepen (Fig. 14.3). The anodic-cathodic differentiation

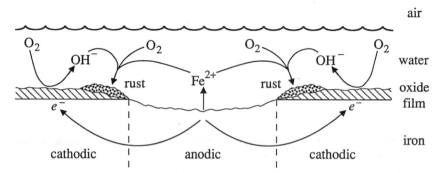

Figure 14.3 Single-metal corrosion of iron by aerated water.

is therefore self-perpetuating and, worse, tends to cause the loss of metal to be locally severe (another point of contrast to "dry" oxidation; Section 7.5).

As in bimetallic corrosion, the Fe^{2+} and the OH^- that constitute the immediate corrosion products are created at quite separate places on the metal surface and must diffuse together over substantial distances in the water phase to complete the electrochemical circuit. Where they come together, oxidation by dissolved O_2 to iron(III) occurs and rust is precipitated in a ring around the anodic spot. The diffusion of the ions towards each other is greatly facilitated if other "inert" ions such as Na^+ and Cl^- are present in the aqueous phase and can move to correct the electrical imbalance caused by the movement of the Fe^{2+} or OH^-. To look at the same problem a little differently, the flow of electrons from the anodic site to the cathodic areas requires that an electrical current flow through the aqueous phase to match it and to complete the electrical circuit. If there are substantial concentrations of ions in addition to Fe^{2+} and OH^- already in the water, they can carry most of this current and so increase the corrosion rate (i.e., the electrical corrosion current). Dissolved electrolytes therefore accelerate corrosion (often greatly), even though they usually do not cause the corrosion themselves—the cause is usually reaction 14.5.

Newly formed Fe^{2+} ions, then, must escape from the electrical field of the electrons they are leaving behind on the metal. In addition to the simple current-carrying function of an "inert" electrolyte in the water (described above), some anions such as chloride can form complexes with the Fe^{2+}, facilitating its departure by reducing its effective charge, i.e., by forming $FeCl^+(aq)$ or even $FeCl_2{}^0(aq)$. The transport of iron(III) as $FeCl^{2+}(aq)$ and $FeCl_2{}^+(aq)$ is similarly enhanced; all these species are more soluble than their hydroxide counterparts. The consequence is that dissolved ionic chlorides such as road salt ($NaCl$, or $CaCl_2$ from Solvay wastes) are especially aggressive in promoting corrosion. Worse yet, they are particularly effective in accelerating crevice corrosion or pitting, as shown in Fig. 14.4.

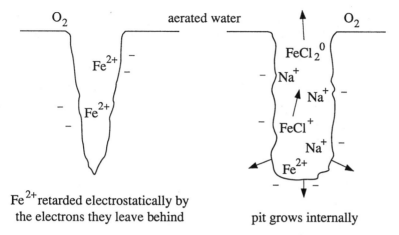

Figure 14.4 Acceleration of corrosion in a crack or crevice (left) by a dissolved electrolyte, NaCl (right).

The result can be the failure of a steel component, weakened by deep pits or crevices, even though the traditional corrosion rate expressed in *mils* per year (1 mil = 0.001 inch or 0.0254 mm), i.e., thickness of metal lost per year if corrosion were uniform across the surface, may have seemed insignificant.

14.3   The Role of Oxide Films

The oxide film which helped create the cathodic areas in the first place may be destroyed by *reductive dissolution,* at least in the case of iron, because of the excess of electrons present at cathodic sites. This is possible for iron because of the accessibility of the iron(II) oxidation state. The extreme insolubility (and hence protective capacity) of the oxide film derives from its iron(III) content, but iron(II) hydroxide has a significant solubility in neutral water, as noted above, and of course is readily soluble in acidic media.

$$Fe_2O_3(s) + 3H_2O + 2e^- \rightleftharpoons 2Fe(OH)_2(s) + 2OH^- \qquad (14.9)$$
$$\rightleftharpoons 2Fe^{2+}(aq) + 6OH^-(aq)$$

Equation 14.9 tells us that the protective oxide film on iron will be preserved in alkaline media, weakened in neutral water, and lost in acidic environments. Indeed, in very acidic solutions, the distinction between extended anodic and cathodic sites will be lost along with the oxide film, and dissolution of iron with accompanying hydrogen evolution becomes general across the surface of the specimen.

The oxide films on the alkali metals and on calcium, strontium, and barium are soluble in water at virtually any pH, and these metals, which also have very negative $E°$ values for reduction of their aqueous ions to the metal, react vigorously—in fact, violently—with water at any pH to liberate hydrogen gas. Sodium, for example, melts from the heat of the reaction, and, being less dense than water, rushes around the water surface on a cushion of hydrogen until the gas ignites with a resounding explosion and the characteristic yellow flame associated with vaporized sodium. It was noted previously (Section 7.5) that the oxide film on these metals is also unprotective in "dry" oxidation, although for an entirely different reason.

At the other extreme, the oxide layers on aluminum, beryllium, titanium, vanadium, chromium, nickel, and tantalum are very insoluble in water at intermediate pH values and do not have easily accessible reduced states with higher solubility. The oxide films on these metals are therefore highly protective against aqueous corrosion.

Aluminum, for example, is a very reactive metal if it is freed of its oxide film:

$$Al^{3+}(aq) + 3e^- \rightarrow Al(s) \quad E° = -1.66\,V \tag{14.10}$$

and, if the oxide film is broken under aerated water, $Al^{3+}(aq)$ forms instantly along with OH^-. Thus, a protective film of $Al(OH)_3$ (etc.) immediately forms again at once on the fresh anodic site, sealing it off and stopping the corrosion reaction at once unless the water is acidic, when $Al^{3+}(aq)$ can escape into bulk solution—or very alkaline, in which case the soluble aluminate ion, $Al(OH)_4{}^-(aq)$, forms:

$$2Al(s) + 2OH^-(aq) + 6H_2O \rightarrow 2Al(OH)_4{}^-(aq) + 3H_2. \tag{14.11}$$

There is *no* stable entity $Al^{2+}(aq)$ to compare with $Fe^{2+}(aq)$; consequently, the mechanism that causes rust to be non-protective because of migration of $Fe^{2+}(aq)$ through the water before precipitation as $FeO(OH)$ does not apply to aluminum, on which $Al(OH)_3$ or $AlO(OH)$ forms, at once, on the anodic site. Conversely, removal of the protective aluminum oxide film cannot occur by the reductive dissolution mechanism described for iron.

The oxide film on chromium, presumably Cr_2O_3, is similarly highly insoluble and not easily reduced to more soluble chromium(II) oxide/hydroxide. Unlike $Al^{3+}(aq)$, $Cr^{3+}(aq)$ *can* be reduced to the sky-blue $Cr^{2+}(aq)$ ion in water, for example by treatment with zinc metal ($E°$ for Zn^{2+}/Zn is $-0.76\,V$):

$$Cr^{3+}(aq) + e^- \rightarrow Cr^{2+}(aq) \quad E° = -0.42\,V. \tag{14.12}$$

However, it is obvious from the $E°$ of reaction 14.12 that $Cr^{2+}(aq)$ is a powerful reductant and is very rapidly re-oxidized to various green chromium(III) species by aerated water. So long as there is *some* oxygen present (more accurately, so long as the E_h of the system is more positive than about $-0.4\,V$), reductive dissolution will not affect the protective Cr_2O_3

film, and the chromium will be very resistant to corrosion. The metal in this condition is said to be *passive*. If, however, the E_h of the environment becomes so negative that reductive dissolution sets in (read Cr for Fe in reaction 14.9), the passivity of the chromium will be lost, and the metal will then become *active* in corrosion.

Stainless steels, all of which contain at least 11% Cr, owe their usual passivity to an oxide layer that can be approximately formulated as $FeCr_2O_4$, or (if nickel is also present) as $(Ni,Fe)Cr_2O_4$. These spinel-type oxides are similar to the mineral chromite (Section 6.4), which is extremely insoluble in aqueous media. Again, however, this film may be weakened or lost under strongly reducing conditions, and the "stainless" steel may then become active. This problem will be analyzed in electrochemical terms in Section 14.6.

14.4   Crevice and Intergranular Corrosion

In the rusting of iron, loss of metal occurs at the anodic sites *only*. Oxygen reduction occurs elsewhere, where the metal is cathodic and therefore protected by the excess of electrons (whether the oxide film survives there or not). Paradoxically, then, locally *high* concentrations of dissolved oxygen tend to create cathodic areas on the metal and so protect it. This phenomenon is referred to as the *principle of differential aeration*. Conversely, anodic sites tend to form in places where the oxygen concentration is locally low—for example, in cracks or crevices where the initially available dissolved oxygen becomes depleted by the first, general oxidation of the iron and cannot be replaced rapidly enough by diffusion.

While some of these crevices occur naturally, others may either be induced by stress or be created by poor engineering practice. For example, the sloppy rivetting of iron plates with an iron rivet may leave confined spaces within the joint (Fig. 14.5) in which the oxygen concentration is soon depleted to well below that in the water outside. The metal inside the joint therefore becomes anodic and, worse yet, the entire outer surface of the rivetted plates becomes an extended cathode where the rate-determining oxygen reduction step in the overall corrosion process occurs. The small anodic region at the rivet therefore supports essentially the whole of a substantial corrosion current, and the rapid metal loss at the rivet will lead to failure of the joint. The deposit of rust will tend to form just outside the crevices where corrosion is occurring, since the $Fe^{2+}(aq)$ will have to diffuse out to where the oxygen concentration and the pH are relatively high to be oxidized to γ-FeO(OH). Of course (as noted earlier in Sections 14.2 and 14.3), the corrosion rate will be even greater if dissolved salts are present in the aqueous phase (e.g., seawater), and worse still if the rivet is of a more reactive metal than the plates.

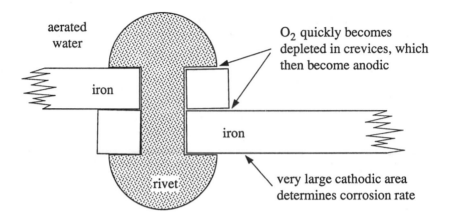

aerated water

iron

iron

rivet

O_2 quickly becomes depleted in crevices, which then become anodic

very large cathodic area determines corrosion rate

Figure 14.5 Differential aeration around a poorly fitted rivet leads to intense local corrosion.

The local accumulation of dirt on a steel structure in a damp environment is enough to set up an anodic area underneath it by excluding air. Similarly, chipped paintwork results in the lateral spreading of anodic areas under the paintwork, radially outward from the chips. At the chipped site, air has relatively free access to the metal, but under the paint the oxygen is excluded and anodic activity becomes intense, spreading under the paint and leaving a trail of rust behind where air has slowly diffused in to oxidize the Fe^{2+} (aq).

Such rusting phenomena as these are distressingly familiar in marine environments or in moderately cold climates where salts are used to de-ice roads. The acceleration of corrosion by seawater or sea spray, or by road salt, has several origins:

(*a*) Salts increase the electrical conductivity of the aqueous phase and so increase the corrosion current (Section 14.2).

(*b*) Some anions such as chloride complex Fe^{2+} and Fe^{3+} and facilitate their dispersal (Section 14.2).

(*c*) Complexing anions like chloride may act as bridging ligands between, say, Fe^{2+} in solution and Fe^{3+} in the oxide film, so facilitating electron transfer into the oxide film (Fe^{III} becoming Fe^{II}) and expediting its breakup:

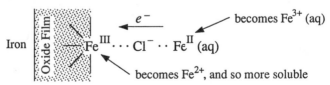

Iron | Oxide Film

$$Fe^{III} \cdots Cl^- \cdots Fe^{II} (aq)$$

e^-

becomes Fe^{3+} (aq)

becomes Fe^{2+}, and so more soluble

(*d*) Dissolved salts depress the freezing point of water, so aqueous corrosion continues below 0 °C.

(*e*) Dissolved salts depress the vapor pressure of liquid water (the ultimate cause of (*d*)) and so retard its evaporation; the metal stays wet longer.

The worst aspect of crevice corrosion is that it weakens the specimen locally, often unseen in a casual inspection, and may lead to failure of the specimen even though only a very small amount of metal has been lost. This type of corrosion can originate where stress has induced defects in the atomic lattice of the metal or even caused cracking (*stress cracking corrosion*).

Even in unstressed metals, however, the boundaries of the grains in the metal structure (Fig. 7.4) have relatively high free energy and thus tend to become anodic, leading to *intergranular corrosion*. The intergranular regions also tend to be the places where impurities, precipitated from solid solution, tend to collect, and this is illustrated by the *weld decay* of austenitic stainless steels, which typically contain 18% Cr and 8% Ni as well as iron and some 0.1 to 0.2% C. At 500 to 800 °C, a chromium carbide phase $Cr_{23}C_6$ separates from the chemically homogeneous stainless steel and precipitates in the grain boundaries, leaving a Cr-depleted zone around the boundaries (Fig. 14.6). The steel depends upon a high Cr content to remain corrosion-resistant (Section 14.3), so that the regions around the grain boundaries become anodic when the cold specimen is left in contact with aerated water, the rest of the steel serving as a much larger cathode. As a result, intense intergranular corrosion will occur in those parts on either side of a weld in a stainless steel specimen where the temperature has been 500 to 800 °C for an extended time during welding (Fig. 14.7).

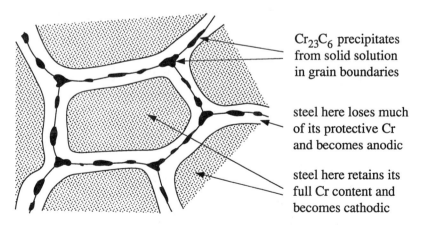

$Cr_{23}C_6$ precipitates from solid solution in grain boundaries

steel here loses much of its protective Cr and becomes anodic

steel here retains its full Cr content and becomes cathodic

Figure 14.6 Creation of microscopic anodic areas in the weld decay of a typical stainless steel.

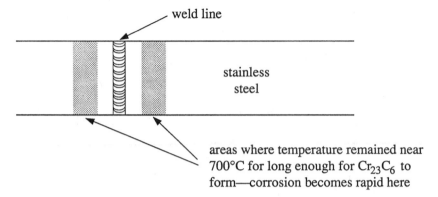

Figure 14.7 Areas susceptible to weld decay (*heat affected zones* or HAZ) in a welded stainless steel article.

There are several ways in which this *weld decay* can be avoided:

(a) High-temperature arc welding helps get the job completed quickly, before much $Cr_{23}C_6$ is precipitated.

(b) The sample might be reheated to about $1120\,^\circ C$, whereupon the $Cr_{23}C_6$ will go back into solid solution. The sample should then be quenched in water.

(c) Elements that form more stable carbides than does chromium can be alloyed into the stainless steel when it is made. The idea is to prevent the chromium from being consumed locally by the carbon by tying up the carbon in another carbide. From the discussion in Section 10.1 it follows that transition metals to the left of chromium in the Periodic Table should be suitable. In practice, titanium is used in AISI type 321 stainless steel,* and niobium ("columbium," in U.S. industrial usage) and tantalum in AISI type 347.

(d) A special low-carbon (< 0.03 weight %) stainless steel can be prepared, e.g., AISI types 304L or 316L, where "L" designates low carbon content. This may, of course, mean some loss of desirable mechanical properties.

14.5   Corrosion by Acids and with Complexing Agents

In general, the susceptibility of metals M to aqueous corrosion is expected to correlate inversely with the E° values for the reduction of $M^{m+}(aq)$ to

---

*The composition of stainless steels is discussed in Section 14.8.

M(s)—the less positive $E°$ is, the greater the tendency of M to corrode in aerated water. Factors that can upset predictions based on $E°$ include the presence of protective films (*passivation*), overpotential effects (Sections 13.4 and 14.6), the effect of complexing agents, and the incursion of a cathode reaction other than O_2 reduction or H_2 evolution.

Complexing agents usually promote corrosion, either by stripping away protective films or metal hydroxide deposits (cf. Section 11.5) or by changing the $E°$ or E_h value itself (Section 13.2). In Exercise 13.2, it was seen that $E°$ for the gold(I)/gold metal couple went from $+1.68$ V for the unstable ion $Au^+(aq)$ to -0.6 V if gold(I) were complexed with cyanide ligands. The latter figure shows that gold can be oxidized by *aerated* aqueous NaCN solution, which dissolves it as $Na[Au(CN)_2]$, and in fact this is used in the hydrometallurgical extraction of elemental gold from its crushed ores (Section 15.2). (Note that pK_a for HCN is about 9, so NaCN solutions are very alkaline, and the lowered E_h for H^+/H_2 plus overpotential effects will prevent the $Au(CN)_2^-/Au$ couple from reducing the solvent to hydrogen.)

Stabilization of nickel(II) and copper(II) by complexing with EDTA or NTA similarly increases the vulnerability of these normally corrosion-resistant metals to corrosion by aerated waters. For this reason, NTA or EDTA solutions should not be used in copper or high-Ni alloy equipment.

The corrosion of metals by aqueous acids with hydrogen evolution is usually rapid and fairly uniform across the surface (*general corrosion*), since the reductive dissolution of the oxide film that helps maintain the distinction between anodic and cathodic sites is favored by low pH (reaction 14.9). In this case, then, pitting is less important than overall loss of metal. If the oxide film is sufficiently insoluble in acids and is also resistant to reductive dissolution, as with titanium or stainless ($> 11\%$ Cr) steels, the metal may remain unaffected by aqueous acids, except at quite negative E_h values. In cases where the cathodic discharge of hydrogen ions

$$H^+(aq) + e^- \rightarrow \dot{H}(aq) \qquad (14.13)$$

does occur but the formation of diatomic hydrogen gas is inhibited (e.g., by the presence of H_2S), the hydrogen atoms may penetrate the metal, giving an interstitial hydride (Sections 10.1 and 13.4). This may lead to *embrittlement* of the metal, or, if the hydrogen atoms diffuse together through grain boundaries or voids to form gaseous H_2 *within* the metal, *hydrogen blistering* of the metal may result. In either case, it is a highly undesirable development from the engineering standpoint.

If oxidizing anions are present, the main cathodic reaction in aqueous corrosion may be neither oxygen absorption nor hydrogen evolution but reduction of the anion. These anions are almost inevitably oxoanions, and their reduction usually involves consumption of several hydrogen ions, so that this type of cathodic reaction is usually encountered only under very acidic conditions. Aqueous nitric acid or concentrated sulfuric acid provide familiar examples:

$$4H^+ + NO_3^- + 3e^- \rightarrow NO(g) + 2H_2O \qquad E^\circ = +0.96\,V \qquad (14.14)$$
$$3H^+ + HSO_4^- + 2e^- \rightarrow SO_2(g) + 2H_2O \qquad E^\circ = +0.17\,V. \qquad (14.15)$$

Copper metal will therefore dissolve readily in relatively dilute aqueous nitric acid (cf. Section 13.2) to give sky-blue $Cu^{2+}(aq)$ and nitric oxide gas:

$$Cu^{2+}(aq) + 2e^- \rightarrow Cu(s) \qquad\qquad E^\circ = +0.34\,V \qquad (14.16)$$

$$3Cu(s) + 8HNO_3(aq) \rightarrow 3Cu(NO_3)_2(aq) + 4H_2O + 2NO(g)$$
$$\Delta E^\circ = +0.62\,V. \qquad (14.17)$$

However, in concentrated (70%) nitric acid the gaseous product is mainly nitrogen dioxide:

$$Cu(s) + 4HNO_3(aq) \rightarrow Cu(NO_3)_2(aq) + 2H_2O + 2NO_2(g). \qquad (14.18)$$

Copper does not dissolve in 1 mol L^{-1} sulfuric acid, since E° for the half-reaction 14.16 is too positive relative to sulfate reduction (reaction 14.15) or hydrogen evolution. It will, however, dissolve in hot, concentrated (18 mol L^{-1}) sulfuric acid, in which the activities of H^+ and HSO_4^- are much higher and the activity of water (now a minor constituent) much lower than in the standard-state conditions for which the E° of half-reaction 14.15 applies:

$$Cu(s) + 2H_2SO_4(\text{hot, concentrated}) \rightarrow$$
$$CuSO_4 \text{ solution} + 2H_2O + SO_2(g). \qquad (14.19)$$

Sometimes it happens that the incursion of oxoanion reduction in place of hydrogen evolution as the cathodic reaction in the corrosion of iron leads, not to an increased rate of corrosion, but to a drastic retardation. This is because strongly oxidizing conditions, e.g., in concentrated nitric acid, can force the immediate oxidation of iron to iron(III), rather than via the persistent iron(II) intermediate (as described in Sections 14.1 and 14.2), so that an insoluble iron(III) oxide layer forms at once on the anodic and cathodic surfaces alike and the iron becomes passivated (Section 14.3). Michael Faraday's demonstration of this phenomenon is instructive:

(a) An iron specimen is placed in dilute nitric acid. It is seen to begin to dissolve with a steady evolution of gas.

(b) The specimen is now transferred to concentrated nitric acid. A few bubbles of gas may escape at first, then no more, and the now passive iron is not attacked further.

(*c*) The passivated iron may be transferred back to the dilute nitric acid. If this is done carefully, the iron remains passive because of its oxide film, and no gas at all is evolved.

(*d*) The specimen, still under dilute HNO_3, is struck lightly. The chemically inert but mechanically fragile film is broken, and gas evolution recommences vigorously, initially where the iron was struck, but subsequently over the whole surface of the specimen as the remnants of the oxide film undergo reductive dissolution. Obviously, such a delicate oxide film has no value in corrosion-proofing iron (contrast the tough "$FeCr_2O_4$" film on stainless steels).

14.6 The Role of Overpotential in Corrosion

Consider first the *polarization curve* (i.e., Tafel plot) for the anodic half-reaction occurring in corrosion of stainless steels (Fig. 14.8). The diagram for the active region is much the same as has been seen for other anodes (Figs. 13.4 to 13.7). As E_h is increased to a certain specific value, however, a sudden and dramatic drop in the anodic current density i occurs, corresponding to the formation of an oxide film. At higher E_h, i remains constant at a very low level (the horizontal scale in Fig. 14.8 is logarithmic), and the metal has become passive, i.e., effectively immune from corrosion.

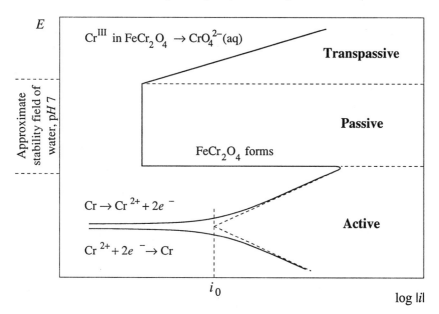

Figure 14.8 Polarization (voltage vs. log current density) curve for a typical stainless steel in water (cf. Fig. 13.5).

The oxide film may be regarded as consisting of chromite ($FeCr_2O_4$), the protective capacity of which is due to the extreme insolubility conferred by its cationic chromium(III) content. Under sufficiently oxidizing conditions, however, chromium(III) is oxidized to chromium(VI), which characteristically forms the soluble oxoanions CrO_4^{2-} (chromate, yellow), $HCrO_4^-$ (hydrogen chromate, $pK_a \approx 6$, orange) and $Cr_2O_7^{2-}$ (dichromate, from condensation of $HCrO_4^-$, orange). There is therefore an upper limit in E_h to the passivity of a stainless steel, beyond which active corrosion again sets in with the release of orange or yellow chromium(VI) into the water phase. This transpassive region is not normally entered in aqueous systems at ordinary temperatures, although in highly oxygenated water far above its normal boiling point (under pressure) chromium(VI) is sometimes seen to form.

The passive range of typical stainless steels conveniently spans most of the E_h stability field of neutral water. This can be appreciated by examination of the $E°$ or E_h values for reactions 14.20 to 14.25, with the caveat that these refer to pure iron and chromium metals rather than to stainless steels and that the conditions are standard ones rather than (for example) the very low $[Cr^{3+}]$ in equilibrium with the "$FeCr_2O_4$" film. The formation of the "$FeCr_2O_4$" film becomes possible when Cr^{II} can be oxidized to Cr^{III}, and when iron metal can be oxidized to Fe^{II}.

$$Cr^{2+}(aq) + 2e^- \rightarrow Cr(s) \qquad E° = -0.91\,V \quad (14.20)$$

$$Fe^{2+}(aq) + e^- \rightarrow Fe(s) \qquad E° = -0.44\,V \quad (14.21)$$

$$Cr^{3+}(aq) + e^- \rightarrow Cr^{2+}(aq) \qquad E° = -0.42\,V \quad (14.22)$$

$$Cr_2O_7^{2-} + 14H^+ + 6e^- \rightarrow 2Cr^{3+} + 7H_2O \qquad E° = +1.36\,V,$$
$$E_{h(pH7)} = +0.39\,V \quad (14.23)$$

$$2H^+(aq) + 2e^- \rightarrow H_2(g) \qquad E° = 0.00\,V,$$
$$E_{h(pH7)} = -0.41\,V \quad (14.24)$$

$$O_2(g) + 4H^+(aq) + 4e^- \rightarrow 2H_2O \qquad E° = +1.229\,V,$$
$$E_{h(pH7)} = +0.81\,V \quad (14.25)$$

A diagram similar to Fig. 14.8 could be constructed for ordinary iron or steel, but the onset of passivity, corresponding to the formation of an Fe_2O_3 film directly over the entire surface of the metal, would occur at a much higher E_h value than for stainless steel ($E°$ for $Fe^{3+}(aq)/Fe^{2+}(aq)$ is $+0.77\,V$). The transpassive region, corresponding to anodic oxidation of the Fe_2O_3 film to soluble FeO_4^{2-}, requires a high impressed EMF and is only feasible in alkaline conditions (cf. the Pourbaix diagram of the iron-water system, Fig. 13.3).

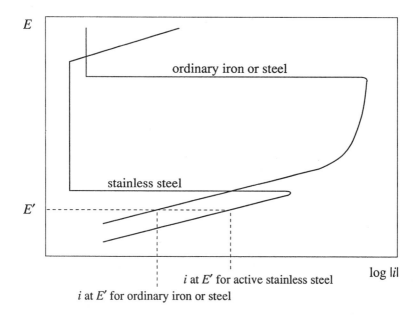

Figure 14.9 Anodic polarization curves for typical stainless and ordinary steels. At low E_h, stainless steel may become active, in which case it corrodes more rapidly than ordinary iron or steels.

In Fig. 14.9, the operationally significant parts of qualitative polarization curves for a typical steel and a stainless steel are superimposed. It will be seen that, for a given E_h value E' in the active range of the stainless steel, the current density will be higher for dissolution of the stainless steel than for corrosion of the iron. It is therefore very important that stainless steels be prevented from becoming active in service, because, if they do, they corrode rapidly—more so than ordinary iron would.

Figures 14.8 and 14.9 show only the anodic polarization curves for corrosion cells. The important question is, where do these curves intersect with the polarization curves for likely cathodic reactions, such as hydrogen evolution or oxygen absorption? The intersection point defines the corrosion current density i_{corr} and hence the corrosion rate per unit surface area. As an illustrative example, let us consider the corrosion of titanium (which passivates at negative E_h) by aqueous acid. In Fig. 14.10, the polarization curves for H_2 evolution on Ti and for the Ti/Ti^{3+} couple intersect in the active region of the Ti anode. To make the intersection occur in the passive region (as in Fig. 14.11), we must either move the H^+/H_2 polarization curve bodily upwards, which means, paradoxically, increasing the hydrogen ion concentration and hence E_h for that electrode, or slide the curve to the right by increasing the exchange current density i_0 for the evolution of

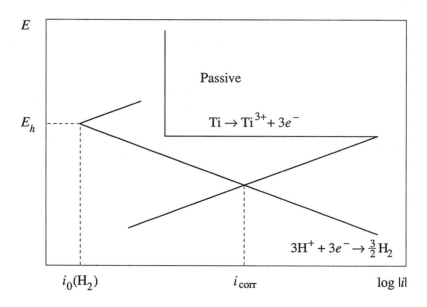

Figure 14.10 Active corrosion of titanium by aqueous acid.

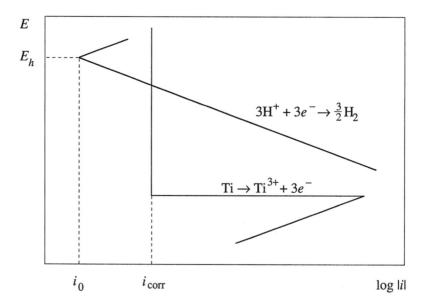

Figure 14.11 Passive titanium-0.2% palladium alloy in aqueous acid. Here, i_0 refers to catalyzed H_2 evolution.

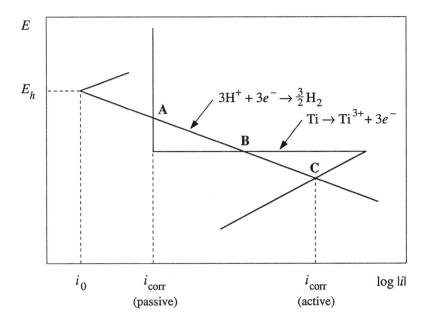

Figure 14.12 Titanium with metastable passivity. If E_h or i_0 are not quite high enough, the metal may go from passive to active behavior without warning. If this happens, the corrosion current will be high (cf. normal active corrosion, Fig. 14.10).

hydrogen on titanium, or both. The exchange current density can be increased by catalyzing hydrogen evolution with some 0.1 to 0.2% palladium, alloyed into the titanium. The resulting alloy costs at least twice as much as ordinary commercial Ti alloys, but may be worth the extra cost if service in acids at low E_h values is required.

We must, however, avoid the situation shown in Fig. 14.12, where the cathodic polarization curve intersects the anodic curve not only in the passive region at point A, but also on the active "nose" of the curve at B and C. The intersection at B is of no real significance, as it is electrically unstable—a slight perturbation in E_h will send the system to situation A or C. Intersection C, however, represents a high rate of corrosion, and although we may seem to have a passive, non-corroding system corresponding to A, this serene situation may change suddenly and without warning to that represented by C. It is therefore very important that E_h and i_0 for the cathode reaction be set high enough to place A well into the passive zone, so that the cathode polarization curve easily clears the "nose" on the anodic polarization curve. The same considerations apply to any metals or alloys (notably stainless steels) that rely upon an oxide film for their corrosion resistance.

14.7   Control of Corrosion

Many ploys to minimize corrosion will suggest themselves from the foregoing—keep steel structures dry and free of dirt and salts, deaerate the water,
etc. It must be kept in mind, however, that steps taken to reduce one technological problem may create another—for example, the softening of water
may, as we shall see, increase corrosion, while the use of toxic chromates
as anodic inhibitors (see below) has obvious environmental limitations. In
corrosion control, as in all applications of chemistry, one has to consider the
broader consequences before taking action, and often one perceived remedy
for corrosion (e.g., removing excess CO_2 from water by thorough aeration)
may accelerate corrosion in another way (in this case, by saturating the
water with oxygen at its atmospheric partial pressure).

(*a*) *Cathodic protection* involves simply making the metal object electrically
negative with respect to its surroundings. This can be done by connecting it
electrically to a more reactive metal or "sacrificial anode," which is allowed
to corrode away while supplying protective electrons to the metal object.
Typical examples are the protection of steel power pylons with magnesium
stakes sunk into the ground, the use of magnesium bars inside hot-water
tanks, and the attachment of magnesium or zinc ingots to the hulls of steel
ships. In all cases, the sacrificial anode will have to be replaced from time
to time. Steel used for tubs or outdoor structures can be coated with zinc
(*galvanized iron*); the idea is the same, except that the lost coating cannot
easily be replaced.   Alternatively, the structure to be protected may be
connected to the negative terminal of a DC source such as a fuel cell; this
is an attractive proposition for natural gas pipelines, since some miniscule
portion of the gas can be bled off to operate the unattended fuel cell.

(*b*) *Protective coatings* such as paint or plastic resins are a familiar means
of corrosion control by cutting off access of water, air, and electrolytes to
the structure. The coating must, however, be complete and remain intact
to be protective—local penetration will generally create an active corrosion site which may be driven by a cathodic reaction involving exposed
areas elsewhere or slow diffusion of oxygen through the large area of surviving coating. Likewise, *plating* with a corrosion-resistant metal such as
chromium, tin, or gold is effective only so long as the plate remains *completely* intact. If the underlying metal (usually steel) is exposed anywhere,
it will immediately become anodic and will have to support a large corrosion current controlled by the much greater cathodic area of intact plate.
The intense localized corrosion so caused will be considerably worse than
in the case of chipped paint or resin.

   If the metal to be corrosion-proofed is one that normally forms a protective oxide film of low electrical conductivity (e.g., aluminum, beryllium,
titanium, or zirconium), this film can be forcibly thickened by making the

metal object the anode of an electrolytic cell until the current it can pass becomes negligible. This process is known as *anodizing* and is extensively applied to aluminum products. Typically, the anodizing of aluminum in 3 mol L^{-1} aqueous sulfuric acid produces an oxide coat 10 μm thick, whereas in air at 400 °C the oxide film thickness would reach only about 2 nm in the same time. The outer layers of oxide produced in anodizing are porous to some degree and can adsorb dyes while they form, giving a lustrous, brightly colored product.

(*c*) *Modification of the environment* may be feasible if one is dealing with a closed system. For example, if oxygen absorption is the main cathodic process driving corrosion, deoxygenation of the water phase by purging with nitrogen or vacuum degassing may be effective in suppressing corrosion. Alternatively, the oxygen in a hot aqueous system (such as boiler water) may be removed chemically by addition of sodium sulfite, which will leave possibly undesirable sodium sulfate in solution, or hydrazine ($H_2N—NH_2$, Section 1.9) which goes to nitrogen and water and so is preferred for "zero dissolved solids" operation. With stainless steel equipment, the E_h of the system must not be taken too low; otherwise, the metal may become active (Section 14.6). Even with ordinary steels it should be remembered that high oxygen concentrations are locally protective, so that deoxygenation is not recommended if the metal to be protected is already cathodic.

Corrosion inhibitors are solutes that blanket the electrochemically active surfaces of the corrosion-prone metal and suppress corrosion by blocking the flow of ions or molecules to or from these surfaces. Many *anodic inhibitors* operate by causing the freshly formed Fe^{2+}(aq) ions at the anodic surfaces of corroding iron to be precipitated immediately on the anode itself (cf. passivation of iron by concentrated HNO_3, which, however, affects *all* surfaces of the iron). *Alkaline solutions* achieve this by causing $Fe(OH)_2$ to precipitate at once, rather than after diffusion of the Fe^{2+}(aq). If dissolved oxygen is present, the $Fe(OH)_2$ oxidizes to insoluble Fe^{III} oxides or hydroxides, while at high temperatures and low oxygen partial pressures (e.g., in deoxygenated boilers, or in the heat transfer circuits of water-cooled nuclear power reactors) $Fe(OH)_2$ decomposes spontaneously to hydrogen and a black, protective film of magnetite (the *Schikorr reaction*):

$$3Fe(OH)_2(s) \xrightarrow{\text{heat}} Fe_3O_4(s) + 2H_2O(l) + H_2(g). \qquad (14.26)$$

Hydrogen could also be generated *at high temperatures* by the direct reaction of iron with water to form $Fe(OH)_2$; in principle, this is marginally possible at high pH even at 25 °C according to Fig. 13.3, but is inhibited by overpotential. Furthermore, in *very* concentrated alkali, the product $Fe(OH)_2$ may form soluble green hydroxo complexes (Section 13.3) and so may not assist in the formation of protective layers. Since hydrogen can cause hydriding of the metal (Sections 10.1 and 14.5), *caustic embrittlement*

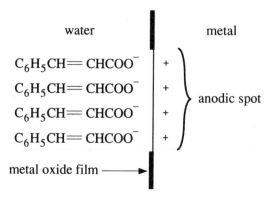

Figure 14.13 Blanketing action of an organic anodic inhibitor—in this case, sodium cinnamate.

of the iron may become locally intense in a boiler or other heat transfer system wherever hot aqueous alkali becomes highly concentrated, e.g., in steam-forming pockets, in stress-induced cracks, or in leaky seams or rivets (cf. Section 12.4), with potentially disastrous consequences. This must be borne in mind whenever alkali is used for corrosion suppression in aqueous systems at high temperatures.

Phosphates act as anodic inhibitors much as do other alkalis, although the iron oxides/hydroxides formed on anodic sites then contain some PO_4^{3-}. Chromates and nitrites oxidize Fe^{2+} (aq) rapidly to insoluble iron(III) oxides on anodic sites. *Dianodic inhibitors* combine complementary inhibition mechanisms; for example, sodium triphosphate may be used with sodium chromate, or sodium molybdate with $NaNO_2$.

A different approach to blanketing anodic areas involves dissolving organic anions with appropriate hydrophobic (i.e., water-repellent) substituents. The anionic heads of the molecules specifically seek out a positively charged anodic spot, and the hydrophobic tails serve to isolate it from the aqueous solution and so block the ionic part of the corrosion circuit (Fig. 14.13). Sodium benzoate or especially cinnamate are particularly effective in this regard.

A serious limitation of the use of anodic inhibitors is that they must be used in sufficiently high concentration to eliminate *all* the anodic sites, otherwise the anodic area that remains will carry the whole corrosion current, which is usually *cathodically* controlled. Intense local corrosion may then result, possibly leading to failure of the specimen. *Cathodic inhibitors,* on the contrary, are helpful in any concentrations e.g., the blanketing of only half the cathodic surface will still roughly halve the corrosion rate. The presence of temporary hardness or magnesium ions can help reduce corrosion through deposition of $CaCO_3$ or $Mg(OH)_2$, specifically upon the cathodic surfaces where OH^- is produced in the oxygen absorption reaction:

$$O_2 + 2H_2O + 4e^- \rightarrow 4OH^- \tag{14.27}$$

$$Ca^{2+} + HCO_3{}^- + OH^- \rightarrow CaCO_3(s) + H_2O \tag{14.28}$$

$$Mg^{2+} + 2OH^- \rightarrow Mg(OH)_2(s). \tag{14.29}$$

It follows that hard water is paradoxically less corrosive than soft, deionized, or distilled water, other things (notably the O_2 concentrations) being equal.

In *acidic* solutions, organic amines protonate to form cations with hydrophobic tails. These ions will seek out and blanket cathodic surfaces, much as carboxylate anions seek out and cover anodic spots in neutral or basic media. The usual choices include amylamine ($C_5H_{11}NH_2$), pyridine (C_5H_5N), cyclohexylamine ($C_6H_{11}NH_2$), and morpholine ($O(CH_2CH_2)_2N$).

Aqueous corrosion can occur even when the metallic object to be protected is ostensibly not immersed in water, if the relative humidity of the atmosphere is more than 60%. In that case, a film of water will in fact be present on the metal surface. Furthermore, if sulfur dioxide is present in the air, corrosion in this thin film of water will be greatly accelerated, partly because the acidity of the dissolved SO_2 facilitates the oxygen absorption reaction:

$$SO_2(g) + H_2O(l) \rightleftharpoons H_2SO_3(aq) \tag{14.30}$$

$$\rightleftharpoons H^+(aq) + HSO_3{}^-(aq) \text{ (etc.)}$$

$$O_2(g) + 4H^+ + 4e^- \rightarrow 2H_2O \tag{14.31}$$

but also because SO_2 will favor reductive dissolution of partially protective Fe^{III} oxide films:

$$Fe_2O_3(s) + H_2SO_3(aq) + 2H^+(aq) \rightarrow$$
$$2Fe^{2+}(aq) + SO_4{}^{2-}(aq) + 2H_2O. \tag{14.32}$$

Dust (especially from industrial activities) and salt spray will also exacerbate atmospheric corrosion (Section 14.4). In enclosed industrial premises, atmospheric corrosion could be minimized by preventing noxious emissions, filtering the air to remove particulate matter, and scrubbing the air with water to remove SO_2 and other objectionable gases, although the humidity should itself be kept as low as possible (e.g., steam leaks should not be tolerated). On the global scale, however, the cost to the public of atmospheric corrosion could be substantially reduced by sharply limiting SO_2 and, to a lesser extent, NO_x emissions from power plants, smelters, automobiles, and other industrial functions. This is an aspect of the acid rain threat (Chapter 2) that is usually overlooked.

(*d*) *Modification of the metal itself,* by alloying for corrosion resistance, is often worth the increased capital cost. *Titanium* has excellent corrosion resistance (even when unalloyed) because of its tough natural oxide film, but it is presently rather expensive for routine use (e.g., in chemical process equipment), unless the increased capital cost is a secondary consideration. *Iron* is almost twice as dense as titanium, which may influence the choice of metal on structural grounds, but it can be alloyed with 11% or more *chromium* for corrosion resistance (see stainless steels, below) or, for resistance to acid attack, with an element such as *silicon* or *molybdenum* which will give a film of an acidic oxide (SiO_2 and MoO_3, the anhydrides of silicic and molybdic acids) on the metal surface. Silicon, however, tends to make steel brittle. Nevertheless, the proprietary alloys *duriron* (14.5% Si, 0.95% C) and *durichlor* (14.5% Si, 3% Mo) are very serviceable for chemical engineering operations involving acids. Molybdenum also confers special acid and chloride resistant properties on type 316 stainless steel (see below). Metals that rely upon oxide films for corrosion resistance should, of course, be used only in E_h conditions under which passivity can be maintained.

Nickel confers excellent resistance to alkaline attack and to stress-cracking corrosion and is also quite resistant to non-oxidizing acids. It is widely used to the extent of a few percent in stainless steels (300-series, see below), and in several high-nickel alloys, including the following proprietary metals:

> Monel 400: 66% Ni, 31% Cu, 1% Fe
> Inconel 600: 76% Ni, 16% Cr, 8% Fe
> Hastelloy C: 56% Ni, 15% Cr, 17% Mo, 5% Fe, 1% Si, 4% W.

As the foregoing suggests, the last-named has superb corrosion resistance over wide ranges of pH and E_h. The main disadvantage of such non-ferrous alloys over stainless steels is usually cost, although the vulnerability of high-nickel and copper alloys to attack by chelating agents (Sections 12.4 and 14.5) should be borne in mind.

Alloying to modify the overpotential of the metal surface for H_2 evolution or O_2 absorption can help control corrosion, although it is not always obvious whether these cathodic processes should be suppressed (i_0 lowered) or stimulated to produce the desired corrosion resistance—in the case of titanium (see Section 14.6), palladium was alloyed in to *catalyze* H_2 evolution and to force the metal into a passive condition.

14.8　Stainless Steels

The common feature of all stainless steels is the presence of at least 11%, and more usually 18%, chromium. This confers corrosion protection through formation of a particularly insoluble and reduction-resistant oxide film. The steels are heat-treatable for improved mechanical properties (Section 10.2) if they can undergo the α-Fe/γ-Fe transition at elevated temperatures. It is useful to distinguish four groups of commonly used stainless steels:

(a) *Martensitic chromium steels* (AISI 400-series) contain no (or very little) nickel, and the chromium content is typically about 12%. These steels can undergo the α-Fe/γ-Fe transition at about 1050 °C, and so can be heat-treated for improved mechanical properties, much as can ordinary carbon steels. Since they have the α-Fe structure at ambient temperatures, they are ferromagnetic in ordinary service. Examples are type 410, which is used for turbine blades, and type 416, which has good machinability.

(b) *Ferritic chromium steels* (also 400 series) again have no nickel content, but the chromium content typically ranges from 12 to 20%, which prevents the α-Fe/γ-Fe transition at any temperature. However, since the structure is basically the α-Fe type, these alloys are ferromagnetic. They have good resistance to oxidizing acids, by virtue of the high Cr content. In an early application of such alloys in the chemical industry, type 430 has been used for fabricating nitric acid tank-cars.

(c) *Austenitic stainless steels* (300 series) usually contain about 18% Cr, and 8 to 12% nickel, which suppresses stress-cracking corrosion and improves general chemical resistance. The nickel content preserves the γ-Fe (austentite) structure at all temperatures, so these austenitic steels are *not* heat-hardenable or ferromagnetic. Type 304 is the widely used "18-8" stainless steel, while type 316, with 2 to 3% molybdenum, has improved resistance to acids and is an excellent choice for chemical process equipment. To avoid weld decay (Section 14.4), types 304L or 316L, with carbon contents less than 0.03%, or types 321 (% Ti not less than 4 × % C) or 347 ((Nb + Ta) at least 10 × % C) may be used. For better high-temperature performance, higher non-ferrous metal contents are required, e.g., 25% Cr and $\approx$20% Ni in type 310 stainless steel.

(d) *200-series stainless steels* are again austenitic, like the 300-series, but usually manganese is used in place of nickel.

References

1. U. R. Evans, "An Introduction to Metallic Corrosion," 3rd edn., Edward Arnold: London, 1981.

2. U. R. Evans, "The Corrosion and Oxidation of Metals," Edward Arnold: London, 1960, and Supplements 1 (1968) and 2 (1978).

3. M. D. Fontana and N. D. Greene, "Corrosion Engineering," 2nd edn., McGraw-Hill: New York, 1978.

4. Anton deS. Brasunas and Norman E. Hamner, "NACE Basic Corrosion Course," National Association of Corrosion Engineers: Houston, Texas, 1970.

5. G. Wranglén, "An Introduction to Corrosion and Protection of Metals," Institut för Metallskydd: Stockholm, 1972.

6. J. O'M. Bockris and A. K. N. Reddy, "Modern Electrochemistry," Vol. 2, Plenum Press: New York, 1970, especially Chapter 11.

7. P. A. Schweitzer (ed.), "Corrosion and Corrosion Protection Handbook," 2nd edn., Marcel Dekker: New York, 1989.

Exercises

14.1 With reference to the discussion of the bimetallic corrosion of iron given in Section 14.1, confirm (*a*) that the solubility of iron(II) hydroxide is 5.8 μmol L^{-1} if OH$^-$ is produced along with iron(II) according to reaction 14.5 and the reverse of reaction 14.6, and (*b*) that the time scale of the oxidation of iron(II) in solution at pH 7 is as given following Eqn. 14.8 (*Hint:* incorporate [O$_2$] and pH into a rate constant k_1 for a first-order process, then calculate the half-period $t_{1/2} = (\ln 2)/k_1$.)

14.2 Express the "mils per year" corrosion rate in terms of electrical current density for iron (density 7.86 g cm^{-3}; one mil = 0.001 inch = 0.00254 cm).

14.3 Provide explanations for the following:

(*a*) Extremely pure metals (e.g., zinc) often corrode less rapidly than the same metals in impure form.

(*b*) If an aluminum sheet (E° for Al^{3+}/Al $= -1.66$ V) is in contact with one of iron (E° for Fe^{2+}/Fe $= -0.44$ V), it is the iron that corrodes, i.e., becomes anodic.

(*c*) A rubber band was slipped over a mild steel plate, which was then left lying in a damp place for several days. A rust deposit formed on the exposed metal (especially on either side of the rubber band), but the metal under the band was rust-free (though markedly etched).

(*d*) Corrosion of steel by a single drop of water produces a ring of rust, the outer diameter of which is somewhat less than the diameter of the drop.

(*e*) A steel pile, partly immersed in still water, tends to corrode mainly at the lower end; there is a protected area under water near the meniscus.

(f) Health authorities have warned against the use of copper or stainless steel fittings in contact with the lead piping that is still found in domestic plumbing in some older houses in Britain.

(g) Water containing high concentrations of dissolved carbon dioxide is very corrosive towards iron.

14.4 Compare and contrast the mechanisms of oxidation of iron by dry air ("scaling"; see Chapter 7) and by aqueous environments.

Chapter 15

Extractive Metallurgy

THE EXTRACTION of metals from minerals[1-7] has many points in common with corrosion control—redox chemistry is applied to wrest metals from Nature, just as it is used to prevent her reclaiming them through corrosion. There is an additional factor to be considered, however. Most ores contain only minor amounts of the metal of interest, and before final reduction or refining, the desired metal or its compounds must be concentrated and separated from other metals that may be present.

15.1   Gravity and Flotation Methods of Ore Concentration

The simplest concentration technique is the use of *gravity* to separate dense metal or ore particles from the much less dense silicate and other rock-forming minerals by suspending the latter, finely divided, in swirling water. Placer gold (density 19.3 g cm^{-3}, cf. about 2.5 for silicates) is the time-honored example. The gold dust that has accumulated in some riverbed silts was concentrated there by the same process, having weathered out of the surrounding rock in which it was present in very small amounts.

Froth flotation[1] is widely used to concentrate sulfidic ores, e.g., galena (PbS). The crushed rock is again suspended in water, but this time the particles of metal sulfide, which may be denser than the unwanted siliceous *gangue*, are nevertheless caught up in a froth generated by blowing air through the mixture (usually after addition of a frothing agent such as pine oil). The froth can then be skimmed off the top and the metal sulfide recovered. This will only work if the metal sulfide particles specifically can be made water-repellent, so that these unwetted particles seek the air–water interface of the froth while the water-wetted gangue particles sink in the aqueous phase. This is achieved by adding to the water a water-soluble

239

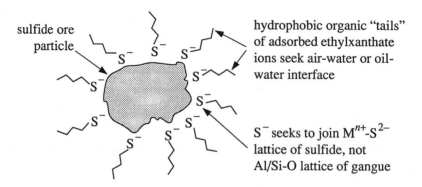

Figure 15.1 Mode of action of sodium ethylxanthate as a froth flotation agent for sulfide minerals.

soap-like salt in which the negative charge of the organic anion is borne by sulfur atoms, which will therefore tend to adsorb on the metal sulfide surfaces but not on the silicate or aluminosilicate gangue surfaces, which have negligible affinity for sulfur. The salt most commonly used is sodium ethylxanthate $(C_2H_5OCS_2^-Na^+)$, which is a sulfur-containing analogue of organic carbonate salts (Fig. 15.1; compare and contrast Fig. 8.7).

15.2   Hydrometallurgical Concentration and Separation

Hydrometallurgical methods[4, 5] are those that use reactions in aqueous solution (often involving metal complex formation) to concentrate and/or to separate the metal ions of interest. Gold particles in crushed rock, for example, can be brought into solution as $Au(CN)_2^-$ ion by leaching the rock with aerated sodium cyanide solution (cf. Exercise 13.2); the aluminosilicates remain undissolved:

$$2Au(s) + O_2(g) + 4CN^-(aq) + 2H_2O(l) \rightarrow$$
$$2Au(CN)_2^-(aq) + H_2O_2(aq) + 2OH^-(aq). \quad (15.1)$$

This reaction is rather slow, requiring several days, after which the gold may be recovered by electrolysis[6] (Section 13.7) or by reaction with zinc:

$$Zn(s) + 2Au(CN)_2^-(aq) \rightarrow Zn(CN)_4^{2-}(aq) + 2Au(s). \quad (15.2)$$

(Gold can also be extracted from crushed rock by dissolving it in mercury, with which it forms an amalgam, and then distilling off the mercury. Amateur gold miners must be aware of the toxicity of cyanides and of mercury vapor before attempting either technique.)

The effectiveness of aqueous ammonia as a complexing agent for Ni^{2+}, Cu^{2+}, and cobalt(III), but not for Fe^{2+}, Fe^{3+}, or Mn ions, is the basis of the *Sherritt Gordon ammonia leach process*, developed in the 1950s. Crushed ore containing small amounts of nickel ($\approx 10\%$), copper ($< 2\%$) and cobalt ($\approx 0.3\%$) sulfides is treated with aqueous $NH_3/(NH_4)_2SO_4$ buffer solution at 85 °C under 10 bars pressure of air, whereupon $Ni(NH_3)_6{}^{2+}$, $Cu(NH_3)_4{}^{2+}$, and $Co(NH_3)_5OH_2{}^{3+}$ form. (The cobalt is present as cobalt(II) originally, but aqueous $Co(NH_3)_6{}^{2+}$ reacts with oxygen to give primarily the aquapentaamminecobalt(III) ion; Section 11.2.) Simultaneously, the sulfide ions are oxidized to sulfate, thiosulfate, and polythionates (Section 1.8), and all these pass into solution, leaving behind a gangue of iron(III) oxides/hydroxides and aluminosilicate minerals. The excess NH_3 is then distilled off, so that the pH falls and the sulfur anions disproportionate giving H_2S, whereupon CuS ($K_{sp} \approx 10^{-36}$) reprecipitates, leaving the Ni (K_{sp} for NiS $= 1 \times 10^{-22}$) in solution along with $Co(NH_3)_5OH_2{}^{3+}$ which is kinetically inert and so does not part with its ammonia ligands at this stage.

The nickel in the aqueous phase is recovered by reduction with H_2 (Section 15.6) until about as much nickel as cobalt remains in rather dilute solution. Both are then precipitated with added H_2S as NiS and CoS, which are redissolved in aqueous H_2SO_4 at pH 2 by oxidation with air, and again converted with ammonia and air to $Ni(NH_3)_6{}^{2+}$ and $Co(NH_3)_5OH_2{}^{3+}$. Now, if this quite concentrated solution is acidified with sulfuric acid, the nickel ammine, being labile (i.e., able to react rapidly), goes to $Ni(H_2O)_6{}^{2+}$, which comes down out of solution as $NiSO_4 \cdot (NH_4)_2SO_4 \cdot 6H_2O$, a rather poorly soluble "double salt" (cf. alums). Cobalt would do the same if it were present as $Co(H_2O)_6{}^{2+}$, but it is not, because $Co(NH_3)_5OH_2{}^{3+}$ is kinetically inert and remains in solution as such. Thus, after a good deal of complicated but instructive aqueous chemistry, we have separated Cu, Ni, and Co from each other. (The reduction of these ions to the metals is described in Section 15.6.) A more recent version of the ammonia leach process uses an ammonia/ammonium carbonate, rather than sulfate, buffer solution.

Acid leaching of metal sulfides such as FeS, CuS, or ZnS with, say, aqueous sulfuric acid might be expected to give H_2S gas and the corresponding aqueous metal sulfates, but in practice many such sulfides are just too insoluble. Treatment with hot aqueous acid under a few bars pressure of air or oxygen, however, can oxidize the sulfide ion to elemental sulfur, polythionates, and eventually sulfate ion (cf. the ammonia leach process), so that the desired metal ions pass into solution. Such hydrothermal processes have an important advantage over sulfide ore roasting (Section 15.8) in that emissions of the notorious air pollutant SO_2 (Section 2.5) can be completely avoided:[4, 8]

$$ZnS(s) + H_2SO_4(aq) + \tfrac{1}{2}O_2(g) \rightarrow ZnSO_4(aq) + S(s) + H_2O. \qquad (15.3)$$

The oxidation of insoluble mineral sulfides to the usually water-soluble sulfates ($PbSO_4$ is an exception) can also be carried out in many cases by *microbial leaching*, that is, by the use of bacteria such as *Thiobacillus ferrooxidans* which can use the sulfide–sulfate redox cycle to drive their metabolic processes. The overall reaction still consumes oxygen:

$$MS(s) + 2O_2(g) \xrightarrow[\text{in water}]{\text{bacteria}} M^{2+}(aq) + SO_4{}^{2-}(aq) \qquad (15.4)$$

and any unwanted iron(II) is usually also oxidized to iron(III) which may then separate as a hydrolytic precipitate—this may, however, be a disadvantage if it impairs the bacterial activity.

Other sulfide-oxidizing processes include the use of nitric acid, a strong oxidant, in place of sulfuric acid:

$$3CuS(s) + 8HNO_3(aq) \rightarrow 8NO + 3CuSO_4(aq) + 4H_2O. \qquad (15.5)$$

Better yet, the metal sulfide may be made the anode of an electrolytic cell, in which the electrolyte is $H_2SO_4(aq)$ and hydrogen evolution is the cathodic reaction:

$$NiS(s) \rightarrow Ni^{2+}(aq) + S(s) + 2e^- \qquad (15.6)$$

$$2H^+(aq) + 2e^- \rightarrow H_2(g). \qquad (15.7)$$

Alkali leach methods are exemplified by the *Bayer process* for the preparation of pure α-Al_2O_3 for electrolysis (Section 15.5) from the mineral bauxite. This consists mainly of α-$AlO(OH)$ (diaspore) and/or γ-$AlO(OH)$ (boehmite), the difference between these being essentially that the oxygen atoms form hcp and ccp arrays, respectively. The chief contaminants are silica, some clay minerals, and iron(III) oxides/hydroxides, which impart a red-brown color to the mineral. Aluminum(III) is much more soluble than iron(III) or aluminosilicates in alkali, so that it can be leached out with 30% aqueous NaOH at 160 °C under $\approx$8 bars pressure, leaving a "red mud" of iron (and other transition metal) oxides/hydroxides and clay minerals:

$$AlO(OH)(s) + OH^-(aq) + H_2O \rightarrow Al(OH)_4{}^-(aq). \qquad (15.8)$$

The hot solution is decanted from the red mud, cooled, and diluted about three-fold, whereupon gibbsite (α-$Al(OH)_3$) will crystallize out if seeded with crystals from a previous batch. The gibbsite is then dehydrated to α-Al_2O_3 by heating to 1250 °C—lower temperatures would give γ-Al_2O_3, which tends to carry absorbed water that would give hydrogen instead of aluminum in the electrolytic reduction to Al that follows. The alkali solution is reconcentrated by evaporation and is recycled.

15.3   Solvent Extraction and Ion Exchange Separations[4, 7, 8]

The formation of mineral deposits occurs through geochemical reactions which will often concentrate and precipitate specific compounds quite locally and separately from others of sufficiently different chemical properties. In such cases, natural processes do much of the work of concentration and separation for us. If, however, the elements of interest have closely similar chemistries, they will tend to occur together and thus will require highly selective chemical techniques to separate them after concentration. Examples include the lanthanides, or *"rare earths"*—this is a misnomer, as they are not so much rare as dispersed. Even when found as a relatively rich deposit such as the phosphate mineral monazite, they are hard to separate from each other. Their chemical properties generally change very slightly but progressively, following trends set by the lanthanide contraction (Section 1.5), as we go from element 57 (lanthanum) to element 71 (lutetium).

A secondary consequence of the lanthanide contraction is that hafnium (element 72), which follows the lanthanides in the Periodic Table, has almost identical atomic and ionic radii and ionization potentials to zirconium, its predecessor in the titanium group, instead of being bigger and more easily ionized than Zr as one might otherwise expect. Indeed, it was not until 1923 that it was realized that the then unknown element 72 was present in samples of zirconium and its compounds prepared up to that time. Yet Zr and Hf differ markedly in their *nuclear* properties; in particular, Zr has a conveniently low *"cross-section"* for neutron absorption, whereas that for Hf is very high. Consequently, in preparing the highly corrosion-resistant zirconium alloy cladding for water-cooled nuclear power reactor fuel elements, it is essential that Hf be eliminated as far as possible.

Such difficult separations can often be effected by *liquid–liquid solvent extraction*,[4, 7, 8] which depends on differences in the distribution of solute species between two immiscible or partially immiscible phases. For a solute species A, this distribution is governed by the *Nernst partition law:*

$$\frac{[A]_e}{[A]_r} = \lambda_A \qquad (15.9)$$

where $[A]_e$ represents the concentration of A in the preferred solvent (the *extract*, usually in an organic solvent) and $[A]_r$ its concentration in the depleted solvent (the *raffinate*, usually the aqueous phase) after the two phases have been thoroughly mixed and allowed to separate out at equilibrium. The *distribution coefficient* λ_A is constant for that particular solvent pair and solute at a given temperature. A similar equation

$$\frac{[B]_e}{[B]_r} = \lambda_B \qquad (15.10)$$

can be written for some other solute B, from which A is to be separated, and so on for solutes C, D, etc. The separation factor

$$\alpha_{AB} = \frac{\lambda_A}{\lambda_B} = \frac{[A]_e[B]_r}{[A]_r[B]_e} \qquad (15.11)$$

should be as large as possible if we want A to be concentrated in the extract and B in the raffinate; if $\alpha_{AB} = 1$, no separation is possible. Often, as in the case of the lanthanide or the zirconium–hafnium separation, the difference between A and B (etc.) in any one chemical property may be small, so we seek an extraction system that compounds differences in several chemical factors, as discussed below. In addition, we rely upon multiple equilibrations of extract and raffinate with fresh solvent portions; it is more effective to make several extractions with small amounts of extractant and then to combine the extracts, than to equilibrate the solution just once with the same total volume of extractant. In practice, one designs a multistage, counter-current extractor to work in the continuous flow mode. Highly efficient separations are possible, but the high cost of extractant that is lost through dissolution or emulsification in the raffinate, and is ultimately run off with the waste, limits the use of solvent extraction technology to extraction of the more expensive metals.

The selective extraction of a metal ion A^{n+} from water into an organic extractant will be facilitated if (*a*) the organic solvent has a significant tendency to solvate that particular metal ion, (*b*) the metal ion can be induced to form a neutral species, e.g., by complexing with the anions that are necessarily present in the solution, and/or (*c*) the metal ion can be made to form a complex of zero net charge with an acidic organic chelating agent when the pH of the aqueous phase is adjusted to an appropriate value.

Factors (*a*) and (*b*) are illustrated by the use of tributyl phosphate (TBP, $(C_4H_9O)_3PO$) in some nuclear fuel reprocessing cycles, in which one wants to separate unburned uranium and plutonium (formed as a by-product of the irradiation of ^{238}U with neutrons in the reactor core) in spent nuclear fuel elements from fission products and then from each other. The spent fuel element is still mainly UO_2 and is dissolved in aqueous nitric acid, which is oxidizing enough to take the U and Pu to the (VI) oxidation state as $UO_2{}^{2+}$(aq) and $PuO_2{}^{2+}$(aq), the *uranyl* and *plutonyl* ions, which can be regarded as hydrolyzed U^{6+}(aq) and Pu^{6+}(aq) (cf. Section 11.6). This solution is then equilibrated with TBP, which is immiscible with water, or TBP in an alkane solvent. The $UO_2{}^{2+}$ and $PuO_2{}^{2+}$ form neutral complexes containing both TBP and the nitrate ions which are present in large excess:

$$MO_2{}^{2+} + 2NO_3{}^- + 2TBP \rightleftharpoons MO_2(NO_3)_2(TBP)_2{}^0 \qquad (15.12)$$

[M = U or Pu]

$$\left[\begin{array}{c} (BuO)_3PO \diagdown \quad O \diagup O \diagdown \\ \overset{\displaystyle O}{\underset{\displaystyle O}{\parallel}} \quad O-N \\ O-M-O \\ N-O \quad OP(OBu)_3 \\ O \diagup \qquad O \end{array} \right]^0$$

These U and Pu complexes, being neutral and already containing TBP solvent ligands, pass preferentially into the TBP solvent phase, leaving almost all the fission products such as Sr^{2+}, I^-, and Cs^+ behind in the aqueous phase. The separation of U from Pu in TBP is easily achieved by reducing one (Pu) but not the other to the (III) oxidation state with an aqueous reductant, whereupon the Pu^{3+} goes into the aqueous phase while $UO_2(NO_3)_2(TBP)_2{}^0$ remains in the TBP. We saw in Exercise 13.5 how the choice of a suitable reductant can be made.

In a zirconium–hafnium separation process developed by Eldorado Nuclear, a mixture of sodium zirconate and hafnate can be obtained by fusing zircon sand with NaOH and dissolving the product in water. Acidification with nitric acid then gives aqueous zirconyl (ZrO^{2+}, or hydrolyzed "Zr^{4+}") and hafnyl (HfO^{2+}) nitrates. Extraction with TBP then gives an extract containing mainly $[HfO(NO_3)_2(TBP)_n]^0$, but most of the ZrO^{2+} remains in the aqueous phase and can be recovered on evaporation as substantially hafnium-free $ZrO(NO_3)_2(s)$. The effectiveness of this process can be ascribed to the compounding of factors (a) and (b) mentioned above to favor extraction of the hafnyl ion.

Factor (c) can be illustrated by considering the formation of complexes between 8-hydroxyquinoline (HQ) and a mixture of metal ions, say, $M^{2+} = Fe^{2+}$, Co^{2+}, Ni^{2+}, and Cu^{2+}. This is in fact the order of increasing stabilities of the complexes $MQ_2{}^0$ (the *Irving–Williams* order; cf. the susceptibility of Ni^{2+} and Cu^{2+} to NTA complexing, Sections 12.4 and 14.5), but M^{2+} must compete with $2H^+$ for Q^-:

$$2HQ + M^{2+} \rightleftharpoons MQ_2{}^0 + 2H^+ \tag{15.13}$$

so we can, by setting the pH low enough, ensure that only copper(II) forms major amounts of an uncharged di(hydroxyquinolinato)-complex, which can then be extracted selectively into, say, chloroform. When this is complete, the pH could be increased enough to permit selective extraction of $NiQ_2{}^0$, etc. Several such schemes have been developed to recover copper from solutions obtained (usually) by acid leaching of low-grade copper ores.

Liquid–liquid solvent extraction is the chief method used at present for separation of the lanthanide(III) ions from one another on the industrial

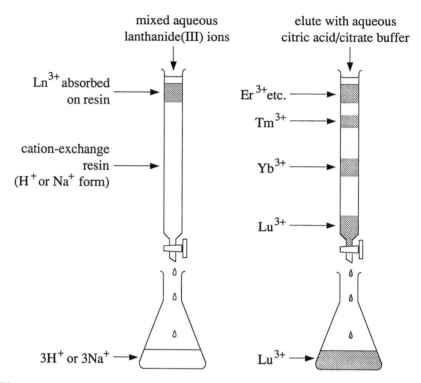

Figure 15.2 Ion-exchange chromatographic separation of mixed aqueous lanthanide(III) ion.

scale (by the Rhône–Poulenc and Molycorp companies),[7] and compounds of the individual lanthanides find various "high-technology" applications such as color video phosphors, lasers (neodymium in yttrium-aluminum garnet), materials with special electrical or magnetic properties for electronic applications (yttrium iron garnet, etc.), superconducting ceramics such as $YBa_2Cu_3O_{7-x}$, and catalysts. Where the highest purity is required, however, the mixed lanthanide(III) ions can be absorbed from solution at the top of a column of a cation exchange resin RSO_3Na (Section 12.4) and then selectively *eluted* from it, i.e., swept down the column in bands by a stream of a solution of a substance that competes with the lanthanide(III) ions Ln^{3+} for sites in the resin. An acidic tribasic chelating agent H_3X (usually citric acid/sodium citrate buffer) is used so that, as in reaction 15.13, the tendency for a specific Ln^{3+} to form neutral LnX^0 and so escape the electrostatic attractions holding it within the resin can be "fine-tuned" by adjusting the pH, i.e., the buffer composition

$$Ln^{3+}(aq) + H_3X(aq) \rightleftharpoons LnX^0(aq) + 3H^+(aq) \qquad (15.14)$$

in competition with

$$Ln^{3+}(aq) + 3RSO_3Na(s) \rightleftharpoons R_3Ln(s) + 3Na^+(aq). \qquad (15.15)$$

The lanthanide contraction determines the order of ease of elution to be $Lu^{3+} > Yb^{3+} > Tm^{3+} >$ etc. (Fig. 15.2), for at least two reasons:

(a) The ions with the largest *hydrated* radii are the least firmly held by electrostatic attractions in the resin, and the extent of *hydration* of Ln^{3+} increases as the core ion becomes smaller. So, paradoxically, $Lu^{3+}(aq)$ is the largest hydrated ion, followed by $Yb^{3+}(aq)$, etc.

(b) The smaller the naked Ln^{3+} ion is, the higher is the stability constant for LnX^0.

15.4 Separations Utilizing Special Properties

The list of special chemical traits that might be exploited in extractive metallurgy is long, but two cases deserve special mention. The only widespread mineral in the Earth's crust that is ferromagnetic (meaning, in effect, that it can be picked up by a powerful magnet) is magnetite (Fe_3O_4). (Maghemite, "γ-Fe_2O_3," can be included with magnetite for the present purpose; Section 6.4.) This means that deposits of this high-Fe-content mineral are readily located in the field through magnetic anomalies, and also that Fe_3O_4 can be picked out selectively from the crushed ore by passing it between the poles of a large magnet.

Most transition metals form complexes known as *carbonyls* with carbon monoxide as ligands. Examples include $Fe(CO)_5$, $Fe_2(CO)_9$, $Cr(CO)_6$, and $Rh_6(CO)_{16}$, in which the metal is ostensibly in the oxidation state of zero, and many mixed-ligand carbonyls such as $Mn(CO)_5I$, $CH_3Mn(CO)_5$, or $(C_6H_6)Mo(CO)_3$ are known. Such compounds have an organic-like chemistry, being essentially covalent (see Section 2.2), and the simple carbonyls such as $Ni(CO)_4$ are volatile liquids that can be purified by fractional distillation. Of all these, however, only $Ni(CO)_4$ forms *rapidly* (and reversibly) from the elemental metal and CO gas at low pressures and temperatures:

$$Ni(s) + 4CO(g) \underset{\approx 200\,^\circ C}{\overset{40-90\,^\circ C}{\rightleftharpoons}} Ni(CO)_4 \text{ (bp } 43\,^\circ C)$$

$$\Delta H^\circ = -155 \text{ kJ mol}^{-1}. \qquad (15.16)$$

Nickel can therefore be separated very specifically from impurities by treating the crude metal with CO gas and pyrolyzing (i.e., decomposing at high temperatures) the $Ni(CO)_4$ vapor so formed to obtain nickel in 99.97% or better purity. This reaction was discovered in 1889 by Mond, and in 1902 the Mond Nickel Company began producing pure nickel by the carbonyl process at Clydach, Wales, using crude nickel obtained by the reduction of nickel oxide with hydrogen.[9] This plant, now greatly modernized, still operates. (Mond amalgamated with Inco in 1929.)

Inco has recently brought on stream a pressure carbonyl plant at Copper Cliff, Ontario, in which crude nickel from the smelter (72% Ni, 18% Cu, 3% Fe, 1% Co, 5% S) is treated with CO at 70 bars and temperatures up to 170 °C. High pressures favor $Ni(CO)_4$ formation because of the large negative molar volume change associated with reaction 15.16, but they also favor formation of small amounts of $Fe(CO)_5$ (bp 103 °C) and $Co_2(CO)_8$ (mp with decomposition, 51 °C) which, fortunately, can easily be removed by fractional distillation. The purified $Ni(CO)_4$ can then be pyrolyzed at 200 °C at *low* pressure.

Tetracarbonylnickel is colorless and odorless but extremely toxic, more so even than CO itself, and no leakage (down to the part-per-billion detection limit) can be tolerated in the plant. This has undoubtedly discouraged more widespread use of carbonyl refining in place of the usual electrolytic processes (see Section 15.5).

15.5   Electrolytic Reduction of Concentrate to Metal

Since chemical reduction means gain of electrons, *electrolysis* is the most direct way of recovering a metal from its ores, so long as these can be handled in a fluid state.[6] Consideration of $E°$ values for reactive metal half-cells such as $Na^+(aq)/Na(s)$, $Mg^{2+}(aq)/Mg(s)$, or $Al^{3+}(aq)/Al(s)$ (-2.71, -2.37, and -1.66 V, respectively) shows that these metals can never be obtained by electrolysis of aqueous solutions of their salts, as H_2 would be produced instead, but they can often be obtained by electrolysis of suitable *molten salts* such as NaCl or $MgCl_2$:

$$2NaCl(l) \xrightarrow[\text{graphite anode}]{\text{steel cathode}} 2Na(l) + Cl_2(g). \qquad (15.17)$$

It is usual to use a mixture of salts (in the case of sodium, 60% NaCl + 40% $CaCl_2$) so as to depress the melting point and permit electrolysis at a lower temperature (600 °C, instead of 801 °C for pure NaCl), but obviously the second metal ion (Ca^{2+}) must be less readily reduced than the one of interest.

In the case of aluminum, the product of the Bayer pre-purification process (Section 15.2) is high-melting α-Al_2O_3 (mp $\approx$ 2015 °C), the direct electrolysis of which is impractical. (In any case, the melt is a non-electrolyte.) It can, however, be dissolved (2 to 8%) in a much lower-melting mixture of cryolite (Na_3AlF_6) with $\approx$10% fluorite; here again, the purpose of the CaF_2 is simply to depress the melting point without itself being preferentially electrolyzed. The electrolysis is carried out at about 970 °C with graphite

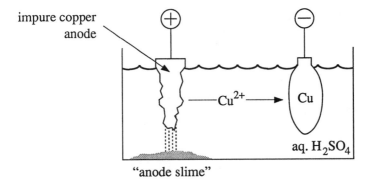

impure copper
anode

Cu²⁺ → Cu

aq. H₂SO₄

"anode slime"

Figure 15.3 Electrorefining of smelter copper.

electrodes, at $\approx 4.5\,\mathrm{V}$ and current density ≈ 1 A cm^{-2}. The graphite anodes, at which the O^{2-} ions are discharged, react to form CO_2 gas and so are consumed as the electrolysis proceeds (the *Hall–Héroult process*):

$$2Al_2O_3\,(\text{in cryolite}) + 3C(s) \xrightarrow{\text{electrolyze}} 4Al(l) + 3CO_2(g). \qquad (15.18)$$

The consumption of the graphite is quite acceptable; inert anode materials for these conditions are hard to find, but in any event the involvement of C in reaction 15.18 reduces the electricity requirement of this energy-intensive electrolysis by nearly half. Other, less commonly used processes that involve lower-temperature electrolysis of molten $AlCl_3$ (mp 183 °C, performed in a closed vessel to prevent sublimation) have been developed. In these, the aluminum is formed as a solid (mp 660 °C), and the anodes are not consumed, but of course one has to convert Al_2O_3 to $AlCl_3$ first, and this is usually done using graphite (coke) anyway:

$$Al_2O_3 + 3C + 3Cl_2 \xrightarrow{900\,°C} 2AlCl_3 + 3CO \qquad (15.19a)$$

$$2AlCl_3(l) \xrightarrow{\text{electrolyze}} 2Al(s) + 3Cl_2(g). \qquad (15.19b)$$

The *electrolysis of aqueous solutions* may be used to obtain less reactive metals, such as Cu, Ni, Zn, or Cr, in high purity, either from aqueous concentrates of the metal salts themselves (*electrowinning*) or from anodes of the crude metal prepared by other, usually pyrometallurgical, techniques (*electrorefining*). For example, crude copper from a smelter can be made the anode in a bath of dilute sulfuric acid; it dissolves when the electric current is passed and is redeposited on the refined Cu cathode in high purity, if the operating conditions of the cell are correctly adjusted (Fig. 15.3). The acid-soluble impurities such as iron also go into solution and remain there so long as the cathode potential is not too negative. Since each metal-ion/metal couple has a characteristic $E°$ value, electrorefining can be highly selective, giving very pure products.

Insoluble impurities fall to the floor of the cell as "anode slime". Despite the derogatory name, this material contains precious metals such as gold, silver, and platinum. The anode slime from the electrorefining of nickel at Sudbury, Ontario, is an important source of the platinum metals, notably palladium.

15.6   Chemical Reduction of Concentrate to Metal

The hydrometallurgical applicability of *gaseous hydrogen* might seem to be limited, since many economically important metals have standard electrode potentials $E°$ for $M^{m+}(aq)/M(s)$ that are negative, that for $H^+(aq)/H_2(g)$ being zero by definition. The conventional values of $E°$, however, refer to standard conditions (in practical terms, 1 mol L^{-1} M^{m+}, 1 mol L^{-1} H^+, 1 bar H_2, 25 °C, etc.), and hydrogen reduction of metal ions can often be made thermodynamically feasible if these conditions are adjusted appropriately—notably, higher hydrogen pressures, higher pH, and, if the reduction is endothermic, higher temperatures.

Nickel(II), for example, would seem an unlikely candidate for H_2 reduction in aqueous solution:

$$Ni^{2+}(aq) + H_2(g) \rightleftharpoons Ni(s) + 2H^+(aq) \quad \Delta E = -0.25\,V,$$
$$\Delta H° = +54 \text{ kJ mol}^{-1}. \text{ (15.20)}$$

In the Sherritt Gordon ammonia leach process, however, aqueous Ni^{2+} is reduced to nickel metal by hydrogen by using a high pressure (30 bars) of H_2, a low acidity (NH_4^+/NH_3 buffer—the pK_a of NH_4^+ is 9.2 at 25 °C and 5.8 at 200 °C), and a high temperature (175 to 200 °C). The kinetics of the reduction are prohibitively slow unless precipitation of the nickel is nucleated (seeded) with pure nickel powder from a previous batch. Cobalt, copper, and several other metals can be produced in this way.

In *pyrometallurgy*[1, 10, 11] (high-temperature dry smelting), however, hydrogen is much less effective as a reductant, but carbon becomes important. For reasons that will emerge later, the concentrate to be reduced is usually in the form of a metal oxide, so the reductant has to be something that has a greater affinity for the available oxygen atoms than the metal itself. In more precise terms, the oxide of the reductant must have a more negative free energy of formation $\Delta G_f°$ *per mole O atoms* than does the metal. The relevant information is usefully summarized in *Ellingham diagrams*, which are plots of $\Delta G_f°$ per mole O against temperature.[10] These plots usually appear almost linear because, although the standard heat capacities $C_p°$

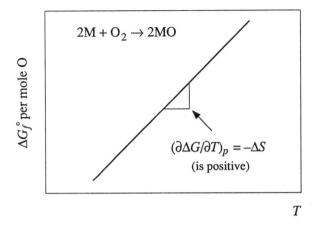

Figure 15.4 Ellingham plot (free energy of formation per mole O atoms versus temperature) for the reaction (metal) + (O_2 gas) → (metal oxide); metal is not necessarily divalent. Plot is rectilinear if heat capacity change of reaction is negligible.

of the individual reactants will be temperature-dependent, the net change ΔC_p° in heat capacity on reaction is usually negligible (Section 1.3).

The standard entropies S° of gases are much larger than those of solids and liquids (Section 1.2). This may be understood on the somewhat simplistic view of S° as a measure of disorder at the molecular level—the molecules of gases have much greater freedom of translational motion, and hence are less ordered, than those of liquids and, particularly, solids. Consequently, for the oxidation of a *solid* metal to a *solid* oxide with the consumption of *gaseous* oxygen

$$M(s) + \tfrac{1}{2}O_2(g) \rightarrow MO(s) \qquad (15.21)$$

the entropy change ΔS_f° will be *negative*. The dependence of ΔG_f° for reaction 15.21 upon temperature, which may be presumed to be linear ($\Delta C_p^{\circ} \approx 0$), will therefore have a *positive* slope, as in Fig. 15.4:

$$\Delta G_f^{\circ} = \Delta H_f^{\circ} - T\Delta S_f^{\circ}. \qquad (15.22)$$

For the oxidation of solid carbon to gaseous carbon monoxide by O_2, however, there is an *increase* in the number of gas molecules, hence an *increase* in ΔS_f° with rising temperature, and so the Ellingham plot has a negative slope (Fig. 15.5):

$$C(s) + \tfrac{1}{2}O_2(g) \rightarrow CO(g). \qquad (15.23)$$

Combining Figs. 15.4 and 15.5 into 15.6, we see that, although MO and C are favored over M and CO at low temperatures, there is a temperature T_r

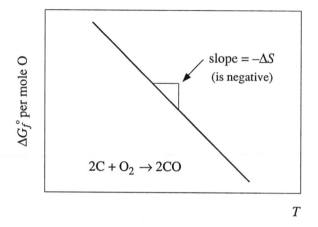

Figure 15.5 Ellingham plot for the combustion of carbon to CO gas.

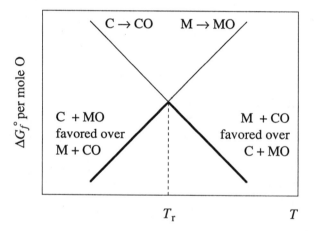

Figure 15.6 Combination of Figs. 15.4 and 15.5, representing the system M + C + O.

above which the reverse becomes true, for the diagram as drawn. In other words, above T_r, carbon (usually supplied as coke, i.e., metallurgical-grade coal from which the volatiles have been distilled) is thermodynamically capable of reducing the metal oxide to the metal.

There are, of course, some complications. Firstly, there may or may not be an intersection as in Fig. 15.6 in the accessible temperature range, which means in practice up to about 2000 °C, above which many refractory materials melt or are themselves reduced. Secondly, the activities of the reactants (particularly the gases) will in general not be the standard ones, so that simple calculations of the reduction temperature T_r based on stan-

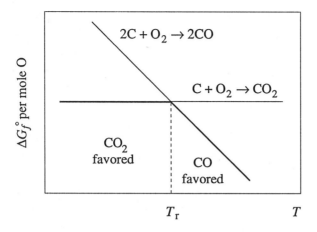

Figure 15.7 Relative stability of carbon monoxide and dioxide in presence of excess carbon.

dard values of ΔH_f° and ΔS_f° should be regarded only as guidelines for real smelting conditions. Thirdly, the reaction *rate* may be prohibitively low at moderate temperatures, even where the reaction is thermodynamically strongly favored; reactions between coarsely mixed solids (C + MO) tend to be inherently slow unless mediated by some fluid component (cf. discussion of iron smelting, below). Fourthly, the predominant product of coke oxidation will be CO_2 at lower temperatures, and it continues to be so at higher temperatures, if the air supply is liberal. Under pyrometallurgical conditions (excess C, limited O_2), however, CO becomes stable with respect to disproportionation above about 700 °C (see Fig. 15.7):

$$2CO(g) \rightleftharpoons C(s) + CO_2(g). \tag{15.24}$$

The entropies of liquids are not strikingly large, compared to those of solids, while those of gases are. Consequently, the Ellingham plot for metal oxides is little affected as we pass the melting point of the metal. However, if the metal *boils* (e.g., magnesium at 1107 °C), the slope of the plot increases sharply since the entropy loss on solid oxide formation is much greater for gaseous than for solid metal:

$$Mg(g) \quad + \quad \tfrac{1}{2}O_2(g) \quad \rightarrow \quad MgO(s) \tag{15.25}$$

1 mole gas $\tfrac{1}{2}$ mole gas   no gas

It is therefore possible to reduce magnesium oxide to the metal with coke above about 1600 °C (Fig. 15.8), even though there would be no intersection of the Ellingham lines for the oxidations of *solid* Mg and coke in an accessible temperature range. The practicability of producing magne-

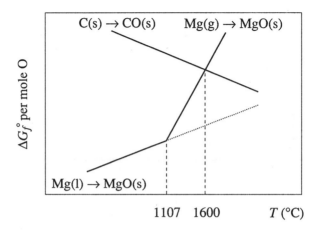

Figure 15.8 Effect of the vaporization of magnesium on the reducibility of magnesium oxide with coke.

sium in this way depends upon the rapid quenching of the Mg vapor below temperatures at which the reverse reaction

$$Mg(g) + CO(g) \rightarrow MgO(s) + C(s) \qquad (15.26)$$

proceeds at a significant rate, since reaction 15.26 again becomes the preferred direction of reaction below 1600 °C. Most magnesium is in fact made by electrolysis of molten $MgCl_2$, as this is currently more economical.

15.7   Pyrometallurgy of Oxides

The combined Ellingham diagram for a selection of oxides is shown in Fig. 15.9. From this it follows that most common metals, even calcium, could, at least in principle, be made by high-temperature reduction of the oxide with coke. In practice, however, this is not always feasible or desirable. For example, Japanese research on the reduction of alumina to aluminum with coke (inspired by the high cost of electricity in Japan) has encountered problems with the volatility of the suboxide Al_2O at the very high reduction temperatures dictated by Fig. 15.9. Titanium dioxide could be reduced by coke, but TiC is a very stable compound (Section 10.1), and this (and possibly TiN from the nitrogen of the air) would be the product of any attempt to make titanium by direct coke reduction. Titanium metal is therefore made by the *Kroll process* (Section 15.8). Copper oxide could be reduced easily by coke, but it turns out that copper can be obtained by the roasting of its sulfide ores without recourse to an added reductant, as described at the end of this chapter. On the other hand, nickel sulfides are

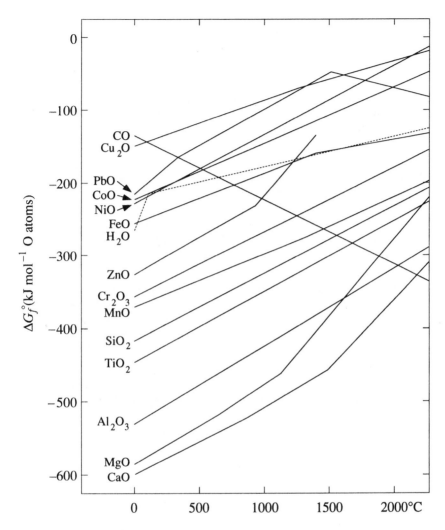

Figure 15.9 Ellingham diagram for the formation of several oxides.

usually converted to the oxide by roasting, and the oxide is reduced with coke to give crude nickel for hydrometallurgical electrorefining.

Figure 15.9 also indicates that hydrogen gas can be an effective reductant for Cu_2O, PbO, NiO, and CoO, but not for FeO, ZnO, etc. Indeed, as we noted in Section 15.4, NiO can be reduced with hydrogen (at 400 °C) to give crude nickel for the Mond process; the gas used is actually water–gas (Section 3.3), and the CO content is used in making the tetracarbonylnickel.

The most powerful reductants for the oxides shown in Fig. 15.9 are aluminum, magnesium, and calcium. Scrap aluminum can be used in the *thermite* reaction to reduce (e.g.) Cr_2O_3:

$$Cr_2O_3 + 2Al \rightarrow Al_2O_3 + 2Cr. \tag{15.27}$$

Much heat is generated, the mixture becoming white-hot, and indeed thermite reactions have been used to generate intense heat as well as the reduced metal for certain welding applications. Chromium, however, is usually made by coke reduction of chromite ($FeCr_2O_4$) in the electric furnace. The product is *ferrochrome*, an Fe-Cr alloy, but this is very acceptable if, as is often the case, the intention is to make stainless steel.

Iron production is the most important application of coke pyrometallurgy in economic terms. Figure 15.10 shows the chemical reactions that occur in the various temperature regimes of the traditional blast furnace. The furnace is charged from the top with a mixture of iron ore (nominally Fe_2O_3), coke (C), and limestone ($CaCO_3$). Air is injected through pipes known as *tuyères*, at the bottom of the furnace, and combustion of coke in this air blast raises the temperature to about 2000 °C, at which temperature the predominant carbon oxide is CO (Fig. 15.7). Thus, the Fe_2O_3 charge at the top of the furnace meets hot CO coming upwards and is reduced, first to magnetite, and then to wüstite (nominally FeO), while the CO goes to CO_2. At the same time, the limestone is converted to lime, adding more CO_2 to the exhaust gases. Any remaining CO will tend to disproportionate to CO_2 and carbon in the cooler, upper part of the furnace (cf. Fig. 15.7), but this process will be relatively slow and incomplete, and in practice CO in the exit gases is burned in heat-exchangers to preheat the air blast before it enters the tuyères.

The critical reduction step involves wüstite and CO at 1000 to 1200 °C:

$$FeO(s) + CO(g) \rightarrow Fe(s) + CO_2(s) \tag{15.28}$$

with the CO arising from the reaction of CO_2 with coke:

$$CO_2(g) + C(s) \rightarrow 2CO(g) \tag{15.29}$$

which is obviously entropy-driven. The net reaction is therefore the one expected on the basis of the Ellingham diagram:

$$FeO(s) + C(s) \rightarrow Fe(s) + CO(g). \tag{15.30}$$

However, the direct reaction of two solids as in reaction 15.30 is almost always slow because of the limited contact between the reactants, whereas in the combined reactions 15.28 and 15.29 the fluid CO/CO_2 phase provides a mechanism for relatively rapid reaction.

As the coke is consumed, allowing the iron and lime to fall towards the hottest part of the furnace, the lime reacts with the silica or silicates that were inevitably present in the iron ore, and a molten calcium silicate *slag* is

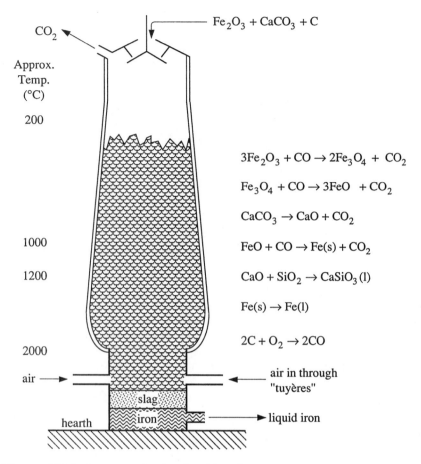

Figure 15.10 Iron ore reduction in a blast furnace.

formed. The iron also melts as much as 400 °C below the melting point of pure iron (1539 °C) because of dissolved carbon (see Fig. 10.1) and collects on the floor of the hearth under a pool of the lighter molten slag. From time to time, molten iron is run out from the hearth, traditionally along narrow channels in sand on the foundry floor into pools called *pigs*, where it solidifies to form *pig iron*. Alternatively, the still-molten iron can be taken directly to steel-making vessels.

Steelmaking. Consideration of the Ellingham diagram shows that, at the high temperatures prevailing near the tuyères, coke may reduce silica to silicon (mp 1410 °C):

$$SiO_2(s) + 2C(s) \rightarrow Si(l) + 2CO(g) \qquad (15.31)$$

which will dissolve in the iron. This is undesirable, as silicon causes iron

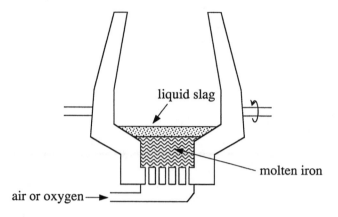

Figure 15.11 Bessemer converter.

to be brittle, as do the other major impurities in pig iron—C, P, Mn, and a minor amount of S. It is the function of the limestone in the blast furnace to minimize this occurrence by converting the SiO_2 to slag, but inevitably the concentrations of impurities, including Si, in pig iron must be reduced to acceptable levels in a separate conversion process, i.e., by oxidation with air or pure oxygen in the presence of lime or limestone, to give serviceable steels. Usually one aims for carbon contents below 1.5% for hard (tool) steels, and not more than 0.3% for mild steels, with Si, P, S, and Mn contents lower still. The reactions occurring in the molten iron/lime mixture are:

$$2C + O_2 \rightarrow 2CO \qquad (15.32)$$

$$Si + O_2 + CaO \rightarrow CaSiO_3(l) \qquad (15.33)$$

$$2P + \tfrac{5}{2}O_2 + 3CaO \rightarrow Ca_3(PO_4)_2(\text{in slag}) \qquad (15.34)$$

$$2Mn + O_2 \rightarrow 2MnO(\text{in slag}) \qquad (15.35)$$

$$Mn + S + CaO \rightarrow CaS(\text{in slag}) + MnO(\text{in slag}). \qquad (15.36)$$

The CO gas is easily removed, while the calcium phosphate and manganese oxide go into the molten slag, which has value in fertilizer manufacture because of its phosphate content. Figure 15.9 shows that Mn, Si, and C will react preferentially with O_2 before oxidation of the molten iron "solvent" begins, but SO_2 and P_2O_5 have less negative ΔG_f° per mole oxygen than does FeO, and it is only because of extraction of $Ca_3(PO_4)_2$ and CaS into the molten slag that P and S can be removed without the oxidation of the liquid iron. The slag, then, has a crucial role to play.

Numerous different steelmaking processes involving these same basic principles are in use today. The bottom-blown *Bessemer converter* (Fig. 15.11), which originated in England in mid-nineteenth century, was largely

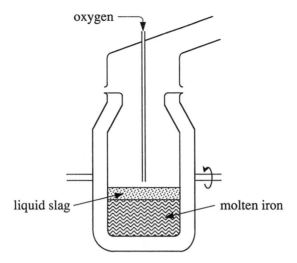

Figure 15.12 The basic oxygen (LD) process.

superseded by the top-blown *Siemens regenerative open hearth furnace*. This in turn has been largely replaced by the *basic oxygen* or *Linz–Donawitz process* (BOP or LD process, Fig. 15.12), in which a powerful jet of oxygen is blown into the molten slag layer from the top, so as to provide better stirring than is possible with the open hearth furnace. This agitation can be improved by rotating the vessel (*KALDO* and *Rotor* processes), in which case a lower-pressure oxygen jet which does not part the slag layer can be used. The exit gases are scrubbed with water to reduce air pollution.

The use of oxygen instead of air is now widespread in the steel industry. This practice has the following advantages which more than compensate for the extra cost:

(*a*) The higher partial pressure of O_2 means that reactions 15.32 through 15.36 proceed more rapidly, so reducing the operating time.

(*b*) The heat loss due to having to heat the nitrogen (79%) and other air components besides O_2 is eliminated. In fact, in most major processes other than the open hearth method, the heat generated by reactions 15.32 through 15.36 is great enough to keep the charge molten, if 100% O_2 is used and the P content is about 2% or more.

(*c*) Nitriding of the steel (Section 10.3) by the N_2 of the air to form small amounts of Fe_4N is eliminated.

Whatever the process, the steelmaking vessel must be lined with a suitable refractory material, usually bricks of calcined dolomite, $(Mg,Ca)O$. Silicate firebricks obviously cannot be used in the presence of lime.

15.8   Pyrometallurgy of Halides and Sulfides

Figure 15.13 shows immediately that carbon is useless as a reductant for chlorides, since CCl_4 is barely thermodynamically stable even at room temperature and becomes endergonic above about $400\,^{\circ}C$. Similarly, CF_4 is the least exergonically formed (per F atom) of all the fluorides covered by Fig. 15.13, and so carbon cannot reduce them. In principle, hydrogen could reduce titanium tetrachloride to titanium, but in reality titanium hydrides (Section 10.1) would be obtained.

The only really satisfactory reductants for metal chlorides and fluorides are magnesium and (less practically on account of their extreme reactivity) sodium or calcium. Titanium (and also zirconium and tantalum) are made industrially by the *Kroll process*, which involves conversion of TiO_2 to $TiCl_4$ by coke reduction in the presence of Cl_2:

$$TiO_2(s) + C(s) + 2Cl_2(g) \xrightarrow{500\,^{\circ}C} TiCl_4(g) + CO_2(g) \qquad (15.37)$$

after which the titanium tetrachloride, which is a liquid at ambient temperature (bp $140\,^{\circ}C$), is condensed, purified by fractional distillation, and then reduced with molten magnesium (mp $649\,^{\circ}C$, bp $1107\,^{\circ}C$) to solid Ti (mp $1660\,^{\circ}C$), giving liquid magnesium chloride (mp $714\,^{\circ}C$):

$$TiCl_4(g) + 2Mg(l) \xrightarrow[850\,^{\circ}C]{Ar} Ti(s) + MgCl_2(l). \qquad (15.38)$$

It is necessary to use argon as the inert atmosphere for reaction 15.38, since even nitrogen will react with Ti to give an interstitial compound (Chapter 10). For the same reason, reduction of TiO_2 by Mg, though theoretically possible according to Fig. 15.9, is not feasible because of retention of interstitial oxygen in the solid titanium. Besides, the other product, MgO, is extremely refractory (mp $2852\,^{\circ}C$) and would be difficult to remove.

In passing, we note that reaction 15.37 is also used in the conversion of natural rutile ore, which is usually red-brown because of its iron(III) content, to pure TiO_2, which is used as a filler for paper and as a white pigment in paints. Its high optical refractive index gives it the best available "covering power" in paints and, unlike the lead carbonate formerly used, it is very inert and non-toxic. The purified $TiCl_4$ from reaction 15.37 is converted to pure TiO_2 (rutile form) by high temperature oxidation:

$$TiCl_4(g) + O_2(g) \rightarrow TiO_2(s) + 2Cl_2(g). \qquad (15.39)$$

(Titanium dioxide, in the form of anatase, can also be obtained from ilmenite by treatment with sulfuric acid, followed by hydrolysis and then thermal dehydration of the resulting $TiO(OH)_2$.)

Crude elemental silicon can be made by the reduction of silica sand with coke in the electric furnace (reaction 15.31) and may be adequate for making

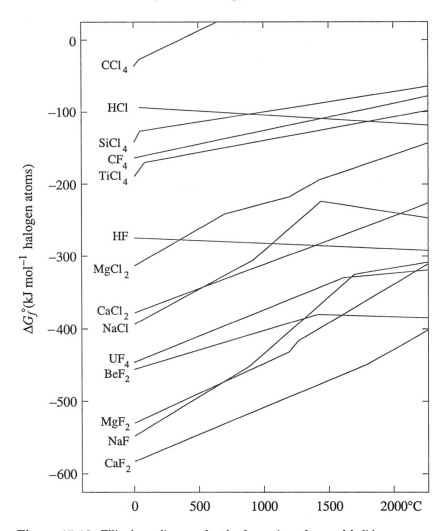

Figure 15.13 Ellingham diagram for the formation of several halides.

ferrosilicon alloys (Section 14.7) or silicones (Section 1.8). The high purity silicon used for electronic "chips," however, is made from silica via silicon tetrachloride, which, like $TiCl_4$, is a volatile liquid (bp 57 °C) susceptible to hydrolysis but readily purifiable by fractional distillation. Indeed, the procedure resembles the Kroll process for titanium, except that an argon atmosphere is not necessary:

$$SiO_2(g) + 2C(s) + 2Cl_2(g) \xrightarrow{\text{red heat}} SiCl_4(g) + 2CO(g) \quad (15.40)$$

$$SiCl_4(\text{redistilled}) + \text{pure Mg} \xrightarrow{\text{heat}} Si(s) + MgCl_2. \quad (15.41)$$

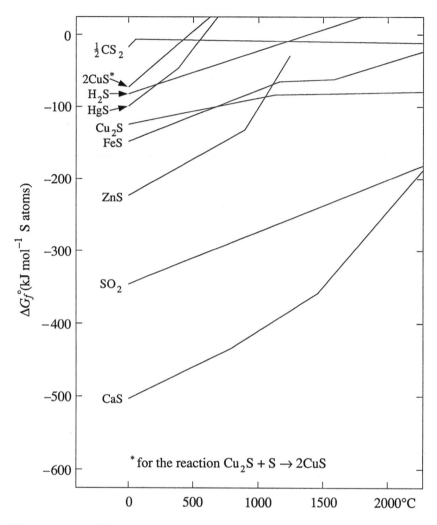

Figure 15.14 Ellingham diagram for the formation of some sulfides.

The magnesium chloride and any excess Mg are then vaporized off, and the remaining silicon is further purified by *zone refining* (in which a band of molten Si is created in a column of the element and is moved slowly along the column, carrying the impurities in it and leaving pure resolidified Si behind) or by *Czochralski crystal growth* (whereby a spinning crystal of pure Si is drawn slowly from a silicon melt, solidifying as it goes). Redistilled SiCl$_4$ is also used in the preparation of pure SiO$_2$ for fiber optics (Section 8.4; compare reactions 8.6 and 15.39).

Figure 15.14 shows that coke cannot serve as a reductant for sulfides, either. Carbon disulfide is seen to be barely stable thermodynamically, while CS, the analogue of carbon monoxide which is so important in oxide pyrometallurgy, is highly unstable and has only recently been characterized in the laboratory. The strongly negative ΔG_f° of SO_2, however, means that oxygen can remove S from many metal sulfides, although the product is usually not the metal but rather an oxide of the metal. This is helpful because these oxides are very often reducible with coke. Thus, coke pyrometallurgy of several sulfides such as ZnS (zinc blende, wurtzite) becomes possible if the ore concentrate is first "roasted" in air:

$$2ZnS + \text{excess } 3O_2 \xrightarrow{\;800\text{–}900\,^\circ C\;} 2ZnO + 2SO_2 \qquad (15.42)$$

$$ZnO + C \xrightarrow{\;\text{blast furnace}\;} Zn + CO. \qquad (15.43)$$

Roasting at a lower temperature in a limited air supply can give the metal sulfate, rather than the oxide:

$$ZnS + 2O_2 \xrightarrow{\;600\text{–}700\,^\circ C\;} ZnSO_4 \qquad (15.44)$$

and the cooled sulfate can be dissolved in water for electrolytic reduction to the metal. For lead, however, such a *sulfatizing roast* is unproductive, since $PbSO_4$ has a very low solubility in water. On the other hand, if the oxygen and SO_2 partial pressures and the temperature are correctly controlled, it is possible to obtain molten lead directly from PbS by the *roast-reduction* reaction:

$$PbS + O_2 \rightarrow Pb + SO_2. \qquad (15.45)$$

Copper, too, can be recovered from its sulfides without resort to a reductant such as coke. The concentrate usually contains iron sulfides (in fact, the most widely occurring ore of copper is $CuFeS_2$, chalcopyrite), as well as the usual silicates. The chemistry of copper smelting is subtle, but the commonly employed two-stage process can be summarized crudely as follows:

(*a*) *Matte smelting* involves roasting the concentrate with lime and recycled converter slag at about $1300\,^\circ C$ to form a slag of molten oxides of iron, silicon, etc., and an immiscible *matte* or molten sulfide layer, which contains Cu_2S along with some FeS.

(*b*) The molten matte layer is transferred to a converter, where air is blown through it and sand is added so that the FeS forms a silicate slag:

$$FeS \xrightarrow{\;O_2 + SiO_2\;} FeSiO_3(l) + SO_2 \qquad (15.46)$$

and then, above $1250\,^\circ C$, the Cu_2S is converted to copper:

$$Cu_2S + O_2 \rightarrow 2Cu + SO_2. \qquad (15.47)$$

This 98% pure *blister copper* may than be refined electrolytically. As far as the copper itself is concerned, the two stages could be represented as:

$$2CuS + O_2 \rightarrow Cu_2S + SO_2 \qquad (15.48)$$

and

$$Cu_2S + O_2 \rightarrow 2Cu + SO_2. \qquad (15.49)$$

However, the role of the slag in what is in effect liquid–liquid solvent extraction is, once again, crucial in the removal of unwanted material such as iron compounds.

References

1. R. H. Parker, "An Introduction to Chemical Metallurgy," 2nd edn., Pergamon Press: Oxford, 1978.

2. J. E. Fergusson, "Inorganic Chemistry and the Earth," Pergamon Press: Oxford, 1982.

3. J. C. Bailar, H. J. Emeléus, R. S. Nyholm and A. F. Trotman-Dickenson (eds.), "Comprehensive Inorganic Chemistry," 5 vols., Pergamon Press: Oxford, 1973.

4. A. R. Burkin, "The Chemistry of Hydrometallurgical Processes," E. and F. N. Spon: London, 1966.

5. F. Habashi, "Hydrometallurgy," *Chemical and Engineering News*, February 8, 1982, 46–58.

6. D. Pletcher, "Industrial Electrochemistry," Chapman and Hall: London, 1982, Chapter 4.

7. R. Thompson (ed.), "Speciality Inorganic Chemicals," Special Publication No. 40, Royal Society of Chemistry: London, 1981, pp. 403–443.

8. (*a*) G. M. Ritchey and A. W. Ashbrook, "Solvent Extraction," Elsevier: New York, 1979. (*b*) Y. Marcus and A. S. Kertes, "Ion Exchange and Solvent Extraction," Wiley-Interscience: New York, 1969.

9. J. R. Boldt, Jr., "The Winning of Nickel," Longmans: Toronto, 1967.

10. C. B. Alcock, "Principles of Pyrometallurgy," Academic Press: London, 1976.

11. O. Kubaschewski and C. B. Alcock, "Metallurgical Thermochemistry," 5th edn., Pergamon Press: Oxford, 1979.

Exercises

15.1 At what pH is hydrogen gas at 1 bar pressure theoretically capable of reducing aqueous Ni^{2+} ion ($E° = -0.25\,V$, assumed 1 mol kg^{-1}) to nickel metal (a) at $25\,°C$; (b) at $150\,°C$ (if $\Delta H = 54$ kJ mol^{-1} and $\Delta C_p° = 0$)? (c) In practice, what factors may prevent the reduction from taking place? (d) How might these be overcome?
[Answers: (a) pH 4.23; (b) pH 2.83.]

15.2 From the following thermodynamic data, with the assumptions that the heat capacities of reaction are negligible and that standard conditions (other than temperature) prevail, calculate the temperatures above which: (a) carbon monoxide becomes the more stable oxide of carbon, in the presence of excess C; (b) carbon is thermodynamically capable of reducing chromia (Cr_2O_3) to chromium metal; (c) carbon might, in principle, be used to reduce rutile to titanium metal; and (d) silica (taken to be α-quartz) may be reduced to silicon in a blast furnace.

Substance	$\Delta H_f°$ (kJ mol^{-1})	$S°$ (J K^{-1} mol^{-1})
C (graphite)	0	5.74
CO	−110.52	197.56
CO_2	−393.51	213.64
Ti	0	30.63
TiO_2 (rutile)	−944.8	50.33
Cr	0	23.77
Cr_2O_3	−1139.72	81.17
Si	0	18.83
SiO_2 (α-quartz)	−910.94	41.84

[Answers: (a) $708\,°C$; (b) $1218\,°C$; (c) $1716\,°C$; (d) $1640\,°C$.]

15.3 On the basis of the following thermochemical data, (a) calculate the equilibrium constant $K°$ for the formation of *gaseous* nickel tetracarbonyl from Ni and Co under standard conditions; (b) estimate the temperature at which $K°$ becomes unity; and (c) explain how this and other related information can be applied in the refining of nickel.

Substance	$\Delta H_f°$ (kJ mol^{-1})	$S°$ (J K^{-1} mol^{-1})
Ni(s)	0	29.87
CO(g)	−110.525	197.674
$Ni(CO)_4$(g)	−602.91	410.6

Note: The heat capacity of reaction is small and may be neglected.
[Answers: (a) 5.7×10^6; (b) $119\,°C$.]

15.4 Magnesium metal can be made industrially by electrolysis of magnesium chloride. Magnesium oxide (which can be obtained from mineral $MgCO_3$ by firing it in a kiln) can be converted to magnesium chloride by passing chlorine gas through a bed of MgO mixed with coke at 600 to 1000 °C:

$$MgO(s) + Cl_2(g) + C(s) \rightleftharpoons MgCl_2(mp\ 714\,°C) + CO(g). \quad (1)$$

Concern has been expressed that the chlorine may react with the carbon monoxide to form the highly toxic gas *phosgene* (carbonyl chloride, $COCl_2$):

$$CO + Cl_2 \rightleftharpoons COCl_2. \quad (2)$$

Are such fears justified? In formulating your answer, you may wish to consider that (*a*) for reaction 2 at 25 °C, $\Delta H° = -108.3$ kJ mol^{-1}, $\Delta S° = -137.2$ J K^{-1} mol^{-1}, $\Delta C_p° = -5.39$ J K^{-1} mol^{-1}; (*b*) chlorine will never be present at the exit from the hot reactant bed unless the supply of one of the other reactants in reaction 1 runs out.

15.5 Trace the steps in the production of high-purity nickel metal from crude sulfide ore: (*a*) by the Sherritt Gordon ammonia leach process (*b*) by the Mond process; and (*c*) by acid leaching.

15.6 Suggest a procedure for the industrial production of hafnium-free zirconium metal from baddeleyite (ZrO_2), on the basis of information given in this chapter and in Appendix C.

Chapter 16

Organometallics and Homogeneous Catalysis

ORGANOMETALLIC COMPOUNDS ("organometallics") are usually defined as those that have at least one direct metal-to-carbon bond. This definition would include metal alkyl compounds such as diethylzinc ($Zn(C_2H_5)_2$) and metal arene complexes like ferrocene ($Fe(C_5H_5)_2$), as well as metal carbonyls such as $Ni(CO)_4$ (Sections 2.2 and 15.4) which do not have organic ligands in the usual sense, but which behave much like organic compounds in terms of their volatility, solubility in non-polar organic solvents, and general properties. On the other hand, lead(IV) acetate ($Pb(OOCCH_3)_4$), a non-polar liquid with organic ligands, is excluded because it contains no Pb—C bonds. Interstitial metal carbides (Section 10.1) might be said to contain metal–carbon bonds, but these refractory solids bear no resemblance at all to organometallics.

The chemistry of organometallics is extraordinarily rich and often complicated. Accordingly, only a brief introduction to the subject is given here, with emphasis on a few aspects of contemporary technological interest. It is, however, an area of continuing intensive research[1–13] and will undoubtedly grow in industrial importance well into the twenty-first century, particularly in the areas of catalysis and in the synthesis of organic compounds.

16.1 Alkyl Compounds of Some Main Group Metals[1–5]

Most truly metallic elements M have electronegativities in the range 0.8 to 1.8 (cf. 2.55 for alkyl carbon), so that M—C bonds can be expected to be strongly polar with a substantial partial negative charge on the carbon. Consequently, we can expect the M—C bond to be highly vulnerable to attack at C by *electrophiles* (i.e., molecules that seek negative charge) such as H^+ and at M by *nucleophiles* such as OH^-, water, or ammonia. The M—C bond is therefore expected to be highly reactive with respect to polar reagents, even though it is often thermochemically quite strong.

Diethylzinc, which was one of the first organometallic compounds to be isolated (Frankland, 1849), epitomizes these characteristics. The compound is easily prepared by heating powdered zinc or zinc–copper alloy with ethyl iodide under an atmosphere of dry nitrogen or CO_2. (The initial product is actually C_2H_5ZnI, which disproportionates on distillation to zinc iodide and diethylzinc.)

$$Zn(s) + 2C_2H_5I(l) \rightarrow Zn(C_2H_5)_2(l) + ZnI_2 \qquad (16.1)$$

Despite a respectable Zn—C bond energy (148 kJ mol^{-1}), however, diethylzinc is immediately hydrolyzed by moisture to form $Zn(OH)_2$ and ethane, catches fire spontaneously in the air to form a ZnO smoke, and reacts readily with many organic compounds as an alkylating or reducing reagent. The hydrolysis reaction is finding application in the preservation of books and documents made with acidic paper; the item to be protected against acid embrittlement is exposed to diethylzinc (DEZ) vapor in a previously evacuated chamber, whereupon hydrogen ions in the paper are mopped up to form ethane and zinc ions, while the deposition of small amounts of $Zn(OH)_2$ within the paper gives lasting protection against re-acidification (see Section 4.3). The procedure cannot, of course, reverse any embrittlement that has already set in.

In the same Periodic group, cadmium forms alkyls much as does zinc, while mercury is readily methylated biologically to form the very toxic methylmercury ion CH_3Hg^+, as discussed in Section 5.3; note that this is a metal alkyl species that resists hydrolysis and is insensitive to the oxygen of the air. In general, however, metal alkyls conform to the zinc stereotype, and indeed it is this reactivity that makes them so important as reagents. As might be expected, alkali metal alkyls are particularly reactive—perhaps too much so, in the case of the reaction of sodium or potassium metals with alkyl or aryl halides RX, which tend to give the alkane R—R and NaX or KX (*Wurtz–Fittig reaction*) rather to follow an analogue of reaction 16.1. On the other hand, lithium metal will react with RX such as *n*-butyl chloride in a suitable solvent such as diethylether or hexane under a nitrogen or (better) argon atmosphere to give synthetically useful organolithium compounds:

$$2Li(s) + RX(\text{in ether}) \xrightarrow[\text{N}_2]{\text{Ar or}} LiR(\text{in ether}) + LiX(s). \qquad (16.2)$$

For example, phenyllithium can be used to make tetraphenyltin:

$$4C_6H_5Li + SnCl_4 \rightarrow (C_6H_5)_4Sn + 4LiCl. \qquad (16.3)$$

For many syntheses that do not require a reagent as aggressive as an organolithium compound, *Grignard reagents* RMgX may be used. These were discovered in 1900 by Victor Grignard, who was awarded the Nobel Prize in 1912 for his work in developing their chemistry. They are invariably prepared in solution in an ether (usually diethyl ether), which serves to stabilize them, probably as complexes of the type $RMgX(OEt_2)_2$, although these exist in solution in equilibrium with other organomagnesium species such as the dimer $RMgX_2MgR$ and dialkylmagnesium R_2Mg (cf. the disproportionation of RZnX, above). The reaction of small magnesium flakes with an organic halide in ether can be carried out in air (contrast the preparation of zinc alkyls), but water must be rigorously excluded—usually, the ether is predried over sodium wire:

$$Mg(s) + RX(\text{in ether}) \rightarrow RMgX(\text{in ether}). \qquad (16.4)$$

Solutions of Grignard reagents have largely replaced organozinc compounds in organic and organometallic syntheses. Unlike the zinc alkyls, they do not require the exclusion of oxygen, but they do react readily with CO_2. In fact, this is a useful way to make unusual carboxylic acids:

$$RMgX + CO_2 \xrightarrow{} RCOOMgX \xrightarrow{HX(aq)} RCOOH + MgX_2. \quad (16.5)$$

Grignard reagents can be used to make organosilicon, -germanium, -tin and -lead compounds by halide abstraction from the corresponding halocompounds (e.g., chlorosilanes, Section 1.8), much as in reaction 16.3. Industrially, however, organosilicon and organotin chlorides are usually obtained by the direct Rochow process (Section 1.8):

$$Sn + 2RX \rightarrow R_2SnX_2 \text{ (etc.)}. \qquad (16.6)$$

Tin alkyls can also be made from the reaction of $SnCl_4$ with aluminum alkyls—highly reactive, oxygen-sensitive materials which are obtained industrially from olefins, hydrogen, and aluminum powder:

$$2Al(s) + 3H_2(g) + 6RCH{=\!=}CH_2 \xrightarrow[60 \text{ bar}]{100\,^\circ C} 2Al(CH_2CH_2R)_3. \qquad (16.7)$$

Industrial organosilicon chemistry centers around silicone production (Section 1.8), while organotin compounds are widely used as stabilizers for polyvinyl chloride plastics (usually as R_2SnX_2, where R is typically *n*-octyl and X may be laurate, maleate, etc.) and as curing agents for silicone rubbers (e.g., di-*n*-butyltin diacetate). Organotin compounds of the type R_3SnX such as *n*-Bu_3SnOH are important biocides and are used as antifouling agents in marine paints, in the suppression of fungal growths in agriculture, and in slime control in the pulp and paper industry. They have the advantage of being very selective, and neither they nor their degradation products are particularly toxic to the higher animals, including humans.

Organolead compounds, notably tetraethyllead, have been extensively used as anti-knock additives for gasoline (see Section 8.3). The Pb—C bond is thermochemically weak, so that PbR_4 molecules decompose readily within the hot cylinders of an automobile engine to give free radicals R that terminate the chain reactions (Section 1.4) that cause explosive rather than smooth burning of the fuel vapor. Thus, it is possible to upgrade low-octane petroleum fractions. However, both tetraalkyllead compounds and their combustion products are dangerously toxic. Lead poisoning is partic-ularly insidious in that its symptoms may not be recognized as such—for example, lead is known to inhibit the mental development of children, and it has been suggested that the phenomenon of steadily falling average scores in certain scholastic aptitude tests may be due in part to ingestion by young children of lead spread in the environment by automobile exhausts (and, to a lesser degree, by lead-carbonate-based paints, which are now largely being phased out in favor of non-toxic TiO_2-containing paints). What is certain is that lead compounds poison the platinum-metal-based catalysts in catalytic converters, so that efforts to abate air pollution by automobiles through the use of catalytic converters (Section 2.4) are dependent upon the use of lead-free fuels. Fuels of sufficiently high octane equivalents can be made from low-octane fractions by shape-selective reforming or by use of additives such as MTBE (Section 8.3) or organometallics of relatively low toxicity and catalyst poisoning power such as methylpentacarbonyl-manganese, $CH_3Mn(CO)_5$.

Several main-group organometallics have significant catalytic activity, e.g., dibutyltin compounds catalyze the formation of polyurethane from organic isocyanates (R—N=C=O) and alcohols, but perhaps the most important are the trialkylaluminums, which, following reaction with tita-nium chlorides, form the famous *Ziegler* and *Ziegler–Natta catalysts* for olefin polymerization. Polymerized olefins such as polyethylene were orig-inally produced from the monomeric olefins only by processes involving elevated temperatures and very high pressures (typically 1000 bars), with the attendant risks and capital expenses. K. Ziegler found that triethyl-aluminum reacts with $TiCl_4$ in inert hydrocarbon solvents to give a brown suspension that causes ethylene (ethene) to polymerize even at room tem-perature and pressure; furthermore, the product is of significantly higher density than polyethylene from the high pressure processes and is ideally suited to molding. Ziegler high density polyethylene has now taken over a large share of the plastics market, although low density polyethylene from the high pressure process is still needed for making plastic film.

The exact nature of the Ziegler catalysts is somewhat obscure. The essential features seem to be that the $TiCl_4$ (a covalent liquid, soluble in hydrocarbon solvents) is alkylated by the trialkylaluminum and also at least partly reduced to titanium(III). Titanium(III) chloride is an ionic solid, insoluble in organic solvents, so it is not surprising that the material so formed is not readily soluble in the reaction medium. This material, which is apparently somewhat variable in structure and composition but which contains Al^{III} as well as Ti^{III}, can add an olefin molecule at a Ti center, presumably in much the same way that ethylene adds to platinum(II) chloride to form *Zeise's salts* (see Section 16.2), and an alkyl group already present on the Ti can then migrate onto the olefin to form a new, longer alkyl group:

$$
\underset{Cl}{\overset{R}{>}\!Ti\!<}
\quad\xrightarrow{H_2C=CHR'}\quad
\overset{H_2C=CHR'}{\underset{Cl}{\overset{R}{>}\!Ti\!<}}
\quad\longrightarrow\quad
\underset{Cl}{\overset{CH_2-CHR'R}{>}\!Ti\!<}
$$

$$
\xrightarrow{H_2C=CHR'}\quad
\overset{H_2C=CHR'}{\underset{Cl}{\overset{CH_2-CHR'R}{>}\!Ti\!<}}
\qquad \text{etc.} \qquad (16.8)
$$

Thus, polymer chains $-(CH_2CHR')_n-$ are created. This is an important route to polypropylene ($R' = CH_3$), which is difficult to make by other methods. Furthermore, it was shown by G. Natta that the heterogeneous (*Ziegler–Natta*) catalysts prepared using $TiCl_3$ in place of $TiCl_4$ produced polymers in which the side-chains R' were arranged in a regular manner in relation to the central carbon chain (Fig. 16.1).

Where the side-chains are all on the same side of the central carbon backbone, the polymer is said to be *syntactic;* where they alternate regularly on either side, it is called *syndiotactic;* and where there is no regular placement of the side-chains, the polymer is *atactic.* Syntactic polypropylene has excellent mechanical properties and is widely used in injection molding and fiber manufacture. By contrast, atactic polypropylene has low strength and has little value. Thus, the work of Ziegler and Natta has assumed huge commercial importance; they were awarded the Nobel Prize for these studies in 1963.

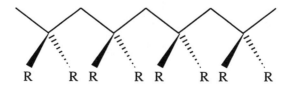

Figure 16.1 A syntactic polymer.

16.2   Organotransition Metal Compounds

The first organometallic compound of the transition metals to be characterized (1827) was *Zeise's salt*, $K[(C_2H_4)PtCl_3] \cdot H_2O$ (Fig. 16.2). It forms
when $K_2[PtCl_4]$ in aqueous ethanol is exposed to ethylene (ethene); a
dimeric $Pt—C_2H_4$ complex with Cl bridges is also formed. In both species,
the ethylene is bonded *sideways* to the platinum(II) center so that the two
carbon atoms are equidistant from the metal. This is called the *dihapto-* or
η^2 mode—a ligand with three adjacent carbons directly bonded to a metal
atom would be *trihapto-* or η^3, and so on.

The nature of the bonding in η^2-olefin complexes is similar in principle
to that explained in Section 2.2 for metal carbonyls, inasmuch as a pair of
electrons (here, the π-bonding electrons of the C═C double bond) gains
donor power through a feedback of d electrons from the metal atom into the
empty *antibonding* π^* orbitals associated with the C═C bond (Fig. 16.3).
Without this "push-pull" mechanism (called *back-bonding*), the π electrons
of the olefin would have negligible tendency to allow themselves to be shared
with the metal. These displacements of electron density in the π and π^*
systems result in the substituents on the carbon atoms (H, in the case of
ethylene) being bent away from the metal atom.

Aromatic hydrocarbons such as benzene (C_6H_6) and the anion $C_5H_5^-$
of cyclopentadiene (C_5H_6) are "side-on bonders" *par excellence*. They
bond the metal atom perpendicularly to the plane of the aromatic ring
by multiple interactions involving the several bonding and antibonding orbitals of their π-electron systems.[1-7] Thus, two cyclopentadienide units can
form a *sandwich complex* with Fe^{II}—the famous *ferrocene*, $Fe(\eta^5\text{-}C_5H_5)_2$
(Fig. 16.4).

This is an orange-brown solid (mp 173 °C) which is easily made by
heating freshly distilled C_5H_6 with an iron(II) salt and solid KOH. It is
extraordinarily inert—it can be heated to 500 °C without decomposition,

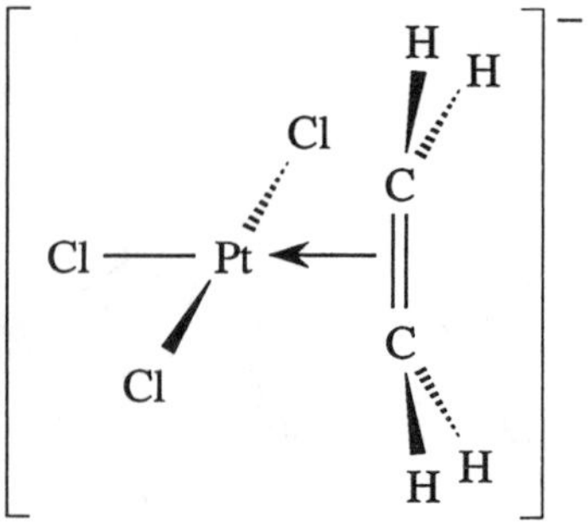

Figure 16.2 Anion of Zeise's salt.

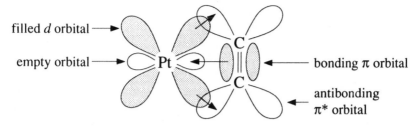

Figure 16.3 Bonding in Zeise's anion.

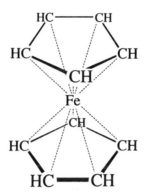

Figure 16.4 Ferrocene.

is stable indefinitely in the air or in contact with water, resists attempts to hydrogenate it, and generally requires rather brutal methods such as Friedel–Crafts syntheses to introduce substituents to the organic rings. It is oxidized by nitric acid or halogens, but even then the Fe—C bonds remain intact, giving the ferricenium ion $Fe^{III}(\eta^5\text{-}C_5H_5)_2{}^+$ or its derivatives. All this stands in sharp contrast to the traditional picture of organometallic reactivity that came from early experience with zinc alkyls, etc., although some other *metal arene* sandwich compounds such as $Cr(\eta^6\text{-}C_6H_6)_2$ and $Ni(\eta^5\text{-}C_5H_5)_2$ (nickelocene) are considerably more reactive than ferrocene and require special handling. Complexes containing just one aromatic ligand with side-on bonding ("open-faced sandwich" or "piano stool" complexes) are also well known. In all cases, the extensive π interactions of the ligand allow it to act as a "sponge" for excess electron density in the complex.

This draining away of excess electron density from the metal atom into the π^* system of olefin, aromatic, or carbonyl ligands helps stabilize low oxidation states of the metal. This is why metal carbonyls such as $Ni(CO)_4$ or $Fe(CO)_5$ are so stable despite the formal oxidation state of zero for the metal. Organophosphines (i.e., organic derivatives of phosphine, PH_3) behave similarly to CO as back-bonding ligands, except that the π-acceptor orbitals are the unoccupied $3d$ orbitals of the P atom. Nitric oxide bonds to metals in the same way as does CO, forming *nitrosyl* complexes. These hardly qualify *per se* as organometallics, but in practice nitrosyl ligands are often associated with carbonyl or organic ligands because of the compatibility of the bonding modes, e.g., in $Co(CO)_3NO$. Even the molecules N_2 and O_2 can bond to transition metal centers using variants of the back-bonding mechanism. G. Kubas has recently shown that H_2 molecules can also bond to metal centers in certain organotransition metal complexes in a dihapto-mode; here, no back-bonding is involved, since H_2 has no π-acceptor system, and the bonding is presumably of the three-center, two-electron type well known in borane (boron hydride) chemistry. More importantly, the presence of π-acceptor ligands on the metal center stabilizes hydrido (H^-) complexes through the "electron sponge" action noted above, and so addition of H_2 to organotransition metal centers usually gives ultimately hydrido- complexes, perhaps via the η^2-H_2 intermediate. In a similar way, alkyl derivatives of the transition metals, which might be expected to be extremely reactive on the basis of electronegativity, are substantially stabilized by the presence of π-acceptor ligands, as in $CH_3Mn(CO)_5$ (see Section 16.1).

Thus, the special bonding characteristics of π-acceptor ligands in organotransition metal compounds enable these complexes to coordinate small molecules such as ethylene, CO, and H_2, and also provide an electronic buffer system to facilitate changes of metal oxidation state and coordination number. These complexes are therefore especially suited to act as catalysts for homogeneous reactions involving these molecules, usually through cycles involving *oxidative addition* and *reductive elimination*, as follows.

16.3   Transition Metal Complexes as Homogeneous Catalysts

Homogeneous catalysts are those that are present in the same phase as the reaction that they are facilitating (e.g., nitric oxide in the lead chamber or Deacon processes, where all the reactants and the catalyst are gaseous). Several transition metal complexes have gained industrial importance as homogeneous catalysts for reactions in solution. In Chapter 9, we considered the many advantages of *heterogeneous* catalysts. Homogeneous catalysts are in many respects less convenient—in particular, one is faced with the problem of separating the catalyst from the products at the end of the reaction—but, as molecular entities, they have precisely reproducible properties and are much more amenable to study and to systematic chemical modification than are catalytic surfaces. Thus, it is possible to *design* homogeneous catalysts for extremely high activity and selectivity through the techniques of synthetic chemistry.

In essence, the function of the catalyst is (a) to bring the reactant molecules together, and (b) to facilitate rearrangement of their chemical bonds by acting as an "electron bank." Transition metal complexes can fulfil function (a) through their ability to accept and exchange various ligands, while the variability of the oxidation states of the central transition metal atoms, particularly where π-acceptor ligands are present, can provide very effectively for function (b). The specificity and effectiveness of the catalytic complex can, in principle at least, be modified at will by altering the "spectator" ligands or by choosing a different central metal atom.

In particular, if we have a complex that normally has n ligands when the oxidation state of the central metal is z, but prefers $(n+2)$ ligands when the oxidation state is increased to $(z + 2)$, we have the prerequisites for facile *oxidative addition* of a polyatomic molecule such as H_2 to form two new ligands (here, hydrido ligands, H^-) by breaking a covalent bond within the molecule and taking two electrons from the metal atom M (reaction 16.9):

$$
\underset{L}{\overset{L}{>}}M\underset{L}{\overset{L}{<}} \;+\; \overset{H}{\underset{H}{|}} \;\longrightarrow\; \underset{L}{\overset{L}{>}}\overset{\overset{\displaystyle H}{|}}{M}\underset{\underset{\displaystyle L}{|}}{\overset{L}{<}}H \tag{16.9}
$$

The reverse process is called *reductive elimination.*

Note that, in reaction 16.9, the H—H bond is broken to form two hydrogen atoms in a reactive form (hydrido ligands) much as H_2 is activated by chemisorption on a nickel metal surface (Section 9.2); these hydrogens can attack appropriate ligands L intramolecularly (hydride transfer). Thus, if one of the ligands L is an olefin (alkene) coordinated to M as in Zeise's anion (Section 16.2), the hydrido ligands can attack the neighboring C=C function to form first an alkyl-M complex and then a free alkane, with concomitant return of the two "borrowed" electrons back to M and resumption of four-coordination in ML_4. The end result is therefore the *hydrogenation* of the olefin by H_2 in homogeneous solution through the action of ML_4, which, as the definition of a catalyst requires, undergoes no net change.

This sequence of events may be illustrated by the homogeneous hydrogenation of ethylene in (say) benzene solution by *Wilkinson's catalyst,* $RhCl(PPh_3)_3$ (Ph = phenyl, C_6H_5; omitted for clarity in cycle 16.10). In this square planar complex, the central rhodium atom is stabilized in the oxidation state (I) by acceptance of excess electron density into the $3d$ orbitals of the triphenylphosphine ligands, but is readily oxidized to rhodium(III), which is preferentially six-coordinate. Thus, we have a typical candidate for a catalytic cycle of oxidative addition and subsequent reductive elimination:

$$\text{(catalytic cycle involving } Cl, P, Rh, H_2, H_2C{=}CH_2, CH_3{-}CH_3 \text{)} \qquad (16.10)$$

The mechanistic details of cycle 16.10 have been represented in a somewhat arbitrary fashion, but the essence of the mode of action of transition metal complexes (in particular, complexes of the Group 9 elements Co, Rh, and Ir in the (I) oxidation state) as homogeneous catalysts for hydrogenation reactions should be clear. Cycles such as 16.10 can be used to catalyze the addition of CO (*carbonylation;* CO readily complexes with transition metal centers) as well as H_2 to a terminal $-CH{=}CH_2$ function, giving an aldehyde:

$$R{-}CH{=}CH_2 + CO + H_2 \xrightarrow{\text{catalyst}} R{-}CH_2{-}CH_2{-}CHO. \qquad (16.11)$$

This reaction is called *hydroformylation* and is typified by the so-called "OXO" process in which the catalyst is $HCo(CO)_4$. In practice, almost any source of cobalt will serve, since it will be converted to $HCo(CO)_4$ in the presence of CO and H_2 under the operating conditions, which are typically 110 to 180 °C and a total gas pressure of 20 to 35 MPa. Evidently, the alkene $R{-}CH{=}CH_2$ adds to the Co complex, as in Zeise's compounds, and is converted by coordinated H to an alkyl group as in cycle 16.10. This alkyl group then migrates to the C atom of one of the carbonyl ligands to form a $M{-}CO{-}CH_2{-}CH_2{-}R$ grouping, which in turn reacts with a further H ligand at the carbonyl C atom and is released as the aldehyde. Hydroformylation is clearly related to the *Fischer–Tropsch reactions*, in which CO/H_2 mixtures (in effect, water–gas; Section 3.3) react over heterogeneous catalysts to give organic compounds such as methanol.

The above examples involve hydrogenation (i.e., reduction) of organic molecules, but transition metal complexes can also catalyze oxidations. For example, the *Wacker process*, which has been widely used to convert ethylene to acetaldehyde, depends on catalysis by palladium(II) in the presence of copper(II) in aqueous HCl. Once again, Zeise-type coordination of the ethylene to the metal center is believed to be involved:

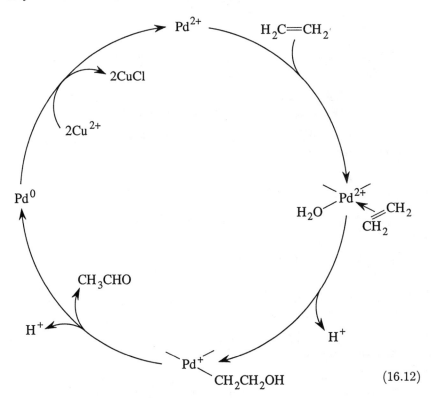

$$2CuCl + 2H^+ + \tfrac{1}{2}O_2 \rightarrow 2Cu^{2+} + 2Cl^- + H_2O. \qquad (16.13)$$

The role of the copper chloride is to provide a means of using air to reoxidize the palladium to palladium(II). The Wacker process itself is becoming obsolete, largely because acetaldehyde is now more economically made by the reaction of methanol with CO via oxidative addition to the homogeneous catalyst $Rh^I(CO)_2I_2{}^-$ (the *Monsanto process*), but it has been seminal in the development of palladium complexes as oxidation catalysts.

In this chapter, only a few simple examples of the mode of action of transition metal complexes and organometallic compounds as catalysts have been considered, by way of illustration of the scope of the field. Much of the chemistry involved in industrial organometallic chemistry and homogeneous catalysis is more complicated than can be effectively presented in a book of this nature. The interested reader is therefore referred to the selected works listed below for further information.

References

1. C. Elschenbroich and A. Saltzer, "Organometallics," VCH Publishers: New York, 1989.

2. I. Haiduc and J. J. Zuckerman, "Basic Organometallic Chemistry," Walter de Gruyter: New York, 1985.

3. G. E. Coates, M. L. H. Green, P. Powell, and K. Wade, "Principles of Organometallic Chemistry," Methuen: London, 1968.

4. N. N. Greenwood and A. Earnshaw, "Chemistry of the Elements," Pergamon Press: Oxford, 1984.

5. F. A. Cotton and G. Wilkinson, "Advanced Inorganic Chemistry," 5th edn., Wiley-Interscience: New York, 1988.

6. A. Yamamoto, "Organotransition Metal Chemistry," Wiley-Interscience: New York, 1986.

7. C. M. Lukehart, "Fundamental Transition Metal Organometallic Chemistry," Brooks/Cole Publishing: Belmont, California, 1985.

8. C. Masters, "Homogeneous Transition-metal Catalysis—A Gentle Art," Chapman and Hall: London, 1981.

9. A. Nakamura and M. Tsutsui, "Principles and Applications of Homogeneous Catalysis," Wiley-Interscience: New York, 1980.

10. J. P. Collman and L. S. Hegedus, "Principles and Applications of Organotransition Metal Chemistry," University Science Books: Mill Valley, California, 1980.

11. G. W. Parshall, "Homogeneous Catalysis," John Wiley & Sons: New York, 1980.

12. R. F. Heck, "Organotransition Metal Chemistry," Academic Press: New York, 1974.

13. R. H. Crabtree, "The Organometallic Chemistry of the Transition Metals," John Wiley & Sons: New York, 1988.

Exercises

16.1 (a) With the aid of Appendix C, calculate the heat of reaction of liquid diethylzinc ($\Delta H_f^\circ = +10.5$ kJ mol^{-1}) with water vapor to form solid zinc hydroxide ($\Delta H_f^\circ = -642.2$ kJ mol^{-1}).

(b) Calculate the heat of reaction of liquid diethylmercury ($\Delta H_f^\circ = +30.1$ kJ mol^{-1}) with water vapor, noting that divalent mercury forms a solid oxide (two forms) but no hydroxide.

(c) Comment on the implications of your results, assuming them to be qualitatively representative of the organometallic chemistries of Zn and Hg.

[*Answers:* (a) -338.4; (b) $+36.6$ (yellow HgO) kJ mol^{-1}.]

16.2 Germanium chloride is a colorless fuming liquid (bp $84\,^\circ$C) that can be prepared simply by distilling a solution of GeO$_2$ in concentrated aqueous HCl. Suggest a means of making tetraethylgermane (bp $161\,^\circ$C) from germanium(IV) oxide.

16.3 It is commonly observed that organometallic, carbonyl, and nitrosyl complexes of the transition metals are usually most stable when the metal has a total of 18 electrons in its "valence shell" (i.e., beyond the preceding noble gas configuration), including electrons donated by the ligands. Ligands such as halide ions, hydride, η^2-ethylene, and carbonyl count as two-electron donors, whereas uncharged NO is a three-electron donor, and aromatic ligands with all six π electrons involved in bonding to the metal (e.g., η^5-C$_5$H$_5$$^-$, η^6-C$_6H_6$^{0}, and the tropylium ion η^7-C$_7$H$_7$$^+$) are six-electron donors. Thus, the Cr in Cr(CO)$_6$ provides six electrons (Cr is the sixth element after Ar in the Periodic Table), and the six CO provide 12 electrons, totalling 18. The iron in ferrocene is formally iron(II), so the metal again provides six electrons and the two C$_5$H$_5$$^-$ ligands give 12 for a total of 18. This is known as the *eighteen electron rule*.

(a) Show that Fe(CO)$_5$, Ni(CO)$_4$, di(η^6-benzene)chromium(0), (OC)$_5$Mn—Mn(CO)$_5$ (note the Mn—Mn single bond), HCo(CO)$_4$, and (η^7-C$_7$H$_7$)Mo(CO)$_3$$^+$ conform to the eighteen electron rule.

(b) Predict x in Mn(CO)$_x$I, Co(CO)$_x$NO, (η^5-C$_5$H$_5$)Fe(η^6-C$_6$H$_6$)$^{x+}$, and Fe(CO)$_4$$^{x-}$.

(c) Explain why the ferricenium ion Fe(η^5-C$_5$H$_5$)$_2$$^+$ behaves as a strong oxidizing agent and cobaltocene Co(η^5-C$_5$H$_5$)$_2$ acts as a strong reducing agent, in suitable solvents, while nickelocene is very difficult to make and handle.

Appendix A

Useful Constants

Ice point temperature	$= 273.1500$ K
Molar gas constant	$R = 8.3143$ J K^{-1} mol^{-1}
Avogadro's number	$N = 6.02252 \times 10^{23}$ elementary entities per mole
Speed of light in a vacuum	$c = 2.997925 \times 10^8$ m s^{-1}
Planck's constant	$h = 6.6256 \times 10^{-34}$ J s^{-1}
Boltzmann's constant	$k_B = 1.38054 \times 10^{-23}$ J K^{-1}
Charge of the electron	$e = -1.60210 \times 10^{-19}$ C
Mass of the electron	$m_e = 9.1091 \times 10^{-31}$ kg
Mass of the proton	$m_p = 1.67252 \times 10^{-27}$ kg
Faraday constant	$F = 9.64870 \times 10^4$ C mol^{-1}
Permittivity of a vacuum	$\varepsilon_0 = 8.8541853 \times 10^{-12}$ A^2 s^4 kg^{-1} m^{-3}

Appendix B

The Chemical Elements: Standard Atomic Masses

RECOMMENDATIONS of the IUPAC Commission on Atomic Weights and Isotopic Abundances, *Pure Appl. Chem.*, *60*, 842 (1988); for elements 107–109, see G. Herrmann, *Angew. Chem. Int. Ed. Engl. 27*, 1417 (1988). Masses are scaled to carbon-12 = 12 exactly. Note that the atomic masses of some elements vary somewhat depending upon their origin, as a result of isotopic fractionation processes or varying contributions from radioactive decay of other nuclides. For elements which have no stable isotopes, the nominal atomic masses given in square brackets refer to the longest-lived isotope. Names of elements given in parenthesis indicate alternative nomenclature or the origin (usually a Latin name) of the chemical symbols.

Element	Symbol	Atomic Number	Atomic Mass
Actinium	Ac	89	[227]
Aluminum	Al	13	26.98154
Americium	Am	95	[243]
Antimony	Sb	51	121.75
(Argentum)	Ag	see Silver	
Argon	Ar	18	39.948
Arsenic	As	33	74.92159
Astatine	At	85	[210]
(Aurum)	Au	see Gold	
Barium	Ba	56	137.33
Berkelium	Bk	97	[247]
Beryllium	Be	4	9.01218
Bismuth	Bi	83	208.98037
Boron	B	5	10.811
Bromine	Br	35	79.904

Element	Symbol	Atomic Number	Atomic Mass
Cadmium	Cd	48	112.41
Cesium	Cs	55	132.9054
Calcium	Ca	20	40.078
Californium	Cf	98	[251]
Carbon	C	6	12.011
Cerium	Ce	58	140.115
Chlorine	Cl	17	35.453
Chromium	Cr	24	51.996
Cobalt	Co	27	58.93320
(Columbium)	(Cb)	see Niobium	
Copper	Cu	29	63.546
(Cuprum)	Cu	see Copper	
Curium	Cm	96	[247]
Dysprosium	Dy	66	162.50
Einsteinium	Es	99	[254]
Erbium	Er	68	167.26
Europium	Eu	63	151.965
Fermium	Fm	100	[253]
(Ferrum)	Fe	see Iron	
Fluorine	F	9	18.998403
Francium	Fr	87	[212]
Gadolinium	Gd	64	157.25
Gallium	Ga	31	69.723
Germanium	Ge	32	72.61
Gold	Au	79	196.96654
Hafnium	Hf	72	178.49
Helium	He	2	4.002602
Holmium	Ho	67	164.93032
(Hydrargyrum)	Hg	see Mercury	
Hydrogen	H	1	1.00794
Indium	In	49	114.82
Iodine	I	53	126.90447
Iridium	Ir	77	192.22
Iron	Fe	26	55.847
(Kalium)	K	see Potassium	
Krypton	Kr	36	83.80
Lanthanum	La	57	138.9055
Lawrencium	Lr	103	[262]
Lead	Pb	82	207.2
Lithium	Li	3	6.941
Lutetium	Lu	71	174.967
Magnesium	Mg	12	24.305

Element	Symbol	Atomic Number	Atomic Mass
Manganese	Mn	25	54.93805
Mendelevium	Md	101	[258]
Mercury	Hg	80	200.59
Molybdenum	Mo	42	95.94
(Natrium)	Na	see Sodium	
Neodymium	Nd	60	144.24
Neon	Ne	10	20.180
Neptunium	Np	93	[237]
Nickel	Ni	28	58.69
Niobium	Nb	41	92.90638
Nitrogen	N	7	14.0067
Nobelium	No	102	[259]
Osmium	Os	76	190.2
Oxygen	O	8	15.9994
Palladium	Pd	46	106.42
Phosphorus	P	15	30.97376
Platinum	Pt	78	195.08
(Plumbum)	Pb	see Lead	
Plutonium	Pu	94	[244]
Polonium	Po	84	[209]
Potassium	K	19	39.0983
Praseodymium	Pr	59	140.90765
Promethium	Pm	61	[145]
Protactinium	Pa	91	[231]
Radium	Ra	88	[226]
Radon	Rn	86	[222]
Rhenium	Re	75	186.207
Rhodium	Rh	45	102.90550
Rubidium	Rb	37	85.4678
Ruthenium	Ru	44	101.07
Samarium	Sm	62	150.36
Scandium	Sc	21	44.95591
Selenium	Se	34	78.96
Silicon	Si	14	28.0855
Silver	Ag	47	107.8682
Sodium	Na	11	22.98977
(Stannum)	Sn	see Tin	
(Stibium)	Sb	see Antimony	
Strontium	Sr	38	87.62
Sulfur	S	16	32.066
Tantalum	Ta	73	180.9479
Technetium	Tc	43	[98]

286 Appendix B.

Element	Symbol	Atomic Number	Atomic Mass
Tellurium	Te	52	127.60
Terbium	Tb	65	158.92534
Thallium	Tl	81	204.3833
Thorium	Th	90	232.0381
Thulium	Tm	69	168.93421
Tin	Sn	50	118.71
Titanium	Ti	22	47.88
Tungsten	W	74	183.85
Unnilennium	Une	109	[266]
Unnilhexium	Unh	106	[263]
Unniloctium	Uno	108	[265]
Unnilpentium	Unp	105	[262]
Unnilquadium	Unq	104	[261]
Unnilseptium	Uns	107	[262]
Uranium	U	92	238.0289
Vanadium	V	23	50.9415
(Wolfram)	W	see Tungsten	
Xenon	Xe	54	131.29
Ytterbium	Yb	70	173.04
Yttrium	Y	39	88.90585
Zinc	Zn	30	65.39
Zirconium	Zr	40	91.224

Appendix C

Chemical Thermodynamic Data

THE FOLLOWING data for substances mentioned in this book are adapted from D. D. Wagman, W. H. Evans, V. B. Parker, R. H. Schumm, I. Halow, S. M. Bailey, K. L. Churney, and R. L. Nuttall, "The NBS Tables of Chemical Thermodynamic Properties," American Chemical Society/American Institute of Physics, Washington, DC, 1982. The tabulated quantities are the enthalpy of formation, entropy, and heat capacity in SI units for 298.15 K and 100 kPa (1 bar) exactly. For solutes, the units are mol per kg of solvent (molal scale). The NBS order of entries is used—oxygen first, then hydrogen, followed by descents of the Periodic Table groupwise, beginning with the noble gases (Group 18) and moving left. Since the thermodynamic properties of ions in solution cannot be separated experimentally from those of the counter-ions which must be present, they are conventionally referred to those for $H^+(aq)$; for example, $S°$ for $Cl^-(aq)$ is conventionally taken to be the same as $S°$ for HCl(aq. ions).

Substance	$\Delta H_f°$ (kJ mol^{-1})	$S°$ (J K^{-1} mol^{-1})	$C_p°$ (J K^{-1} mol^{-1})
O(g)	249.170	161.055	21.912
O$_2$(g)	0	205.138	29.355
O$_3$(g)	142.7	238.93	39.20
H(g)	217.965	114.713	20.784
H$^+$(aq)	0	0	0
H$_2$(g)	0	130.684	28.824
OH(g)	38.95	183.745	29.886
OH$^-$(g)	−143.5	—	—
OH$^-$(aq)	−229.994	−10.75	−148.5
HO$_2$(g)	10.5	229.0	34.89

Substance	ΔH_f° (kJ mol^{-1})	S° (J K^{-1} mol^{-1})	C_p° (J K^{-1} mol^{-1})
HO$_2$$^-$(aq)	−160.33	23.8	—
H$_2$O(g)	−241.818	188.825	33.577
H$_2$O(l)	−285.83	69.91	75.291
H$_2$O$_2$(aq)	−191.17	143.9	—
F(g)	78.99	158.754	22.744
F$^-$(g)	−255.39	—	—
F$_2$(g)	0	202.78	31.30
HF(g)	−271.1	173.779	29.133
HF(aq., undissoc.)	−320.08	88.7	—
Cl(g)	121.679	165.198	21.840
Cl$^-$(g)	−233.13	—	—
Cl$^-$(aq)	−167.159	56.5	−136.4
Cl$_2$(g)	0	223.066	33.907
Cl$_2$(aq)	−23.4	121.	—
ClO(g)	101.84	226.63	31.46
ClO$^-$(aq)	−107.1	42.	—
ClO$_2$(g)	102.5	256.84	41.97
ClO$_2$(aq)	74.9	164.8	—
ClO$_2$$^-$(aq)	−66.5	101.3	—
ClO$_3$$^-$(aq)	−103.97	162.3	—
ClO$_4$$^-$(aq)	−129.33	182.0	—
HCl(g)	−92.307	186.908	29.12
HClO(aq., undissoc.)	−120.9	142.	—
Br(g)	111.884	175.022	20.786
Br$^-$(g)	−219.07	—	—
Br$^-$(aq)	−121.55	82.4	−141.8
Br$_2$(l)	0	152.231	75.689
Br$_2$(g)	30.907	245.463	36.02
BrO$^-$(aq)	−94.1	42.	—
BrO$_3$$^-$(aq)	−67.07	161.71	—
HBr(g)	−36.40	198.695	29.142
I(g)	106.838	180.791	20.786
I$_2$(s)	0	116.135	54.438
I$_2$(g)	62.438	260.69	36.90
I$_2$(aq)	22.6	137.2	—

Substance	ΔH_f° (kJ mol^{-1})	S° (J K^{-1} mol^{-1})	C_p° (J K^{-1} mol^{-1})
I$^-$(aq)	-55.19	111.3	-142.3
I$_3{}^-$(aq)	-51.5	239.3	—
HI(g)	26.48	206.594	29.158
α-S	0	31.80	22.64
S^{2-}(aq)	33.1	-14.6	—
SO$_2$(g)	-296.830	248.22	39.87
SO$_2$(aq)	-322.980	161.9	—
β-SO$_3$(s)	-454.51	70.7	—
SO$_3$(g)	-395.72	256.76	50.67
SO$_3{}^{2-}$(aq)	-635.5	$-29.$	—
SO$_4{}^{2-}$(aq)	-909.27	20.1	$-293.$
S$_2$O$_3{}^{2-}$(aq)	-648.5	67.	—
S$_2$O$_8{}^{2-}$(aq)	-1344.7	244.3	—
S$_4$O$_6{}^{2-}$(aq)	-1224.2	257.3	-67.8
HS$^-$(aq)	-17.6	62.8	—
H$_2$S(g)	-20.63	205.79	34.23
H$_2$S(aq., undissoc.)	-39.7	121.	—
H$_2$SO$_3$(aq., undissoc.)	-608.81	232.2	—
HSO$_3{}^-$(aq)	-626.22	139.7	—
HSO$_4{}^-$(aq)	-887.34	131.8	$-84.$
H$_2$SO$_4$(l)	-813.989	156.904	138.91
N(g)	472.704	153.298	20.786
N$_2$(g)	0	191.61	29.125
N$_3{}^-$(aq)	275.14	107.9	—
NO(g)	90.25	210.761	29.844
NO$_2$(g)	33.18	240.06	37.20
NO$_2{}^-$(aq)	-104.6	123.0	-97.5
NO$_3{}^-$(aq)	-205.0	146.4	-86.6
N$_2$O(g)	82.05	219.85	38.45
N$_2$O$_4$(g)	9.16	304.29	77.28
N$_2$O$_4$(l)	-19.50	209.2	142.7
NH$_3$(g)	-46.11	192.45	35.06
NH$_3$(aq)	-80.29	111.3	—
NH$_4{}^+$(aq)	-132.51	113.4	79.9
N$_2$H$_4$(l)	50.63	121.21	98.87
HN$_3$(l)	264.0	140.6	—
HNO$_2$(aq)	-119.2	135.6	—

Substance	ΔH_f° (kJ mol^{-1})	S° (J K^{-1} mol^{-1})	C_p° (J K^{-1} mol^{-1})
HNO$_3$(l)	−174.10	155.60	109.87
NH$_4$NO$_2$(s)	−256.5	—	—
NH$_4$NO$_3$(s)	−365.56	151.08	139.3
NCl$_3$(l)	230.	—	—
NOCl(g)	51.71	261.69	44.69
NH$_4$Cl(s)	−314.43	94.6	84.1
NH$_4$ClO$_4$(s)	−295.3	186.2	—
(NH$_4$)$_2$SO$_4$(s)	−1180.85	220.1	187.49
P(s, white)	0	41.09	23.840
P(red, triclinic)	−17.6	22.80	21.21
PO$_4^{3-}$(aq)	−1277.4	−222.	—
P$_4$O$_{10}$(s, hexagonal)	−2984.0	228.86	211.71
HPO$_4^{2-}$(aq)	−1292.14	−33.5	—
H$_2$PO$_4^-$(aq)	−1296.29	90.4	—
C(graphite)	0	5.740	8.527
C(diamond)	1.895	2.377	6.113
C(g)	716.682	158.096	20.838
CO(g)	−110.525	197.674	29.142
CO$_2$(g)	−393.509	213.74	37.11
CO$_2$(aq)	−413.80	117.6	—
CO$_3^{2-}$(aq)	−677.14	−56.9	—
CH$_4$(g)	−74.81	186.264	35.309
HCO$_3^-$(aq)	−691.99	91.2	—
CH$_3$OH(l)	−238.66	126.8	81.6
CH$_3$OH(g)	−200.66	239.81	43.89
CF$_4$(g)	−925.	261.61	61.09
CCl$_4$(l)	−135.44	216.40	131.75
CCl$_4$(g)	−102.9	309.85	83.30
COCl$_2$(g)	−218.8	283.53	57.66
CHCl$_3$(l)	−134.47	201.7	113.8
CF$_3$Cl(g)	−695.	285.29	66.86
CF$_2$Cl$_2$(g)	−477.	300.77	72.26
CS$_2$(l)	89.70	151.34	75.7
CS$_2$(g)	117.36	237.84	45.40
COS(g)	−142.09	231.57	41.51
CN$^-$(aq)	150.6	94.1	—
HCN(l)	108.87	112.84	70.63

Substance	ΔH_f° (kJ mol^{-1})	S° (J K^{-1} mol^{-1})	C_p° (J K^{-1} mol^{-1})
HCN(g)	135.1	201.78	35.86
HCN(aq. undissoc.)	107.1	124.7	—
CO(NH$_2$)$_2$(s)	−333.51	104.60	93.14
NH$_4$CO$_2$NH$_2$(s)	−645.05	133.5	—
NCS$^-$(aq)	76.44	144.3	−40.2
C$_2$H$_2$(g)	226.73	200.94	43.93
C$_2$H$_4$(g)	52.26	219.56	43.56
C$_2$H$_6$(g)	−84.68	229.60	52.63
$\frac{1}{n}$(C$_2$F$_4$)$_n$(Teflon)	−820.5	—	—
C$_2$F$_4$(g)	−650.6	300.06	80.46
Si(s)	0	18.83	20.00
SiO$_2$(α-quartz)	−910.94	41.84	44.43
SiO$_2$(α-cristobalite)	−909.48	42.68	44.18
SiO$_2$(α-tridymite)	−909.06	43.5	44.60
SiO$_2$(amorphous)	−903.49	46.9	44.4
H$_4$SiO$_4$(aq)	−1468.6	180.	—
SiH$_4$(g)	34.3	204.62	42.84
SiCl$_4$(l)	−687.0	239.7	145.31
SiCl$_4$(g)	−657.01	330.73	90.25
SiC(s, cubic)	−65.3	16.61	26.86
[(CH$_3$)$_3$Si]$_2$O(l)	−815.0	433.84	311.37
[(CH$_3$)$_3$Si]$_2$O(g)	−777.72	535.06	238.49
Sn(s, white)	0	51.55	26.99
Sn(s, gray)	−2.09	44.14	25.77
SnO(s)	−285.8	56.5	44.31
SnO$_2$(s)	−580.7	52.3	52.59
SnCl$_4$(l)	−511.3	258.6	165.3
SnCl$_4$(g)	−471.5	365.8	98.3
Pb(s)	0	64.81	26.44
Pb^{2+}(aq)	−1.7	10.5	—
PbO(s, yellow)	−217.32	68.70	45.77
PbO(s, red)	−218.99	66.5	45.81
PbO$_2$(s)	−277.4	68.6	64.64
Pb$_3$O$_4$(s)	−718.4	211.3	146.9
PbS(s)	−100.4	91.2	49.50
PbSO$_4$(s)	−919.94	148.57	103.207

Substance	ΔH_f° (kJ mol^{-1})	S° (J K^{-1} mol^{-1})	C_p° (J K^{-1} mol^{-1})
Pb(N$_3$)$_2$(s)	478.2	148.1	—
B(s)	0	5.86	11.09
B$_2$O$_3$(s)	−1272.77	53.97	62.93
B(OH)$_3$(s)	−1094.33	88.83	81.38
BF$_3$(g)	−1137.00	254.12	50.46
BF$_3$NH$_3$(s)	−1353.9	—	—
Al(s)	0	28.33	24.35
Al(g)	326.4	164.54	21.38
Al^{3+}(g)	5483.17	—	—
Al^{3+}(aq)	−531.	−321.7	—
Al$_2$O(g)	−130.	259.35	45.69
Al$_2$O$_3$(s, α)	−1675.7	50.92	79.04
Al$_2$O$_3$(s, γ)	−1656.9	—	—
Al$_2$O$_3$·H$_2$O(boehmite)	−1980.7	96.86	131.25
Al$_2$O$_3$·H$_2$O(diaspore)	−1998.91	70.67	106.19
Al$_2$O$_3$·3H$_2$O(gibbsite)	−2586.67	136.90	183.47
Al$_2$O$_3$·3H$_2$O(bayerite)	−2576.5	—	—
Al(OH)$_4^-$(aq)	−1502.5	102.9	—
AlF$_3$(s)	−1504.1	66.44	75.10
AlCl$_3$(s)	−704.2	110.67	91.84
Al$_2$Cl$_6$(g)	−1290.8	490.	—
Zn(s)	0	41.63	25.40
Zn(g)	130.729	160.984	20.786
Zn^{2+}(aq)	−153.89	−112.1	46.
ZnO(s)	−348.28	43.64	40.25
ZnS(zinc blende)	−205.98	57.7	46.0
ZnS(wurtzite)	−192.63	—	—
ZnSO$_4$(s)	−982.8	110.5	99.2
Hg(l)	0	76.02	27.983
Hg(g)	61.317	174.96	20.786
Hg^{2+}(aq)	171.1	−32.2	—
Hg$_2^{2+}$(aq)	172.4	84.5	—
HgO(s, red)	−90.83	70.29	44.06
HgO(s, yellow)	−90.46	71.1	—
HgCl$_2$(s)	−224.3	146.	—

Substance	ΔH_f° (kJ mol^{-1})	S° (J K^{-1} mol^{-1})	C_p° (J K^{-1} mol^{-1})
$Hg_2Cl_2(s)$	−265.22	192.5	—
$HgS(s, red)$	−58.2	82.4	48.41
$HgS(s, black)$	−53.6	88.3	—
$Cu(s)$	0	33.150	24.435
$Cu^+(aq)$	71.67	40.6	—
$Cu^{2+}(aq)$	64.77	−99.6	—
$CuO(s)$	−157.3	42.63	42.30
$Cu_2O(s)$	−168.6	93.14	63.64
$CuS(s)$	−53.1	66.5	47.82
$Cu_2S(s)$	−79.5	120.9	76.32
$CuSO_4(s)$	−771.36	109.	100.
$Cu(NH_3)_4{}^{2+}(aq)$	−348.5	273.6	—
$Ag(s)$	0	42.55	25.351
$Ag^+(aq)$	105.579	72.68	21.8
$Ag_2O(s)$	−31.05	121.3	65.86
$AgCl(s)$	−127.068	96.2	50.79
$AgBr(s)$	−100.37	107.1	52.38
$Ag_2S(s, \alpha)$	−32.59	144.01	76.53
$Au(s)$	0	47.40	25.418
$Au(CN)_2{}^-(aq)$	242.3	172.	—
$Ni(s)$	0	29.87	26.07
$Ni^{2+}(aq)$	−54.0	−128.9	—
$NiO(s)$	−239.7	37.99	44.31
$Ni(OH)_2(s, \alpha)$	−529.7	88.	—
$NiS(s)$	−82.0	52.97	47.11
$NiSO_4(s)$	−872.91	92.	138.
$NiSO_4 \cdot 7H_2O(s)$	−2976.33	378.94	364.59
$Ni(NH_3)_6{}^{2+}(aq)$	−630.1	394.6	—
$Ni(CO)_4(l)$	−633.0	313.4	204.6
$Ni(CO)_4(g)$	−602.91	410.6	145.18
$Co(s)$	0	30.04	24.81
$Co^{2+}(aq)$	−58.2	−113.	—
$CoO(s)$	−237.94	52.97	55.23
$Co(OH)_2(pink, pptd.)$	−539.7	79.	—
$Co(NH_3)_6{}^{3+}(aq)$	−584.9	146.	—

Substance	ΔH_f° (kJ mol^{-1})	S° (J K^{-1} mol^{-1})	C_p° (J K^{-1} mol^{-1})
$Co(NH_3)_5Cl^{2+}$ (aq)	−628.0	341.4	—
Fe(s)	0	27.28	25.10
Fe^{2+} (g)	2749.93	—	—
Fe^{3+} (g)	5712.8	—	—
Fe^{2+} (aq)	−89.1	−137.7	—
Fe^{3+} (aq)	−48.5	−315.9	—
$FeO_{0.947}$ (s, wustite)	−266.27	57.49	48.12
Fe_2O_3 (s, hematite)	−824.2	87.40	103.85
Fe_3O_4 (s, magnetite)	−1118.4	146.4	143.43
$FeOH^{2+}$ (aq)	−290.8	−142.	—
$Fe(OH)_2$ (s, pptd.)	−569.0	88.	—
$Fe(OH)_3$ (s, pptd.)	−823.0	106.7	—
$Fe_{1.000}S$ (s)	−100.0	60.29	50.54
FeS_2 (s, pyrite)	−178.2	52.93	62.17
Fe_7S_8 (s)	−736.4	485.8	398.57
$FeSO_4$ (s)	−928.4	107.5	100.58
$Fe(CO)_5$ (l)	−774.0	338.1	240.6
$Fe(CO)_5$ (g)	−733.9	445.3	—
$Fe(CN)_6^{3-}$ (aq)	561.9	270.3	—
$Fe(CN)_6^{4-}$ (aq)	455.6	95.	—
Mn(s, α)	0	32.01	26.32
Mn^{2+} (aq)	−220.75	−73.6	50.
MnO(s)	−385.22	59.71	45.44
MnO_2 (s)	−520.03	53.05	54.14
MnO_4^- (aq)	−541.4	191.2	−82.0
Mn_2O_3 (s)	−959.0	110.5	107.65
Mn_3O_4 (s)	−1387.8	155.6	139.66
$Mn(OH)_2$ (s, pptd.)	−695.4	99.2	—
Cr(s)	0	23.77	23.35
CrO_4^{2-} (aq)	−881.15	50.21	—
Cr_2O_3 (s)	−1139.7	81.2	118.74
$Cr_2O_7^{2-}$ (aq)	−1490.3	261.9	—
$HCrO_4^-$ (aq)	−878.2	184.1	—
$FeCr_2O_4$ (s)	−1444.7	146.0	133.64
Mo(s)	0	28.66	24.06

Substance	ΔH_f° (kJ mol^{-1})	S° (J K^{-1} mol^{-1})	C_p° (J K^{-1} mol^{-1})
$MoO_2(s)$	−588.94	46.28	55.98
$MoO_3(s)$	−745.09	77.74	74.98
$MoO_4^{2-}(aq)$	−997.9	27.2	—
$MoS_2(s)$	−235.1	62.59	63.55
$W(s)$	0	32.64	24.27
$WO_3(s)$	−842.87	75.90	73.76
$V(s)$	0	28.91	24.89
$VO^{2+}(aq)$	−486.6	−133.9	—
$VO_2^+(aq)$	−649.8	−42.3	—
$VO_3^-(aq)$	−888.3	50.	—
$V_2O_5(s)$	−1550.6	131.0	127.65
$Ti(s)$	0	30.63	25.02
$TiO(s, \alpha)$	−519.7	50.	39.96
$TiO_2(s, anatase)$	−939.7	49.92	55.48
$TiO_2(s, rutile)$	−944.7	50.33	55.02
$Ti_2O_3(s)$	−1520.9	78.78	97.36
$TiH_2(s)$	−119.7	29.7	30.1
$TiCl_4(l)$	−804.2	252.34	145.18
$TiCl_4(g)$	−763.2	354.9	95.4
$TiN(s)$	−338.1	30.25	37.07
$TiC(s)$	−184.5	24.23	33.64
$Zr(s)$	0	38.99	25.36
$ZrO_2(s, \alpha)$	−1100.56	50.38	56.19
$ZrH_2(s)$	−169.0	35.02	30.96
$ZrCl_4(s)$	−980.52	181.6	119.79
$ZrCl_4(g)$	−870.3	368.3	98.28
$U(s)$	0	50.21	27.665
$U^{4+}(aq)$	−591.2	−410.	—
$UO_2(s)$	−1084.9	77.03	63.60
$UO_2^{2+}(aq)$	−1019.6	−97.5	—
$U_3O_8(s, \alpha)$	−3574.8	282.59	238.36
$UF_4(s)$	−1914.2	151.67	116.02
$UF_6(g)$	−2147.4	377.9	129.62
$Be(s)$	0	9.50	16.44

Substance	ΔH_f° (kJ mol^{-1})	S° (J K^{-1} mol^{-1})	C_p° (J K^{-1} mol^{-1})
BeO(s)	−609.6	14.14	25.52
BeCl$_2$(s)	−490.4	82.68	64.85
Be$_2$SiO$_4$(s)	−2149.3	64.31	95.56
Mg(s)	0	32.68	24.89
Mg(g)	147.70	148.650	20.786
Mg^{2+}(g)	2348.504	—	—
Mg^{2+}(aq)	−466.85	−138.1	—
MgO(s, periclase)	−601.70	26.94	37.15
Mg(OH)$_2$(s)	−924.54	63.18	77.03
MgCl$_2$(s)	−641.32	89.62	71.38
MgCO$_3$(s)	−1095.8	65.7	75.52
MgSiO$_3$(s, enstatite)	−1549.0	67.74	81.38
Mg$_2$SiO$_4$(forsterite)	−2174.0	95.14	118.49
Mg$_3$Si$_2$O$_5$(OH)$_4$(s, chrysotile)	−4365.6	221.3	273.68
Mg$_3$Si$_4$O$_{10}$(OH)$_2$(talc)	−5922.5	260.7	321.7
MgAl$_2$O$_4$(spinel)	−2299.9	80.63	116.19
Ca(s)	0	41.42	25.31
Ca(g)	178.2	154.884	20.786
Ca^{2+}(g)	1925.90	—	—
Ca^{2+}(aq)	−542.83	−53.1	—
CaO(s)	−635.09	39.75	42.80
Ca(OH)$_2$(s)	−986.09	83.39	87.49
CaF$_2$(s)	−1219.6	68.87	67.03
CaCl$_2$(s)	−795.8	104.6	72.59
CaSO$_3 \cdot \frac{1}{2}$H$_2$O(s)	−1311.7	121.3	—
CaSO$_4$(s, anhydrite)	−1434.11	106.7	99.66
CaSO$_4 \cdot \frac{1}{2}$H$_2$O(s)	−1576.74	130.5	119.41
CaSO$_4 \cdot 2$H$_2$O(gypsum)	−2022.63	194.1	186.02
Ca$_3$(PO$_4$)$_2$(s, β)	−4120.8	236.0	227.82
Ca$_5$(PO$_4$)$_3$OH(apatite)	−6738.5	390.35	384.95
Ca$_5$(PO$_4$)$_3$F(s)	−6872.	387.85	375.95
CaC$_2$(s)	−59.8	69.96	62.72
CaCO$_3$(calcite)	−1206.92	92.9	81.88
CaCO$_3$(aragonite)	−1207.13	88.7	81.25
CaTiO$_3$(perovskite)	−1660.6	93.64	97.65
CaMg(CO$_3$)$_2$(dolomite)	−2326.3	155.18	157.53

Substance	ΔH_f° (kJ mol^{-1})	S° (J K^{-1} mol^{-1})	C_p° (J K^{-1} mol^{-1})
Ba(s)	0	62.8	28.07
Ba^{2+}(aq)	−537.64	9.6	—
BaO(s)	−553.5	70.42	47.78
BaSO$_4$(s)	−1473.2	132.2	101.75
BaCO$_3$(s)	−1216.3	112.1	85.35
Li(s)	0	29.12	24.77
Li$^+$(aq)	−278.49	13.4	68.6
Li$_2$O(s)	−597.94	37.57	54.1
LiOH(s)	−484.93	42.80	49.66
LiCl(s)	−408.61	59.33	47.99
LiI(s)	−270.41	86.78	51.04
Li$_2$CO$_3$(s)	−1215.9	90.37	99.12
LiAlH$_4$(s)	−116.3	78.74	83.18
Na(s)	0	51.21	28.24
Na(g)	107.32	153.712	20.786
Na$^+$(g)	609.358	—	—
Na$^+$(aq)	−240.12	59.0	46.4
Na$_2$O(s)	−414.22	75.06	69.12
Na$_2$O$_2$(s)	−510.87	95.0	89.24
NaH(s)	−56.275	40.016	36.401
NaOH(s)	−425.609	64.455	59.54
NaF(s)	−573.647	51.46	46.86
NaCl(s)	−411.153	72.13	50.50
NaClO$_3$(s)	−365.774	123.4	—
NaClO$_4$(s)	−383.30	142.3	—
NaBr(s)	−361.062	86.82	51.38
NaI(s)	−287.78	98.53	52.09
Na$_2$S(s)	−364.8	83.7	—
Na$_2$SO$_4$(s)	−1387.08	149.58	128.20
NaHSO$_4$(s)	−1125.5	113.0	—
NaN$_3$(s)	21.71	96.86	76.61
NaNO$_2$(s)	−358.65	103.8	—
NaNO$_3$(s)	−467.85	116.52	92.88
Na$_3$PO$_4$(s)	−1917.40	173.80	153.47
Na$_5$P$_3$O$_{10}$(s)	−4399.1	381.79	327.02
Na$_2$CO$_3$(s)	−1130.68	134.98	112.30
Na$_2$CO$_3 \cdot$10H$_2$O(s)	−4081.32	562.7	550.32

Substance	ΔH_f° (kJ mol^{-1})	S° (J K^{-1} mol^{-1})	C_p° (J K^{-1} mol^{-1})
NaHCO$_3$(s)	−950.81	101.7	87.61
NaCN(s, cubic)	−87.49	115.60	70.37
Na$_2$SiO$_3$(cryst.)	−1554.90	113.85	—
Na$_2$B$_4$O$_7$·10H$_2$O(borax)	−6288.6	586.	615.
NaBH$_4$(s)	−188.61	101.29	86.78
Na$_3$AlF$_6$(s)	−3301.2	238.5	215.89
NaAlSi$_2$O$_6$·H$_2$O(s, analcite)	−3300.8	234.3	209.91
NaAlSi$_3$O$_8$(albite)	−3935.1	207.40	205.10
Na$_2$CrO$_4$(s)	−1342.2	176.61	142.13
K(s)	0	64.18	29.58
K(g)	89.24	160.336	20.786
K$^+$(g)	514.26	—	—
K$^+$(aq)	−252.38	102.5	21.8
KO$_2$(s)	−284.93	116.7	77.53
K$_2$O(s)	−361.5	—	—
K$_2$O$_2$(s)	−494.1	102.1	—
KOH(s)	−424.764	78.9	64.9
KF(s)	−567.27	66.57	49.04
KCl(s)	−436.747	82.59	51.30
KClO$_3$(s)	−397.73	143.1	100.25
KClO$_4$(s)	−432.75	151.0	112.38
KBr(s)	−393.798	95.90	52.30
KI(s)	−327.900	106.32	52.93
K$_2$SO$_4$(s)	−1437.79	175.56	131.46
K$_2$S$_2$O$_8$(s)	−1916.1	278.7	213.09
KNO$_2$(s)	−369.82	152.09	107.40
KNO$_3$(s)	−494.63	133.05	96.40
K$_2$CO$_3$(s)	−1151.02	155.52	114.43
KHCO$_3$(s)	−963.2	115.5	—
KCN(s)	−113.0	128.49	66.27
KNCS(s)	−200.16	124.26	88.53
KAl(SO$_4$)$_2$·12H$_2$O(s)	−6061.8	687.4	651.03
KAlSi$_3$O$_8$(orthoclase)	−3959.7	232.88	204.51
KAl$_3$Si$_3$O$_{10}$(OH)$_2$(muscovite)	−5984.4	306.3	—
KMnO$_4$(s)	−837.2	171.71	117.57
K$_2$CrO$_4$(s)	−1403.7	200.12	145.98
K$_2$Cr$_2$O$_7$(s)	−2061.5	291.2	219.24

Appendix D

Standard Electrode Potentials for Aqueous Solutions

FOR MORE comprehensive information, see A. J. Bard, R. Parsons and J. Jordan, "Standard Potentials in Aqueous Solution," Marcel Dekker, Inc., New York, 1985, from which the following $E°$ values are taken.

Acidic Solutions ($[H^+] = 1.0$ mol kg^{-1})

Half Reaction	$E°$(V)
$Li^+ + e^- \rightleftharpoons Li$	-3.045
$K^+ + e^- \rightleftharpoons K$	-2.925
$Na^+ + e^- \rightleftharpoons Na$	-2.714
$La^{3+} + 3e^- \rightleftharpoons La$	-2.37
$Mg^{2+} + 2e^- \rightleftharpoons Mg$	-2.356
$\frac{1}{2}H_2 + e^- \rightleftharpoons H^-$	-2.25
$Be^{2+} + 2e^- \rightleftharpoons Be$	-1.97
$Zr^{4+} + 4e^- \rightleftharpoons Zr$	-1.70
$Al^{3+} + 3e^- \rightleftharpoons Al$	-1.67
$Ti^{3+} + 3e^- \rightleftharpoons Ti$	-1.21
$Mn^{2+} + 2e^- \rightleftharpoons Mn$	-1.18
$V^{2+} + 2e^- \rightleftharpoons V$	-1.13
$SiO_2(glass) + 4H^+ + 4e^- \rightleftharpoons Si + 2H_2O$	-0.888
$Zn^{2+} + 2e^- \rightleftharpoons Zn$	-0.763
$U^{4+} + e^- \rightleftharpoons U^{3+}$	-0.52
$Fe^{2+} + 2e^- \rightleftharpoons Fe$	-0.44
$Cr^{3+} + e^- \rightleftharpoons Cr^{2+}$	-0.424

Half Reaction	$E°$ (V)
$Cd^{2+} + 2e^- \rightleftharpoons Cd$	−0.403
$PbSO_4 + 2e^- \rightleftharpoons Pb + SO_4^{2-}$	−0.351
$Eu^{3+} + e^- \rightleftharpoons Eu^{2+}$	−0.35
$Co^{2+} + 2e^- \rightleftharpoons Co$	−0.277
$H_3PO_4 + 2H^+ + 2e^- \rightleftharpoons H_3PO_3 + H_2O$	−0.276
$Ni^{2+} + 2e^- \rightleftharpoons Ni$	−0.257
$V^{3+} + e^- \rightleftharpoons V^{2+}$	−0.255
$2SO_4^{2-} + 4H^+ + 4e^- \rightleftharpoons S_2O_6^{2-} + 2H_2O$	−0.253
$N_2 + 5H^+ + 4e^- \rightleftharpoons N_2H_5^+$	−0.23
$CO_2 + 2H^+ + 2e^- \rightleftharpoons HCOOH$	−0.16
$AgI + e^- \rightleftharpoons Ag + I^-$	−0.152
$Sn^{2+} + 2e^- \rightleftharpoons Sn$	−0.136
$Pb^{2+} + 2e^- \rightleftharpoons Pb$	−0.125
$2H^+ + 2e^- \rightleftharpoons H_2$	0.000
$HCOOH + 2H^+ + 2e^- \rightleftharpoons HCHO + H_2O$	+0.056
$AgBr + e^- \rightleftharpoons Ag + Br^-$	+0.071
$TiO^{2+} + 2H^+ + e^- \rightleftharpoons Ti^{3+} + H_2O$	+0.100
$S + 2H^+ + 2e^- \rightleftharpoons H_2S$	+0.144
$Sn^{4+} + 2e^- \rightleftharpoons Sn^{2+}$	+0.15
$SO_4^{2-} + 2H^+ + 2e^- \rightleftharpoons H_2SO_3 + H_2O$	+0.158
$Cu^{2+} + e^- \rightleftharpoons Cu^+$	+0.159
$AgCl + e^- \rightleftharpoons Ag + Cl^-$	+0.222
$HCHO + 2H^+ + 2e^- \rightleftharpoons CH_3OH$	+0.232
$UO_2^{2+} + 4H^+ + 2e^- \rightleftharpoons U^{4+} + 2H_2O$	+0.27
$VO^{2+} + 2H^+ + e^- \rightleftharpoons V^{3+} + H_2O$	+0.337
$Cu^{2+} + 2e^- \rightleftharpoons Cu$	+0.340
$Fe(CN)_6^{3-} + e^- \rightleftharpoons Fe(CN)_6^{4-}$	+0.361
$2H_2SO_3 + 2H^+ + 4e^- \rightleftharpoons S_2O_3^{2-} + 3H_2O$	+0.400
$H_2SO_3 + 4H^+ + 4e^- \rightleftharpoons S + 3H_2O$	+0.500
$2H_2SO_3 + 4H^+ + 6e^- \rightleftharpoons S_4O_6^{2-} + 6H_2O$	+0.507
$Cu^+ + e^- \rightleftharpoons Cu$	+0.520
$I_2 + 2e^- \rightleftharpoons 2I^-$	+0.5355
$I_3^- + 2e^- \rightleftharpoons 3I^-$	+0.536
$MnO_4^- + e^- \rightleftharpoons MnO_4^{2-}$	+0.56
$S_2O_6^{2-} + 4H^+ + 2e^- \rightleftharpoons 2H_2SO_3$	+0.569
$CH_3OH + 2H^+ + 2e^- \rightleftharpoons CH_4 + H_2O$	+0.59

Half Reaction	E° (V)
$HN_3 + 11H^+ + 8e^- \rightleftharpoons 3NH_4^+$	+0.695
$O_2 + 2H^+ + 2e^- \rightleftharpoons H_2O_2$	+0.695
$Rh^{3+} + 3e^- \rightleftharpoons Rh$	+0.76
$(NCS)_2 + 2e^- \rightleftharpoons 2NCS^-$	+0.77
$Fe^{3+} + e^- \rightleftharpoons Fe^{2+}$	+0.771
$Hg_2^{2+} + 2e^- \rightleftharpoons 2Hg$	+0.796
$Ag^+ + e^- \rightleftharpoons Ag$	+0.799
$2NO_3^- + 4H^+ + 2e^- \rightleftharpoons N_2O_4 + 2H_2O$	+0.803
$Hg^{2+} + 2e^- \rightleftharpoons Hg$	+0.911
$NO_3^- + 3H^+ + 2e^- \rightleftharpoons HNO_2 + H_2O$	+0.94
$NO_3^- + 4H^+ + 3e^- \rightleftharpoons NO + 2H_2O$	+0.957
$HNO_2 + H^+ + e^- \rightleftharpoons NO + H_2O$	+0.996
$N_2O_4 + 4H^+ + 4e^- \rightleftharpoons NO + 2H_2O$	+1.039
$Br_2 + 2e^- \rightleftharpoons 2Br^-$	+1.065
$N_2O_4 + 2H^+ + 2e^- \rightleftharpoons 2HNO_2$	+1.07
$H_2O_2 + H^+ + e^- \rightleftharpoons OH + H_2O$	+1.14
$ClO_4^- + 2H^+ + 2e^- \rightleftharpoons ClO_3^- + H_2O$	+1.201
$O_2 + 4H^+ + 4e^- \rightleftharpoons 2H_2O$	+1.229
$MnO_2 + 4H^+ + 2e^- \rightleftharpoons Mn^{2+} + 2H_2O$	+1.23
$N_2H_5^+ + 3H^+ + 2e^- \rightleftharpoons 2NH_4^+$	+1.275
$Cl_2 + 2e^- \rightleftharpoons 2Cl^-$	+1.358
$Cr_2O_7^{2-} + 14H^+ + 6e^- \rightleftharpoons 2Cr^{3+} + 7H_2O$	+1.36
$PbO_2 + 4H^+ + 2e^- \rightleftharpoons Pb^{2+} + 2H_2O$	+1.468
$2BrO_3^- + 12H^+ + 10e^- \rightleftharpoons Br_2 + 6H_2O$	+1.478
$Mn^{3+} + e^- \rightleftharpoons Mn^{2+}$	+1.51
$Au^{3+} + 3e^- \rightleftharpoons Au$	+1.52
$NiO_2 + 4H^+ + 2e^- \rightleftharpoons Ni^{2+} + 2H_2O$	+1.593
$2HBrO + 2H^+ + 2e^- \rightleftharpoons Br_2 + 2H_2O$	+1.604
$2HClO + 2H^+ + 2e^- \rightleftharpoons Cl_2 + 2H_2O$	+1.630
$PbO_2 + SO_4^{2-} + 4H^+ + 2e^- \rightleftharpoons PbSO_4 + 2H_2O$	+1.698
$MnO_4^- + 4H^+ + 3e^- \rightleftharpoons MnO_2 + 2H_2O$	+1.70
$Ce^{4+} + e^- \rightleftharpoons Ce^{3+}$	+1.72
$H_2O_2 + 2H^+ + 2e^- \rightleftharpoons 2H_2O$	+1.763
$Au^+ + e^- \rightleftharpoons Au$	+1.83
$Co^{3+} + e^- \rightleftharpoons Co^{2+}$	+1.92
$HN_3 + 3H^+ + 2e^- \rightleftharpoons NH_4^+ + N_2$	+1.96

Half Reaction	E° (V)
$S_2O_8{}^{2-} + 2e^- \rightleftharpoons 2SO_4{}^{2-}$	+1.96
$O_3 + 2H^+ + 2e^- \rightleftharpoons O_2 + H_2O$	+2.075
$\dot{O}H + H^+ + e^- \rightleftharpoons H_2O$	+2.38
$F_2 + 2H^+ + 2e^- \rightleftharpoons 2HF$	+3.053

Basic Solutions ($[OH^-] = 1.0$ mol kg^{-1})

Half Reaction	$E°$ (V)
$Ca(OH)_2 + 2e^- \rightleftharpoons Ca + 2OH^-$	-3.026
$Mg(OH)_2 + 2e^- \rightleftharpoons Mg + 2OH^-$	-2.687
$Al(OH)_4^- + 3e^- \rightleftharpoons Al + 4OH^-$	-2.310
$SiO_3^{2-} + 3H_2O + 4e^- \rightleftharpoons Si + 6OH^-$	-1.7
$Mn(OH)_2 + 2e^- \rightleftharpoons Mn + 2OH^-$	-1.56
$2TiO_2 + H_2O + 2e^- \rightleftharpoons Ti_2O_3 + 2OH^-$	-1.38
$Cr(OH)_3 + 3e^- \rightleftharpoons Cr + 3OH^-$	-1.33
$Zn(OH)_4^{2-} + 2e^- \rightleftharpoons Zn + 4OH^-$	-1.285
$Zn(NH_3)_4^{2+} + 2e^- \rightleftharpoons Zn + 4NH_3$	-1.04
$MnO_2 + 2H_2O + 4e^- \rightleftharpoons Mn + 4OH^-$	-0.980
$Cd(CN)_4^{2-} + 2e^- \rightleftharpoons Cd + 4CN^-$	-0.943
$SO_4^{2-} + H_2O + 2e^- \rightleftharpoons SO_3^{2-} + 2OH^-$	-0.94
$2H_2O + 2e^- \rightleftharpoons H_2 + 2OH^-$	-0.828
$HFeO_2^- + H_2O + 2e^- \rightleftharpoons Fe + 3OH^-$	-0.8
$Co(OH)_2 + 2e^- \rightleftharpoons Co + 2OH^-$	-0.733
$CrO_4^{2-} + 4H_2O + 3e^- \rightleftharpoons Cr(OH)_4^- + 4OH^-$	-0.72
$Ni(OH)_2 + 2e^- \rightleftharpoons Ni + 2OH^-$	-0.72
$FeO_2^- + H_2O + e^- \rightleftharpoons HFeO_2^- + OH^-$	-0.69
$2SO_3^{2-} + 3H_2O + 4e^- \rightleftharpoons S_2O_3^{2-} + 6OH^-$	-0.58
$Ni(NH_3)_6^{2+} + 2e^- \rightleftharpoons Ni + 6NH_3$	-0.476
$S + 2e^- \rightleftharpoons S^{2-}$	-0.45
$O_2 + e^- \rightleftharpoons O_2^-$	-0.33
$CuO + H_2O + 2e^- \rightleftharpoons Cu + 2OH^-$	-0.29
$Mn_2O_3 + 2H_2O + 2e^- \rightleftharpoons 2Mn(OH)_2 + 2OH^-$	-0.25
$2CuO + H_2O + 2e^- \rightleftharpoons Cu_2O + 2OH^-$	-0.22
$O_2 + H_2O + 2e^- \rightleftharpoons HO_2^- + OH^-$	-0.065
$MnO_2 + H_2O + 2e^- \rightleftharpoons Mn(OH)_2 + 2OH^-$	-0.05
$NO_3^- + H_2O + 2e^- \rightleftharpoons NO_2^- + 2OH^-$	$+0.01$
$Co(NH_3)_6^{3+} + e^- \rightleftharpoons Co(NH_3)_6^{2+}$	$+0.058$
$HgO(\text{red form}) + H_2O + 2e^- \rightleftharpoons Hg + 2OH^-$	$+0.098$
$N_2H_4 + 2H_2O + 2e^- \rightleftharpoons 2NH_3 + 2OH^-$	$+0.1$
$Co(OH)_3 + e^- \rightleftharpoons Co(OH)_2 + OH^-$	$+0.17$
$HO_2^- + H_2O + e^- \rightleftharpoons H\dot{O} + 2OH^-$	$+0.184$

Half Reaction	$E°(V)$
$O_2^- + H_2O + e^- \rightleftharpoons HO_2^- + OH^-$	+0.20
$ClO_3^- + H_2O + 2e^- \rightleftharpoons ClO_2^- + 2OH^-$	+0.295
$Ag_2O + H_2O + 2e^- \rightleftharpoons 2Ag + 2OH^-$	+0.342
$Ag(NH_3)_2^+ + e^- \rightleftharpoons Ag + 2NH_3$	+0.373
$ClO_4^- + H_2O + 2e^- \rightleftharpoons ClO_3^- + 2OH^-$	+0.374
$O_2 + 2H_2O + 4e^- \rightleftharpoons 4OH^-$	+0.401
$NiO_2 + 2H_2O + 2e^- \rightleftharpoons Ni(OH)_2 + 2OH^-$	+0.490
$FeO_4^{2-} + 2H_2O + 3e^- \rightleftharpoons FeO_2^- + 4OH^-$	+0.55
$BrO_3^- + 3H_2O + 6e^- \rightleftharpoons Br^- + 6OH^-$	+0.584
$MnO_4^{2-} + 2H_2O + 2e^- \rightleftharpoons MnO_2 + 4OH^-$	+0.62
$ClO_2^- + H_2O + 2e^- \rightleftharpoons ClO^- + 2OH^-$	+0.681
$BrO^- + H_2O + 2e^- \rightleftharpoons Br^- + 2OH^-$	+0.766
$HO_2^- + H_2O + 2e^- \rightleftharpoons 3OH^-$	+0.867
$ClO^- + H_2O + 2e^- \rightleftharpoons Cl^- + 2OH^-$	+0.890
$ClO_2 + e^- \rightleftharpoons ClO_2^-$	+1.041
$O_3 + H_2O + 2e^- \rightleftharpoons O_2 + 2OH^-$	+1.246
$\dot{O}H + e^- \rightleftharpoons OH^-$	+1.985

Appendix E

Nomenclature of Coordination Compounds

THE CONVENTIONS for the naming of inorganic compounds are established by the International Union of Pure and Applied Chemistry (IUPAC), and are reviewed on a continuing basis. The following rules, which cover the more commonly encountered complexes, were current in 1989.

Ligands

For *neutral ligands*, there are four special names that have survived from the early days of coordination chemistry: *aqua* for coordinated water, *ammine* for ammonia (not to be confused with *amine*, meaning organic compounds RNH_2, R_2NH, etc., which can also act as ligands), *carbonyl* for complexed carbon monoxide, and *nitrosyl* for bound nitric oxide (NO). For all other neutral ligands, the ordinary name of the molecule is used without modification. For *anionic* molecules as ligands, the final "e" of the anion name is replaced with "o", but there are irregular cases involving *some* anions which end in "ide": chloride (Cl^-) becomes chloro (and the other halides likewise); oxide (O^{2-}), hydroxide (OH^-), peroxide (O_2^{2-}) and superoxide (O^{2-}) become oxo, hydroxo, peroxo and superoxo; cyanide (CN^-) becomes cyano. Most others follow the regular rule:

nitride (N^{3-})	becomes	nitrido
sulfide (S^{2-})		sulfido
azide (N_3^-)		azido
amide (NH_2^-)		amido
carbonate (CO_3^{2-})		carbonato
nitrate (NO_3^-)		nitrato
nitrite (NO_2^-)		nitrito (if bonded through an O)
		nitro (if bonded through N)
thiocyanate (NCS^-)		thiocyanato-*S* (if S-bonded)
		thiocyanato-*N* (if N-bonded)

Complexes

The ligands are named first, starting with any anionic ones, and the name of the metal is followed without a space by its oxidation state in Roman numerals (or the Arabic 0, for zero-valent metal centers) in parentheses. If the complex as a whole is anionic, the metal name is made to end in -*ate*, which replaces endings such as -ium or -um (nickelate, chromate, tantalate) and is followed by the oxidation state. Where the chemical symbol is derived from a Latin name, the anion name is usually also Latinized: cuprate, argentate, aurate, ferrate, stannate, plumbate—but mercurate is an exception.

The numbers of ligands are indicated by the Greek prefixes di, tri, tetra, penta, hexa, hepta, octa, nona (ennea), deca, etc. If, however, the names of the ligands themselves already contain these prefixes (e.g., *di*ethylene*tri*amine), the ligand name is placed in parenthesis and the prefix outside becomes bis, tris, tetrakis, pentakis, hexakis, etc.

$[Co(NH_3)_5Cl]Cl_2$	chloropentaamminecobalt(III) chloride
$Co(CO)_3NO$	nitrosyltricarbonylcobalt(0)
$K_4[Fe(CN)_6]$	potassium hexacyanoferrate(II)
$Co(en)_3{}^{3+}$	tris(ethylenediamine)cobalt(III) ion

(Strictly speaking, names such as ethylenediamine are not systematic and should not be used, but tradition dies hard.)

Isomers and Bridged Complexes

As explained in Section 11.2, geometrical isomers are designated by the prefixes *cis*- and *trans*-, or, for octahedral complexes ML_3X_3, *fac* (meaning facial—all ligands L adjacent to each other, defining one octahedral face) or *mer* (for meridional—the three Ls occupy a north–south "meridian").

Bridging ligands are indicated by a prefix μ-.

$(H_2O)_4Fe\underset{\underset{H}{O}}{\overset{\overset{H}{O}}{<\quad>}}Fe(OH_2)_4{}^{4+}$ di(μ-hydroxo)octaaquadiiron(III) ion

The use of the η^n- prefix in organometallic complexes is explained in Chapter 16.

Index

Solubility product (K_{sp})
  defined, 149, 150
Solvation, 137–39
Solvay process, 59, 61–62, 216
Solvent extraction, 243–46, 264
Sorbite, 134
Sphalerite, 76
Spinel, 84
Spinel structure, 108
  inverse, 85
  normal, 84–85
Square planar complexes, 141
Stability constants, 147–50
Stability fields, 185–87, 190
Stabilizers for plastics, 269
Stacking faults, 97
Stambaugh, E. P., 34
Standard state, 2–3
Steam turbines, 161
Steatite, 110
Steels
  case hardening of, 135
  metallurgy of, 132–35
  nitriding of, 135, 259
  stainless steels, 256
    AISI designations of stainless
      steels, 235
    austenitic, 235
      weld decay in, 221–22
    composition and properties of,
      234–35
    corrosion of, 225–27
    corrosion resistance of, 219,
      223, 234–35
    ferritic, 235
    heat treatment, 134, 234–35
    low carbon, 222
    martensitic, 235
    oxide film on, 219
    protective oxide film on, 225–26
    uses of, 55
  steelmaking, 257–59
  tempering of, 134
STM (scanning tunneling
  microscopy), 124
Stoichiometry, 98
Stratosphere, 28–29, 36, 69
Strecker synthesis, 145

Stress cracking corrosion, 215, 221
Sublimation, 13
  heat of, 13
Sulfanes, 16–17
Sulfatizing roast, 263
Sulfide ores
  acid leaching of, 241
  leaching with nitric acid, 242
  microbial leaching, 242
  roasting, 33, 51, 241
Sulfides
  organic, 53–54
Sulfite process, 53, 55–57
Sulfonates, 55
Sulfur, 16, 49–57, 170
  in coals, 34
  dioxide, 258
    atmospheric pollution by,
      33–35, 241
    in the Claus process, 50
    environmental damage by, 31
    in the Hargreaves process, 45
    kinetics of oxidation of, 8–9
    liquid, 138
    as a mild reductant, 66
    in pulp- and papermaking,
      55–57
    from roasting of sulfide ores,
      263
    role in metallic corrosion, 233
    in sulfuric acid production,
      51–52
  occurrence, 49
  production, 49–51
  sulfates
    in agriculture, 39
    in ammonia leach process, 241
    in natural waters, 161
    reduction by bacteria, 170
  sulfide ion, 34
  sulfite ion, 33
    in corrosion suppression, 231
  tetrathionate ion, 17, 179
  thiosulfate ion, 17, 151, 179, 241
  trioxide, 33–34, 51–52, 123
Sulfurcrete, 51
Sulfuric acid, 8–9, 25, 34, 165
  in acid rain, 33–35

Tom Swaddle was born in Newcastle upon Tyne, England, in 1937, and read chemistry at University College London (B.Sc. 1958). At the University of Leicester he studied organogermanium chemistry under Prof. C. Eaborn and was awarded the Ph.D. degree in 1961. After post-doctoral studies with Profs. John P. Hunt (Washington State University) and Edward L. King (Universities of Wisconsin and Colorado), he joined the academic staff of The University of Calgary in 1964 and has been Professor of Chemistry there since 1974. He has also held visiting appointments at the Universities of Adelaide (1971), Lausanne (1980), and Melbourne (1988), and at the Tokyo Institute of Technology (1984). He is the author of over 90 papers and reviews in inorganic chemistry, mainly in the areas of reaction kinetics, high temperature aqueous chemistry, and the effects of high pressures. He is a Fellow of The Chemical Institute of Canada, and a member of The Royal Society of Chemistry, The American Chemical Society, AAAS, The Royal Astronomical Society of Canada, and Sigma Xi. He and his wife, Shirley, live in Calgary. He enjoys mountaineering, skiing, fly-fishing, and classical music.